Strategies of the
Major Oil Companies

Research for Business Decisions, No. 70

Richard N. Farmer, Series Editor

Professor of International Business
Indiana University

Other Titles in This Series

Strategies of the Major Oil Companies

by
William N. Greene

Manager
Management Advisory Services
Deloitte, Haskins & Sells
Los Angeles, California

UMI Research
Press

Ann Arbor, Michigan

Produced and distributed by
UMI Research Press
an imprint of
University Microfilms International
A Xerox Information Resources Company
Ann Arbor, Michigan 48106

Library of Congress Cataloging in Publication Data

Greene, William N. (William Noel)
Strategies of the major oil companies.

(Research for business decisions ; no. 70)
Revision of thesis (Ph.D.)—Harvard University, 1982.
Bibliography: p.
Includes index.
1. Petroleum industry and trade—History.   I. Title.
II. Series.
HD9570.5.G673   1985   338.7'6223382'09   84-24076
ISBN 0-8357-1606-6 (alk. paper)

Contents

Preface

This book is an analytical and historical study of the strategies and policies of the seven largest oil companies in the world, which are collectively called the "Majors." It attempts to identify, document, and analyze the strategies and policies of the Majors over a period from their origins in the late nineteenth and early twentieth centuries through the mid-1970's. The Majors are Exxon, Gulf (which was acquired by Chevron in mid-1984), Mobil, Standard Oil of California ("Socal," which changed its name to Chevron on July 1, 1984), Texaco, Royal Dutch/Shell ("Shell"), and British Petroleum ("BP"). Data sources and references for the company histories include previous historical studies as well as industry documents, annual reports, and more than one hundred interviews with senior executives of these companies.

The Major oil companies collectively form the largest group of leading industrial firms in any industry and their actions are of great interest to scholars of industrial organization. They form an easily identifiable group on the basis of size, geographical and vertical spread, and importance in industry history. Individually, these companies are among the most different from the entrepreneurial micro-competitors which have been the focus of traditional economic theory.

Given its focus on major firms in the important international oil industry, the book has three primary purposes. It will develop an analytical history of each of the Majors to provide a body of comparable information which will be used to identify the critical events in their development and to document the components of the strategy of each company. My first purpose is thus to identify and document a record of the policies which constitute the strategies of each of the Majors in a form which may be used to analyze the determinants of these policies. Chapters 4 through 10 address this first purpose.

My second purpose is to contribute to the theory of strategy formation by analyzing how these policies resulted from conscious decisions and natural events in the oil industry, its technology, and its politics. This is thus a study of corporate strategy over time which uses the largest firms in one industry as empirical data to illustrate—and analyze the underlying reasons for—the similarities and differences among strategies of firms in the same business facing a common environment.

My third purpose is to analyze the importance of strategy as a determinant of firm and industry structure, conduct, and performance. I thus attempt to synthesize the fields of industrial organization and microeconomics with the field of business policy. The second and third purposes are addressed in Chapter 11.

The approach of this book differs from that of previous historical, industry, and strategic studies in several respects. Although I have used standard types of historical source material, my purpose is analytical rather than documentary. This study also covers a much longer time period and a full range of strategic decisions, rather than a single type of strategy. I emphasize the interdependence of the strategic choices made by each of the firms I study, at any one time as well as over time, and I also examine the cross-sectional differences in the major policies and strategic choices of firms at different periods in the history of the oil industry.

The introductory chapter previews methods which will be used to identify different strategies and policies of the major firms in the oil industry. It closes with some tables which summarize critical dimensions of the position of each of the Majors in a recent year and the evolution of their positions over time.

Following this chapter are two background chapters on the petroleum industry and the strategy of Standard Oil, the parent of three of the Majors. Then there are seven chapters which attempt to identify and document the strategies of each of the Majors. The final chapter presents an analysis of how policies were formed and changed and some conclusions about strategies from the date in this study.

No one completes a task of this size without accumulating large debts to others for their help along the way. My special thanks go to Professors Robert Stobaugh and Alfred Chandler, who guided this research; to Professor John Lintner for wise commentary; to Bob Gordon of Chevron, Alan Peters of Shell, and Warren Davis of Gulf for their invaluable assistance; to Baker Library at Harvard for providing a marvelous tool for research; to Courtenay Eversole, my wife, Nonette, my family, and my father, who never had the chance to go to college but gave me the aspiration and indispensable encouragement to complete this book.

1

A Method for Identifying Different Strategies in the Oil Industry

The business policy model suggests that behavior is the result of the interaction of strategy and changing opportunities and risks in the environment. (See Appendix A for a summary of the relevant academic literature.) By identifying the behavior and the environmental factors from the historical record, one should be able to deduce the strategy. As Andrews noted, "The current strategy of a company may almost always be deduced from its behavior."[1] From the historical record of a firm's behavior, its environment, and its resources, one should be able to deduce a set of policies which are the components of its strategy.

This will be the process occurring in the historical narratives to follow. Statements of the policies of each company will be made on the basis of its historical behavior. After the date of origin, subsequent behavior will be observed to confirm the continuation, modification, or termination of each policy. The strategic behavior of each company is classified into six historically developed categories of strategic change. (See Appendix B for the development of these categories.) An explanation of each policy will normally be given in terms of the business policy variables (i.e., environmental conditions, resources and handicaps, personal values, and social responsibilities) which appear to have affected the origin of the policy. At the end of each company chapter are exhibits which list the policies stated, the type of strategic decisions they represent (see below), their dates of origin and termination, and the environmental condition, resource, handicap, personal value, and social responsibility variables involved. Figure 1.1 may be helpful in summarizing this process.

The strategy of each company at any point in time may be viewed as the sum of the policies in effect at that date (Figure 1.2).

Figure 1.1
The Process of Policy Determination

Policies (PL#) result from:	Behavior measured by significant changes in:	Business Policy Variables caused by specific:
	Size or Scale (S)	Environmental
	Geographic Expansion (G)	Conditions (EC#)
	Horizontal Combination (H)	Resources and
	Vertical Integration (V)	Handicaps (R#, H#)
	Product and Industry	Personal Values (PV#)
	Diversification (P)	Social Responsibilities
	Administrative Structure (A)	(SR#)

Figure 1.2

$$\text{Strategy}_t \text{ (Co. A)} = \sum \text{Policies in Effect at t of Co. A}$$

At any moment in any particular economic, technical, or political environment, some policies were always more visible than others. Some were in ascendance; some, in decline. Some were advantageous, and some were handicaps. But all those in effect played some part in shaping the issues with which senior management had to deal and knowledge and control of them was invaluable.

While this method for identifying corporate strategy was developed for the oil industry, there is no reason to believe that it might not be applicable elsewhere.

For this analysis to be credible and consistent between companies and over time, there must be objective measures and classifications of what decisions are strategic as opposed to merely transitory. The preceding paragraphs developed a classification of major strategic decisions that will be used in this study. Behavior that will be considered strategic is that which produces a significant change in the company (as measured in the categories below):

(1) choice of size or scale,
(2) geographic expansion,
(3) horizontal combination (within the industry),
(4) vertical integration,

(5) product and industry diversification, and

(6) administrative structure and policies.

These categories provide one basis for classifying strategic decisions by variables readily recognizable to economists. They have the further advantage of being directly derived from historical changes in the firm and industries which have occurred in the last century with the rise of the large-scale firm. A summary of these changes is presented in Appendix B. For these reasons these six categories will be used in this study to measure the behavior resulting from the strategies, and from the changes in strategies, of the firms described in the following chapters.

1.1 Measures Used in This Book

The measures for size of scale will be volumes (generally thousands of barrels per day, abbreviated TBD hereafter) or dollars (usually in millions) or relative percentages of one or the other. For geographical diversification the measures will be countries, continents, hemispheres, or areas where the firm has activities. For vertical integration it will be an index, specifically devised for this study, of net production to refining volume to marketing volume, expressed as a percentage of the largest of the three. For horizontal concentration the measure will be market share. For product and industry diversification there will be two measures: (1) the balance among petroleum products, usually given as percentages or volumes, and (2) the number, type and financial size of non-petroleum operations. The administrative structure measures will include a description of the method of organization, measures of financial and administrative policies, and consideration of the centralization/decentralization of authority.

1.2 Classification of Strategic Options

Because of the size and complexity of the petroleum industry it is useful to identify the options for each type of strategic choice by the major functional areas of the industry: upstream activities, downstream activities, transportation of crude and products, non-petroleum activities, and organizational activities. Upstream activities include exploration, production, and purchase of crude, crude reserves, leases, and concessions. Downstream activities are crude sale, refining, and marketing. Transportation includes marine and pipeline transport of crude and products. Non-petroleum activities are those which do not produce petroleum fuels or refined products. They include byproducts like petrochemicals, sulfur, real estate ventures, coal, uranium, mining, and other ventures related to petroleum technology or management plus a variety of others in unrelated areas ranging from Exxon's venture capital subsidiary to Mobil's retailing and paper companies. Administrative structure and policies are those affecting the firm as a whole entity rather than any single part. A summary of strategic options is given in Exhibit 1-1.

1.3 Preview of Results

A preview of the results of applying this method to the Majors is given in Tables 1-1 and 1-2. One of the conclusions of this book is that, in addition to strategic similarities, permanent and systematic differences have arisen among the positions and strategies of each of these companies as a competitor. Measurements of these similarities and differences are shown in the comparison Tables 1-1 and 1-2. There are differences between Majors in geographic, product and industry markets, degrees of vertical balance of production, refining, and marketing activities, market shares, administrative policies, and philosophy.

Table 1-1 presents a detailed comparison of the positions of the Majors in 1976 and Table 1-2 presents abbreviated comparisons for 1970, 1960, 1950, 1938, 1927, 1918, and 1912. Among the similarities and differences shown are the participation of all seven companies in nearly all petroleum markets worldwide (Table 1-1.2, 1-1.3, 1-1.5). Exxon is the largest of the companies in most categories except in chemicals and tankers, where Shell is largest (Table 1-1.1). BP is the least profitable, was the only Major without a strong position in the U.S. gasoline market until its acquisition of Sohio in 1969, and is the most committed to the Eastern Hemisphere (Table 1-1.1, 1-1.2b, 1-1.3, 1-1.5D). Gulf is the most committed to the U.S. market (Table 1-1.2). It is also third largest in petrochemicals though it ranks much lower in refining capacity (Table 1-1.1). Mobil is the most diversified; BP and Texaco are the least (Table 1-1.5). Socal, BP, and Gulf were the largest crude sellers (Table 1-1.1). All the Majors are crude short today and are large purchasers from OPEC (Table 1-1.4). But prior to 1973 their vertical positions (volume of production to refining and marketing volumes) were much different (Table 1-2). Exxon, Mobil, and Shell historically were crude short, Texaco was close to evenly balanced, and BP, Gulf and Socal were long on crude.

Exhibit 1.1
Strategic Options for the Major Oil Companies

A. Upstream Policies
1. Size/scale: (a) many small sources; (b) few large sources.
*   2. Geographic Diversification: (a) home market; (b) foreign imports; (c) limited foreign production; (d) worldwide production.
3. Horizontal Concentration (Market Share): (a) total; (b) dominant; (c) significant; (d) minor interests. Does not apply to exploration.
*   4. Vertical integration: (a) spot purchase; (b) long-term purchase; (c) purchase of proved reserves or producing property; (d) new field (wildcat) exploration.
5. Product Diversification: Applicable mainly to purchases (a) light/heavy crude; (b) high/low impurities (sulfur, metals, nitrogen, etc.); (c) paraffinic/asphaltic crude; (d) lubricant/specialty feedstock; (e) gas/condensate/ natural gas liquids
6. Administration: (a) centralization (b) decentralization of authority.

B. Downstream Policies

*   1. Size/scale
a.   Refining: (1) small, local market; (2) large, export
b.   Marketing: (1) bulk sales—high volume, low margin
(2) specialty products—low volume, high margin
*   2. Geographic Diversification:
a.   Refining: (1) near production; (2) near markets; (3) offshore
b.   Marketing: (1) home market; (2) export; (3) multinational; (d) worldwide.

*   3. Horizontal Concentration:
a.   Refining: total, dominant, significant, minor market shares.
b.   Marketing: leading, top few, fringe/specialty marketer.
*   4. Vertical Integration: (1) sell crude; (2) refine and sell at refinery gate; (3) wholesale (bulk) product marketing; (4) retail sales.
*   5. Product Diversification: (1) crude; (2) fuel oil; (3) middle distillates, home heating oil; (4) kerosene, diesel fuel, jet fuel; (5) gasoline, naphtha; (6) lubricants, wax, asphalt, specialties.
6. Administration: (1) central coordinating; (2) decentralized management, local autonomy.

C. Transportation

*   1. Size/scale; (a) low cost, large scale; (b) higher cost, smaller scale, more flexible delivery.
2. Geography : (a) local, shorthaul, coastal; (b) worldwide, longhaul, high seas.
*   3. Horizontal Concentration:
a.   dominant transporter—sell transportation service to others;
b.   carry own products only;
c.   joint ownership of transportation;
d.   contract outside for transportation.
*   4. Vertical Integration
a.   Pipelines: (1) wholly owned; (2) joint venture.
b.   Tankers: (1) owned; (2) bareboat lease; (3) long-term charter; (4) spot/short-term/voyage charter.

(Note: * indicates most important strategic options)

    5.   Product Diversification:
        a.   Mode of transport: (1) pipeline; (2) tanker; (3) rail; (4) barge; (5) truck.
        b.   Product transported: (1) crude oil; (2) refined products.
    6.   Administration: (a) profit center; (b) cost/service center.

D. Non-petroleum

    1.   Size/scale: (a) significant (b) insignificant portion of assets, sales, income, potential liability, or management time requirements.
*   2.   Geographic Diversification: (a) U.S. or home country; (b) foreign.
    3.   Horizontal Concentration: dominant, significant, minor market share.
    4.   Vertical Integration: (a) byproducts, (b) other.
    5.   Product Diversification: (a) related to petroleum products, technology or management skills; (b) unrelated
    6.   Administration: (a) autonomous; (b) integrated into parent

E. Organization

    1.   Size/scale: NA
    2.   Geographic Diversification (a) home country of parent; (b) shareholder distribution.
    3.   Horizontal Concentration: NA
    4.   Vertical Integration: relative importance of production, refining, marketing.
    5.   Product Diversification: identity of central business
    6.   Administration
        a.   centralization/ decentralization of authority;
        b.   finance: dividend, debt, lease, exploration expense, and capital expenditure allocation policies.
        c.   joint venture policy.

Table 1-1.1
Comparisons of the Resources of the Major Oil Companies in 1976

	Exxon	Shell	Gulf	Mobil	Socal	Texaco	BP	Industry
Comparisons of Size								
Assets ($ billion)	36.3	29.6	13.4	18.8	13.8	18.2	14.9	
Net Income ($ million)	2641	2210	816	943	880	870	306	
Capital Expenditures ($ million)	5100	3701	1178	1380	1863	1505	1660	
Employees (000)	126	153	53	200	38	73	137	
Production (000 b/d)	2256	2119-E	703	1123	1945	1911-E	600	59555
Refinery Runs (000 b/d)	4359	4191	1697	2049	2258	2860	1780	56700
Products Marketed (000 b/d)	5353	4642	1609	2264	2339	3277	1820	58790
Crude Purchases (000 b/d)	2839	2613	1319	1033	1596	2104	2900	
Crude Sales (000 b/d)	NA	564	326-E	NA	1385	738-E	1580	
Tankers—# Ships	214	287	76	127	90	161	141	
Tonnage (Mdwt)	NA	32.7	8.2	14.2	11.8	17.3	15.6	320.7
Chemicals Sales ($ million)	3238	3745	1062	1027	798	NA	8529	

Notes: b/d - barrels per day
000 - thousands
Mdwt - millions of deadweight tons
NA - not available
E - estimate by author
F - includes processing by others for Mobil

Sources: Annual reports and statistical summaries of the companies. Industry data is from *BP Statistical Review of the World Oil Industry 1976.*

Table 1-1.2a
1976 Comparisons of Geographical Diversification (% Total Dollars)

	Exxon A/I/C	Shell A/I/C	Gulf A/I/C	Mobil A/I/C	Socal A/I/C	Texaco A/I/C	BP A/I/C
Western Hemisphere	51/63/63	35/40/47	84/ /89	/73	70/68/58	61/59/88	NA
USA	35/49/50	20/22/38	68/54/68	51/60/68	57/52/49	46/55/65	Europe
Canada		/5.5	15/14/20	3P/ /3			Canada
Latin America		/3.5	.9/ /.5	.8P/ /1.2			U.S.A.
Eastern Hemisphere	43/34/37	65/58/46	10/ /9	34P/ /27	30/32/42	39/41/12	Australia
Europe	/23	/37	8.2/ /5.6	19P/ /16			Middle East
Africa			1.4/ /.2	8P/ /7			
Asia & Australia			0.4/ /1.3	6P/ /4			
Middle East			/0.2	1P/ /.5			
Tankers	6/ 3/	5/ /7	5.6/ /1.5				
World	100/100/100 for all seven companies						

Notes:
A—Assets
I—Income
C—Capital Expenditure
P—Petroleum operations only

Table 1-1.2b
1976 Comparisons of Geographical Diversification (% Company Total)
of Oil Supply (including Purchases), Refinery Runs, and Marketing Volumes

	Exxon	Shell	Gulf	Mobil	Socal	Texaco	BP	World
	S / R / M	S / R / M	S / R / M	S / R / M	S / R / M	S / R / M	S / R / M	S / R / M
Western Hemisphere	44/48/52	33/40/40	30/69/71	22/38/41	15/55/56	25/51/55	6/ 5	27/37/38
USA	36/29/34	20/25/25	20/49/51	17/38/38	12/47/50	17/34/40		16/24/29
Canada	7/ 9/ 9	7/ 7/ 8	5/18/16	5/ 0/0.4	3/ 3/NA	3/ 3/ 6	x / 6/ .5	3/ 3/ 3
Venezuela	4/	5	6	0	0	1		
Caribbean		7	2	.1	5	9.3		
Eastern Hemisphere	56/52/48	67/60/60	67/31/29	78/63/59	85/45/44	79/49/45	94/85	73/63/62
Europe	3/40/36	2/40/38	e/22/20	1.2/34/31	e/20/NA	.6/23	5/82/71	2/24/24
U.K. & Ireland		10/ 9	3	6-B	0	5	24	3/ 3
France		11/ 8	3	7-B	3	2	16	4/ 4
Germany		6/ 9	0	11-B	2	6	15	4/ 5
Italy		0/ 0	4	5-B	2	3	15	5/ 3
Netherlands		8/ 3	4	5-B	7	4	15	2/ 1
Africa	5	13/ 2/NA		10/ 5/ 7	0.2	.3/1.3	12	10/ 2/ 2
Nigeria	5/2.9/	9		7	0.2	.2		3
Libya		2		3	13			3
Middle East	39	42/.05/NA	48/ 0/ .3	65/ 4/.8	73	64/10	72	37/ 4/ 2

Table 1-1.2b (Cont.)

1976 Comparisons of Geographical Diversification (% Company Total)
of Oil Supply (including Purchases), Refinery Runs, and Marketing Volumes

	Exxon	Shell	Gulf	Mobil	Socal	Texaco	BP	World
Saudi Arabia	37	2		48	65			14
Iran		16		13	7		50	10
Iraq	2	3		NA				4
Kuwait		7		NA			14	3
Far East	10/ 9/	4/17	9/ 9	1/20/20	12/12/NA	10/15	9/ 8	3/13/14
Japan		5/ 9	2	7-B	7			8/ 9
Australia		3/ 3		4-B	2			1/ 1
Singapore		6		6-B				
Intl. Marine Bunkers							10	

Notes: S - oil supply including purchases
          R - refinery runs
          M - marketing volumes
          B - refinery capacity
          x - some, amount unknown
          e - less than 0.05%

Table 1-1.3a
Comparisons of Horizontal Concentration (% World Market) in Net Production in 1976

	Exxon	Shell	Gulf	Mobil	Socal	Texaco	BP	World (000b/d)
World	3.8	3.6-E	1.2	1.9	3.3	3.2	1.0	59555
Western Hemisphere	6.2	5.9	3.0	2.7	2.8	5.4		15905
USA	8.3	6.2	3.5	3.7	3.8	6.3		9725
Canada	9.6	4.6	4.4	4.2	3.9	7.0		1605
Venezuela (Industry nationalized January 1, 1976.)								
Eastern Hemisphere	6.3	2.9	0.5	1.7	3.4	2.5	1.4	43650
Europe		8.5	0.2	2.7	0.1	2.9	20.0	905
Middle East		6.2-E		2.3	12.3	2.7	0.8	22175
Saudi Arabia	14.0			5.3	17.8	6.4-E		8525
Iran								5920
Iraq	2.2							2280
Kuwait								1950
Africa	1.9	9.1-E	3.8	3.2	0.1	0.2	4.9	5850
Nigeria		21.1	6.4	7.3	0.4	0.4	14-E	2065
Libya	10.1	5.1		4.6				1930
Algeria								1075
Far East	15.5	10.3		0.8	22.5	21.3		1865
Indonesia				0.9	27.9	25.8		1505
China								1500
USSR								10315

Note: E—estimate by author

Figures are percentages calculated by dividing company production data from annual reports by world production (righthand column) taken from *BP Statistical Review of the World Oil Industry 1976*.

Table 1-1.3b

Comparisons of Horizontal Concentration (Market Share)
in Refining and Marketing in 1976

	Exxon R / M	Shell R / M	Gulf R / M	Mobil R / M	Socal R / M	Texaco R / M	BP R / M	World (000 b/d) R / M
World	7.7/9.1	7.4/7.9	2.9/2.7	3.6/3.9	4.0/4.0	5.0/5.6	3.1/3.1	56700/58790
Western Hemisphere	10/12.2	8.0/8.2	5.5/5.1	3.8/4.1	6.0/5.8	7.0/8.0	0.5/0.4	20805/22610
USA	9.5/10.8	5.0/6.7	6.0/4.9	3.7/5.0	7.9/6.9	7.2/7.9		13425/16980
Canada	24/25.5	16.5/19.7	17.2/14.4	0.4	4.6/NA	8.5/NA	5.8/5.6E	1710/1790
Eastern Hemisphere	9.4/10.2	10.4/11	2.1/1.8	5.2/5.3	4.2/4.0	5.7/5.8	6.9/6.8	24320/25330
Europe	12.7/13.5	12.5/12.2	2.7/2.2	5.1/4.9	3.3/NA	4.7/NA	10.6/9.3	13625/14340
U.K. & Ireland		21.4/20.4	2.6/NA	5.8B/NA		6.7/NA	21.7/15.2	1933E/1975
France		20.8/15.8	2.3/NA	5.6B/NA		2.4/NA	12.6/11.7	2227E/2385
West Germany		12.9/14.7		9.8B/NA		8.1/NA	13/9.7	1994E/2885
Italy			2.2/NA	3.4B/NA		0.4/NA	NA/NA	2854E/1970
Netherlands		27.9/18.5	5.2/NA	6.4B/NA		6.7/NA	20.8/NA	1248/800
Africa	3.8/NA	7.6/NA		9.2/14.8	8.7/NA	3.7/NA	2.4/4.7E	1020/1105
Middle East	3.8/NA	0.9/NA	0.3	3.5/1.3	8.7/NA	12.2/NA	2.4/4.7E	2300/1445
Far East	5.6/NA	9.9/NA	2.0/1.6	5.4/5/4	3.7/NA	5.9/NA	2.2/1.9	7375/8440
Japan	5.6/NA	4.8/8.0	0.7/NA	5.2/NA	3.6/NA	NA/NA	NA/NA	4305/1595
Australia		20.6/18.7		17.6/NA		NA/NA	NA/NA	655/765
USSR, E. Europe, & China								11575/10850

Notes: R- refinery runs

M- products marketed

E - estimate by author

B - refinery capacity

Figures are calculated as in Table 1-1.3a above.

Table 1-1.4
Comparisons of Vertical Integration in 1976

	Exxon	Shell	Gulf	Mobil	Socal	Texaco	BP	World
World	42:81:100	52E:90:100	42:100:97	50:91:100	83:97:100	58:87:100	33:98:100	100:95:99
Western Hemisphere	36:75:100	37:90:100	41:100:100	47:85:100	34:94:100	47:81:100	NA:100:100	70:92:100
USA	44:70:100	53:91:100	41:98:100	42:91:100	32:91:100	46:73:100		57:79:100
Canada	52:90:100	21:80:100	24:100:88	100:0:12	81:100:NA	77:100:NA	NA:100:100	90:96:100
Latin America	4:84:100	0:96:100	100:62:90	0:5:100	0:100:NA	44:100:NA		81:100:68
Eastern Hemisphere	58:88:100	62E:90:100	45:100:91	54:95:100	100:68:68	74:95:100	35:98:100	100:56:58
Europe	3:89:100	4:97:100	0:100:87	3:99:100		4:100:NA	13:100:93	6:95:100
Africa		100:15:NA	100:0:NA	100:50:87		32:100:NA		100:17:19
Middle East	100:80:NA	100:2:NA	0:0:100	100:16:4		100:46:NA	100E:19:29	100:27:17
Far East	57:100:NA	26:100:NA	0:100:96	3:87:100		91:100:NA	0:100:100	22:87:100
Japan	0:100:NA	0:94:100	0:100:NA	0:100:NA				0:83:100
Australia	0:100:NA	0:94:100	0:0:NA	0:100:NA				56:86:100
USSR, E. Europe, China								100:95:89

$$\text{Vertical Integration Index} = \frac{P : R : M}{\max(P, R, M)}$$

E—estimate
P—net production
R—refinery runs
M—marketing volumes

Table 1-1.5

Comparisons of Product Diversification in 1976

	Exxon	Shell	Mobil	Socal	Texaco	BP	Gulf
A. Revenues by Source (% total)							
Oil and Gas	91.5	83	78	97	NA	94	91
Chemicals	6.2	11	4	3		5	6
Associated Companies		0.7	0.4		2		0.2
Interest & Dividends	1.8	1.1	0.3				1
Other Non-Petroleum	0.5	4.8	11.2				1.5
Major source		Metals	Marcor			Animal Feed	Minerals
Percent		2.3	11.2			0.9	0.9
B. Net Income by Source							
Petroleum	70 Ex&P 18 R&M	NA	82	94	91	87E	84
Chemicals	8.1		9	Marcor	9	13E	18
Nuclear			8	6			
Minerals	(1.3)						
Coal & Nuclear	(1.6)						
Intl. Marine	3.5						
Non-operating	2.5						(0.6)

Notes: Ex—exploration
P—production
R—refining
M—marketing
E—estimate

Table 1-1.5 (Cont.)
Comparisons of Product Diversification in 1976

	Exxon	Shell	Gulf	Mobil	Socal	Texaco	BP
	A / C	A / C	A / C	A / C	A / C	A / C	A / C
C. Fixed Assets (A) and Capital Expenditures (C) by Function (% total)							
Petroleum							
Exploration	89.5/90.1	82.2/80.1	83.8/80.5	81.5/81.6	NA/NA	95.9/89.9	94.2/90.1
Oil Rights & Concessions	/8.3	/16	/20.9	/13.4	/29.7	/17.4	
Production	/8.3	6.2/7.0		/9.5			
Refining	33.8/41	23.2/29.3	40.6/37	33.8/27	41/24.1	52.5/37.8	52.3/69
Marketing	19/6.8	23.8/11	14.2/5.8	19.6/3.8	29/26.2	19.2/16.7	14.4/6.3
Pipelines	14.4/4.0	20.5/9.1	14.4/4.8	16.5/4.3	15/4.8	13.5/5.1	21.5/13
Tankers	/13E	NA/NA	9.6/6.9	4.6/0.1	12/6.2	2.4/1.6	
Natural Gas & Liquids	/5.2	8.5/5.9		5.7/4.0		6.5/7.6	6.0/1.2
Tar Sands - Canada	NA/1.0		3.0/NA				
Oil Shale	NA/3.7		/5.7				
Other Petroleum	2.3/10.7		/0.2				
Non-Petroleum							
Chemicals	10.5/7.6+X	13.6/19.1	14.9/19.5	18.5/18.4	NA	4.1/10.1	5.8/9.9
Metals/Minerals	7.4/7.6	12.9/18.6	10.7/8.6	4.2/7.2		4.1/10.1	5.8/9.9
Coal	x/x	0.7/0.5	4.2/				
Uranium	x/x	x/x	/2.4				
Nuclear	x/x		/2.6				
Paper	x/x						
Retailing				/5.4			
Other	3.1/			14.3/5.8			

Notes: x—some, but amount unknown

Table 1-1.5 (Cont.)
Comparisons of Product Diversification in 1976
Percent of Total Volume by Area

	Exxon T/US/Cn/WH/Eur/EH	Shell T	Gulf T/US/Cn/Other	Mobil T/US/Foreign	Socal US	Texaco T/US/Other WH/EH	BP NA
D. Product Mix (% total)							
Gasoline	27/31/35/34/21/22	32.1	42/58/41/15	34/51/24	35	30/42/40/21	
Naphtha	.9						
Jet Fuel	5.9/7.5/5.7/6.5/4.7/4.8	9.4	4.5/5/1/6		12	7.4/10/5/5	
Kerosene	27/16/33/25/38/25	25.3					
Gas/Diesel Oil				32/27/35		25/20/29/29	
Distillates			28/26/32/30		21		
Residual	28/30/15/30/29/32	24.4	19/7/15/38	21/8/29	22	27/19/22/39	
Lubes	1.5/0		1.3/1.3/.8/1.6			1.1/1.3/1.3/.7	
Other		8.8			10	Subsidiaries only—excludes Caltex	
Supply Sales (wholesale)	20						
Specialty Petroleum Products	11/12/12/5/8/17						

E. Principal Non-Petroleum Activities

Exxon	Shell	Gulf	Mobil	Socal	Texaco	BP
chemicals	chemicals	chemicals	chemicals	chemicals	chemicals	chemicals
minerals	metals	tar sands	retailing	asphalt	tar sands	animal feed
uranium	agriculture	oil shale	paper	oil shale	oil shale	coal
mining	nuclear reactors	coal	solar	gilsonite	real estate	uranium exploration
nuclear fuel fabricating		uranium	coal exploration	metals & minerals	coal	metals exploration
coal	coal	solvent refined coal	uranium exploration	land development	agriculture	
venture capital	exploration	nuclear reactors	real estate			
tar sands	asphalt	nuclear fuel reprocessing	phosphate rock			
		real estate				

Notes: T - total
US - United States
Cn - Canada
WH - Western Hemisphere
EH - Eastern Hemisphere
Eur - Europe

Table 1-1.6
Comparisons of Administrative Policies

	Exxon	Shell	Gulf	Mobil	Socal	Texaco	BP	CMB Group	Non-Majors
A. Finance									
Debt-Equity Ratio	.20	.39	.17	.21	.22	.29	.43	.33	.49
Payout Ratio	.46	.29	.41	.38	.42	.62		.41	.36
Return on Assets	7.3	6.5	6.1	5.9	6.3	4.8	1.8	5.8	5.5
Return on Equity	14.3	18.7	11.7	12.9	12.6	9.7	6.3	12.5	12.2

B. Structure
Principal Divisions of each Organization

Exxon	Shell	Gulf	Mobil	Socal	Texaco	BP
chemicals	oil	energy & minerals	paper	chemicals	chemicals	trading
research & engineering	chemicals	refining & marketing	retailing	asphalt	coal	coal
exploration	metals	chemicals	chemicals	Intl. sales	gas	Ex. & P.
USA	coal	trading & transport	oil	U.S.	marine	research
Canada	nuclear	science & technology	Ex. & P.	W. Opns.	USA	minerals
Middle East	marine	real estate	U.S. R&M	E.& S.Opns.	Canada	U.K.
Intls. (whsl & marine)	research	Canada	Intl.	*interests*	U.K.	Europe
Far East	gas		Middle East	AMOSEAS (expl.)	*interests*	Africa
Europe	USA			AMAX (metals)	AMOSEAS	Middle East
Latin America	Canada			Aramco	Aramco	Far East
	others by country			Caltex	Caltex	W. Hemisphere
						chemicals

	Exxon	Shell	Gulf	Mobil	Socal	Texaco	BP
C. Degree of Autonomy of Subsidiaries	regionally & functionally decentralized	decentralized by country	highly centralized	intermediate	intermediate	notoriously centralized	unknown

Notes: Ex. & P.—exploration and production
R & M—refining and marketing
Intl.—international
whsl—wholesale

Table 1-2
Prior Year Comparisons

	Shell	Exxon	Gulf	Mobil	Socal	Texaco	BP	CMB Group	Industry
A. Total Assets ($ million not adjusted for inflation)									
1970	14289	19242	8672	7921	6571	9923	6319	110237	
1960	8898	10090	3843	3455	2782	3647	2016	41649	
1950	2380	4187	1344	1610	1232	1449	750	18219	
1938	2116	2045	547	923	605	605	255	7690	
1927	1900	1426	553	678S 159V	579	325	220	NA	
1918	816	691	173	234S 75V	145	175	56	NA	
1912	434	369	32	91S	67	60	12	NA	
B. Net Income ($ million not adjusted for inflation)									
1970	890	1312	550	483	455	822	274	6943	
1960	498	689	330	183	266	392	174	2905	
1950	143	408	111	128	150	149	95	1739	
1938	160	76	13	40	29	25	53	300	
1927	254	40	14	11S	40	20	15	NA	
1918	163	57	15	29S 4.9V	15	7	6.4	NA	
1912	65	35	4.5	15S 4.2V	7	6.6	(1.3)	NA	

Notes: S—Socony   V—Vacuum

Table 1-2 (Cont.)

C. Total Capital and Exploration Expenditures ($ million unadjusted for inflation). For 1938 and earlier figures are additions to property, plant and equipment (ΔPPE), which exclude exploration and non-physical spending.

	Shell	Exxon	Gulf	Mobil	Socal	Texaco	BP		Industry
1970	1878	2116	1149	880	900	1148	774	12865	21465
1960	1061	946	453	320	370	627	305	5610	11610
1950	238	385	137	145	183	175	157	1894	5325
ΔPPE 1938	NA	169	52	86	27.1	49	NA	668	NA
1927	NA	NA	67	NA	27.5	47	NA	NA	NA
1918	NA	NA	19	NA	10.8	NA	NA	NA	NA
1912	NA	NA	5.6	NA	NA	NA	NA	NA	NA

ΔPPE - Change in property, plant, and equipment.

D. Worldwide Net Production (000 b/d)

	Shell	Exxon	Gulf	Mobil	Socal	Texaco	BP	Industry
1970	3244	4665	2934	1573	2404	2987	3980	46130
1960	1752	2196	1464	713	928	1234	1523	21348
1950	1005	1289	558	339	461	490	824	10407
1938	598	561	173	159	127	144	234	5446
1927	328	190	150	82	63	128	112	3456
1918	83	26	51	NA	62	36	18	1379
1912	40	11	20	0	11	18	0	966

E. Worldwide Refinery Runs (000 b/d)

	Shell	Exxon	Gulf	Mobil	Socal	Texaco	BP	Industry
1970	5067	5270	1591	1884	1741	2719	2080	45625
1960	2549	2871	849	948	836	1327	894	21308
1950	1266	1640	448	583	428	509	630	10537
1938	640	818	209	308	149	272	230	5368
1927	290	434	144	145	106	159	114	NA
1918	80	163	47	NA	82	64	16	
1912	30	101	NA	6	51	27	0	

Table 1-2 (Cont.)

F. Worldwide Product Sales (000 b/d)

	Shell	Exxon	Gulf	Mobil	Socal	Texaco	BP	
1970	5246	4521	1663	2145	2045	2917	2118	NA
1960	2459	2608	842	1129	1001	1368	860	
1950	1320	1659	478	603	484	574	NA	
1938	780	799	193	374	163	222	200	
1927	400	430	120	NA	NA	167	NA	
1918	125	62	45	NA	85	NA	NA	
1912	75	43	NA	NA	61	NA	0	

G. Vertical Integration

	Shell	Exxon	Gulf	Mobil	Socal	Texaco	BP
1970	62:97:100	89:100:88	100:54:57	73:88:100	100:72:85	100:91:98	100:52:53
1960	69:100:96	76:100:91	100:58:58	63:84:100	93:84:100	90:97:100	100:59:56
1950	76:96:100	78:99:100	100:80:86	56:97:100	95:98:100	85:89:100	92:100:NA
1938	77:82:100	69:100:98	83:100:92	43:82:100	78:91:100	53:100:82	98:100:NA
1927	82:73:100	44:100:99	100:96:80	57:100:NA	59:100:NA	77:95:100	NA
1918	66:64:100	16:100:38	100:92:88	NA	73:98:100	56:100:NA	NA
1912	53:40:100	11:100:43	NA	0:100:NA	18:84:100	67:100:NA	NA

Notes: P—worldwide net production from D above
R—worldwide refinery runs from E above
M—worldwide product sales from F above
NA—not available
*—Index = $\dfrac{P:R:M}{\max (P,R,M)}$

Data Sources: Company data are from annual reports; industry data are from *BP Statistical Review* (various years); CMB Group Data are from Chase Manhattan Bank. *Capital Investments of the World Petroleum Industry* and *Financial Analysis of a Group of Petroleum Companies*, both printed annually.

2

Before Oil

This chapter serves as a background for the stage upon which the seven major companies will shortly act out their separate roles. In it are outlined important influences from outside the industry which affected its origin and early years along with brief descriptions of the major changes in each functional area of the industry since that time. Both these outside influences and internal changes form the common environment which all the companies share. They are some of the environmental variables described in the previous chapter as essential to the understanding of industry structure, conduct, and performance.

2.1 The Origins of the Industry

The oil industry did not arise spontaneously. It had a beginning rooted in the past from which it grew. The business policy model asserts a strong explanatory role for environmental influences external to the company or industry. This chapter traces four of these environmental conditions and suggests that they are the origin of several of the significant characteristics of the early years of the petroleum industry.

The first environmental condition, EC1,* is that the rise of the petroleum industry around the demand for kerosene, as compared to the many other products ranging from fuel to lubricants which could have been made from petroleum with the technology of the time, occurred because of the search for a substitute for whale oil for lighting. A corollary is that the growth of the demand for kerosene was due to its superiority to oil in lamps.

The second is that the initial technology for the industry was borrowed (EC2). Refining technology for petroleum was borrowed from the coal industry's recently developed technology to refine kerosene from shale and coal. Drilling technology came from the drillers of salt and artesian water wells. Storage, handling, and transportation technology came from the open tanks, barrels, and tins that were used for storing, moving, and handling water and dry goods.

---

*Environmental conditions are referenced in the text as (EC#) and listed in Exhibit 2-1.

The third is that the business practices and ethics of the industry, which emphasized secrecy and Darwinian competiton—that is, the survival of the fittest by any means—were the business practices of the day, which extolled the pioneering virtues of individual freedom, self-reliance, privacy, and triumph over any obstacles (EC3). A corollary is that the lack of government interference in the industry reflected this same laissez faire attitude toward minimum interference with individual freedom.

The fourth is that the overdrilling, waste, and boom and bust cycles of the industry resulted from the inappropriate adaptation of the property laws of surface and mineral rights to petroleum in the form of the rule of capture (EC4).

These explanations are set forth, not because there is any doubt about their validity, but because their explanatory power over the course of the development of the industry is enormous. Some, like the rule of capture, directly affect the industry today. The peculiar structure of the U.S. producing industry reflects the rule of capture. Others triggered changes which affect the present. The ethics and practices of the nineteenth century produced the antitrust laws and the Standard Oil divestiture of 1911, the single most important determinant of the structure of the industry today. The public reaction to those ethics and practices produced a public image and congressional perception of the industry that is in times of stress little changed since the turn of the century. Still other environmental conditions, such as the borrowed technology and the demand for kerosene, were merely jumping off points for the development of new technologies and products. They remain as memories (oil for the lamps of China), units of measure (the barrel), and important influences on the early development of the strategies of the seven major companies (the commitment of Standard Oil to kerosene allowed Texaco to gain a grip on the gasoline market; Mobil sold lubricants to Europe and kerosene to the Orient). The subsequent sections of this chapter discuss each of these four environmental conditions.

2.2 Kerosene and its Sources

The technical revolution in lighting came early in the Industrial Revolution. High temperature distillation of coal produced manufactured gas, which was a much larger industry in the first half of the nineteenth century than kerosene.[1] High temperature distillation preceded, and inspired, low temperature distillation, which produced kerosene and other petroleum products. Refining was essentially the treatment of these new materials and products with heat and chemicals.

Manufactured gas (EC1a) was introduced commercially in 1802. By 1812 the streets of the city of London were the first in the world to be lighted by gas, courtesy of the recently formed London and Westminster Gas Light and Coke Company.[2] In 1850 fifty or more cities in the U.S. were illuminated by gas, but distribution was limited almost entirely to streetlights, public buildings, and factories.[3] For homes and rural areas better lamplight was needed to counter the bright lights of the cities.

The home and rural markets in 1859 were clearly ready for a new discovery. For a generation men had been searching for better illuminants than whale oil, tallow candles, camphene and other animal and vegetable oils. Whale oil was too expensive and the other sources gave poor light, smoked, had disagreeable odors, and sometimes exploded. An illuminating oil was needed that burned brightly and was safe and inexpensive. Various substitutes had been devised for whale oil, but none of them were very successful until in 1850 James Young patented a process for extracting a crude liquid hydrocarbon by low temperature distillation from coal and oil shale (EC2a).[4]

By distilling the liquid he produced four basic petroleum products: naphtha (gasoline), the top fraction; coal oil or kerosene, the desired product which was the second fraction; a fraction suitable for making lubricants; and paraffin wax.[5]

All of these products immediately found markets. Naphtha was used as an industrial solvent, for example, to dissolve rubber. Coal oil was used for illumination, replacing whale oil, camphene, and lard oils. Lubricating oils also replaced whale oil and other oils in machinery and leather dressings. Paraffin replaced tallow candles.

Coal oil gave a brilliant light—five candlepower compared to 1.4 for the best camphene lamp (EC2b).[6] It was safe, odor free, and cheaper than camphene, the best common alternative. As with many new products, quality was the early problem. Impure or improperly refined or improperly burned products performed poorly. Variation in raw material sources was also a source of quality problems. Refining was in its infancy. Scientists had not yet discovered organic chemistry. Distillation produced a wide spectrum of exciting new products, some more dangerous and explosive than others. Yields of products from different hydrocarbon sources varied widely.

The technical revolution in lighting had begun with new processes, high and low temperature distillation, which produced a variety of new and useful products. The two most important of these, manufactured gas and kerosene, revolutionized lighting throughout the world. In just another generation a still greater invention, electricity, would replace them both. But in the middle of the nineteenth century, they were huge achievements and demand was enormous. As W.H. Libby of Standard Oil wrote to the governor general of India about kerosene:

> I may claim for petroleum that it is something of a civilizer, as promoting among the poorest classes of these countries a host of evening occupations, industrial, educational, and

recreative, not feasible prior to its introduction; and if it has brought a fair reward to the capital ventured in its development, it has also carried more cheap comfort into more poor homes than almost any discovery of modern times."[7]

In the 1850's scores of coal oil manufacturers sprang up in the cities and towns of the United States, using partly Young's techniques and partly those of others.[8] While these innovations developed the market for kerosene made from coal, other entrepreneurs began to produce and refine petroleum gathered from natural seepages and salt water wells. Demand for kerosene in 1859 was so great that the product sold for over $42 per barrel more than the crude from which it was refined.[9] The market was clearly ready to reward anyone who could find a supply of inexpensive crude petroleum (EC1b).

Inspired by an optimistic report by Benjamin Silliman, professor of chemistry at Yale in 1855, on the economic possibilities of refining crude oil into valuable products, investors in New Haven were susceptible to an issue of shares for $300,000 by the Pennsylvania Rock Oil Company of Connecticut. Silliman was its first president. The company proposed to drill for petroleum on land leased at oil springs on the land of a lumber company in Titusville, Pennsylvania. A man was needed to go to Titusville, perfect the title to the lease, and conduct the operation. Edwin L. Drake, a conductor on the New York and New Haven Railroad since 1849, was residing in the Tontine Hotel in New Haven, recovering from an illness during the summer of 1857 when he was hired at $1000 a year. At 38 he possessed no drilling or business experience but had two visible assets: he was immediately available and, since he could obtain a railroad pass to get to Titusville, his services would be cheap.[10]

With Drake's discovery of the first commercial oil well (EC1c) on August 27, 1859 in Titusville, Pennsylvania, the petroleum industry was born. Drake was not the discoverer of petroleum. It had been known to exist and had been used for medicinal and caulking purposes worldwide as early as 3000 B.C. In the U.S. it was often found as an exudate in the brine wells around Kanawha, West Virginia and Tarentum, Pennsylvania. Brine was found by boring wells into underground reservoirs. Wells as deep as 1000-2000 feet had been drilled in the 1820's and 1830's.[11] Drake's discovery was that large quantities of petroleum existed in underground reservoirs which could be tapped by drilling similar to the boring for salt water wells.

2.3 Borrowed Technology

Most of the technology for the new industry was borrowed (EC2). Drillers of wells supplied the tools, boring techniques, methods of power generation, and storage facilities for petroleum drilling and production. Barrel makers provided the first container for storage facilities for petroleum drilling and production (EC2d). From the coal industry refiners got equipment, methods of distillation, and treatment of distillates with sulfuric acid and caustic soda (EC2c). Petroleum products were marketed at first through the established general purpose channels of trade (EC2e)—along with shoes and ships and sealing wax, cabbages and kings.[12] Lamps used for whale oil, camphene, and other oils needed only minor modifications to accommodate the new fuel.

The technology borrowed from others was rapidly developed beyond the recognition of its predecessors. By the turn of the century exploration was becoming more scientific. Geologists would begin to appear after World War I. Drilling tools were heavier and more effective. Wells were cased with pipe. Iron replaced wood in tanks and elsewhere. Refiners improved stills and condensers, learned to crack petroleum to increase kerosene yield, and discovered the use of charcoal to deodorize lubricants. Products were packaged in cans and cases. Brand names and trademarks appeared by the score. Tank and barrel makers and repairers gave way to tank cars, pipelines, barges, and steamers, leaving the industry with memories and a unit of measure (the 42-gallon barrel) that is a reminder of the rustic past.

2.4 Laissez Faire Competitors

After the Civil War the nation got back to the business of pioneering, both geographically as settlers flooded westward and industrially as the manufacturing sector of the economy rapidly expanded. The spirit of laissez faire individualism was never higher. Thousands of new businesses were born to serve rapidly expanding markets. It was the age of the entrepreneur. Few businesses were publicly owned and even among those the constraints on management were minimal. Businesses were managed by their owners for profit as they saw fit.

The Industrial Revolution was something new in the United States (EC3a), having been imported from Britain only two or three decades before. The country had passed from the colonial era to the age of the pioneer, but it would not begin to think of itself as an industrial nation until the late 1880's and 1890's. Procedures and norms for behavior in these newly emerging patterns of business were far from worked out. The country was, compared to the European nations, still quite new. Large-scale nationwide industry was even newer; it had just begun to appear with the railroads (EC3a). Entrepreneurial businesses were new. On top of all this, the people in this new and exciting petroleum industry were young. Many were veterans of the Civil War seeking their fortunes (EC3b). With its risk, danger, and arduous labor, petroleum production, more than refining or marketing, was a young man's business.

It is not, therefore, surprising that businessmen were concerned primarily with the short run, with the commerce of the moment, and not with the long-term return on large fixed investments. The idea was to convert the relatively small amounts of investment required into cash as quickly as possible. It was a commercial merchant's view of business: charge what the market will bear and turn over the inventory as rapidly as possible (EC3c, d).

In the absence of nationwide markets (EC3e) and central exchanges which balanced supply and demand at a single price for all buyers and sellers, each transaction between buyer and seller depended on local conditions and information at the moment, producing violent price movements, especially in the producing industry (EC3f). Monthly average prices for crude oil in 1861 ranged from 10¢ to $2.25 and from $4.00 to $12.25 in 1864.[13] Annual averages from 1860-1906 are presented in Exhibit 2-2. Due to these two factors, local markets (EC3e) and a short-term view of the business (EC3d), and a third, the rule of capture (EC4) which will be described below, market conditions in the early days of the industry were chaotic as inventories were marked down and sold or bid up to the sky (EC3f).

This vigorous price competition was accompanied by a lack of restraint in other business practices (EC3). Competitors were obstacles to be overcome by any means at hand, including some practices construed today as deceptive, fraudulent, monopolistic, or criminal (EC3g). Many practices were simply hard fought competition with the spoils going to the victor. Nineteenth-century business ethics allowed any competitive practice not clearly prohibited by law, and some that were.[14] Often there was little protection in the courts or the law. For example, many approved the circumvention of patent rights. Col. E.A.L. Roberts, the inventor of the torpedo, an explosive device for expanding the yield of wells, instituted hundreds of suits to protect the monopoly granted in his patent.[15] As applied to commercial practice, laissez faire individualism endorsed freedom of entry into any occupation, the sanctity of private property, the obligation of an owner to manage, and the right to secrecy.[16] With inadequate legal protection, often the only way to protect one's commercial advantages from competitors was secrecy (EC3h).

This was especially true in the producing industry, where the rule of capture allowed any leaseholder to share in a bonanza beneath his property and where discovery and lease sale information had high commercial value. Secret companies, nominees, silent partners, and other subterranean practices served to protect (sometimes) the actions and intentions of a buyer or seller (EC3h). This was especially true of Standard Oil, whose presence or interest could send bids soaring overnight. Secrecy in business practice carried over into the legal arena, where businessmen testified to the legal truth, and nothing more, a practice which surely did as much damage to the public image of business as any (EC3i).

The response to secrecy and absence of information, of course, was to spy upon one's competition and to install what today may be euphemistically called primitive management information systems to keep track of competitors' shipments and sales (EC3j). Standard's manager in Fresno, California once reported, "Our Assistant Cashier goes to the Railroad Depot every day at half past 4 o'clock to pay freight billings, and while there he makes a pretense of checking our freight.... The shipments of our competitors... are all put into the same Railroad Warehouse, and while checking our freight it is an easy matter for him to secure the names to whom our competitors are shipping and the product shipped."[17] Every week the local managers sent the district manager an itemized statement of his own sales and an estimate of his competitors.[18]

These characterisics of the business environment in the 1860's (EC3-3j)—the pioneering spirit, entrepreneurial sovereignty, the newness of everything, the short run view, the absence of national markets, the lack of restraint and legal protection—plus the risky nature of the discovery business and the rule of capture produced the chaos, the ruthlessness, and the secrecy of the early years in the petroleum industry which have had such a lasting effect upon its public image.

2.5 The Rule of Capture

A final factor whose contribution to the structure of the U.S. industry is difficult to overstate is the rule of capture (EC4). This application of the common law to the production of petroleum is, more than anything, responsible for the peculiar atomistic structure of the U.S. producing industry, its waste (suboptimal recovery) and high cost from overdrilling, and the chaotic boom and bust cycles that prevailed until proration in the 1930's.

The rule of capture was based upon two legal premises. The first was that the owner of land also owned the rights to minerals beneath the land, a concept long accepted in Anglo-Saxon countries (EC4a). Because oil is a fugacious mineral which can move across property boundaries (EC4b), judges sanctioned the practice of oilmen that the landowner or leaseholder who brought it to the surface owned it.[19] In other words, it was like a fish or animal with the captor receiving title (the second legal premise)—hence, the rule of capture. The rule of capture meant that oil in the ground had no economic value to a landowner as a mineral asset. Title came only after it was brought to the surface (EC4c).

This encouraged drilling as many wells as possible to capture production before one's competitors did (EC4d). The results were evanescent flush production, rapid exhaustion of reservoirs, waste of gas, and minimum recovery of oil (EC4e). Competition led to maximum inefficiency in production and the boom and bust syndrome of wildly fluctuating volumes and prices. The average life expectancy of a producing well in 1865 was about 18 months.[20] Like gold mining claims oil leases were abandoned when the vein played out. Aptly named Pithole City was the typical boomtown. It grew from nothing when oil was discovered in the spring of 1864 to 15,000 in September 1865, including 3,000 teamsters and thousands of Civil War veterans. Its mail volume of 3,000 letters per day was exceeded only by Pittsburgh and Philadelphia in Pennsylvania. It was called "the sewer city" by the *Nation* (a newspaper). Speculators, stock schemes and scoundrels abounded. Output peaked in late 1865 at 6,000 barrels per day. One year later production had slumped to 2,000 b/d. The exodus of people began in earnest in 1866. Population dropped to 3,000 at the end of 1867 as production had all but ceased. In June 1968 the rest of the population moved on to new fields at Shamburg and Pleasantville and Pithole City was no more.[21]

The effects of the rule of capture were immediately recognized, even though reservoir engineering techniques necessary to define the boundaries of reservoirs did not yet exist (EC4b). But in more than a century of petroleum industry experience, an acceptable alternative to the rule of capture has not been found. The motivation of the landowner is to subdivide his property as small as he possibly can to maximize lease revenue. Interests in a single lease as small as 1/128 have been recorded. The higher mathematics of fractions have sometimes been a source of difficulty for oilmen. It was not unknown for the overzealous to sell fractions totaling more than 100% of a lease—for example, 17 shares or more of 1/16 each.[22] In addition each well may tap three or four producing zones at different depths. Reservoirs in different zones may have varying perimeters, types of crude, and rates of flow. Thus allocation of reservoir ownership according to surface rights has frequently proved impossible. Also, after flush production is gone, the damage to the reservoir may be irreparable, requiring unitized management of a field before producing wells are drilled.

The goal of unitized management in advance of drilling has proved elusive. Unitization of already producing fields in the U.S. producing industry has for the most part proved impossible, except through sale of the lease. For existing fields only compulsory unitization seems to offer hope of increasing the efficiency of recovery. For prospective producing areas the only solutions have been to award leases large enough so that overlapping claims in a discovery are minimal and, where overlaps occur, to encourage unitized field management through joint producing ventures. Critics suggest that the worldwide web of joint ventures in production have been used by the industry to restrict production and, as a consequence, price competion.

2.6 Evolution of the Environment

The reader may wish to reflect for a moment on the appeal of the propositions he has just been offered.

The theory implicit in the narrative above is that the state of the industry at any time owes a great deal to singular historical events which may have both large and lasting influence, the rule of capture for example, or in other cases, large but transitory influence, as did borrowed technology. Because the causative connection of these events to the consequences cited is empirically verifiable, unless their influence can be refuted, it cannot be ignored. Models which seek to explain observed outcomes without the aid of these events are likely to be misspecified and their policy prescriptions, erroneous. This implicit theory does not suggest that these variables or events alone are sufficient to explain observed behavior, but merely that they are necessary.

Examples of misspecification would be attributing the large kerosene component of the product mix to the physical composition of paraffinic crude, ignoring the demand for kerosene as an illuminant; or attributing the excesses of Standard Oil to a group of evil men or to monopoly power, without attention to the ethical norms of the day; or attributing the atomistic structure of the producing industry to large numbers of Civil War veterans who wished to enter the industry, ignoring the rule of capture.

If the effects of these variables are plausible and significant in producing the characteristics of the industry observed at the time of its origin (i.e., kerosene dominance in product mix, initial technology, business ethics and distribution channels, and industry structure), then the next step is to examine the evolution of the environment of the industry for subsequent events of singular significance in explaining what one observes about the industry at later times.

A set of such events or environmental conditions is presented in Exhibit 2-3. They mark the relatively small number of major changes in the environment of the industry external to the firms affected and common to the industry as a whole from the time of its origin to the present. They have in common the effect of causing the business strategies of a number of firms in the industry to change, as each adapts to the new environment. By no means do they explain all of the changes in the industry or completely explain even one. But they are the most important changes affecting the whole industry in the past century. As such they cannot be ignored, as they have frequently been in industry analyses, particularly in the antitrust area.

Exhibit 2-1
Environmental Conditions Before 1885

EC 1	Worldwide demand for a substitute for whale oil for lighting.
EC 1a	Introduction of manufactured gas for lighting in cities, 1802.
EC 1b	Demand for source of inexpensive crude petroleum to produce kerosene, 1850's.
EC 1c	Drake's discovery of the first commercial oil well, August 27, 1859: i.e., that large quantities of petroleum existed in underground reservoirs which could be discovered by drilling.
EC 2	Initial petroleum technology borrowed: refining from shale and coal; drilling technology from salt and water wells; storage, handling, and transportation technology from tanks, barrels, and tins used for water and dry goods.
EC 2a	Low temperature distillation of shale by James Young in 1850, producing gasoline, kerosene, lubricants and wax.
EC 2b	Brilliant light produced by kerosene refined from coal.
EC 2c	Treatment of distillates with sulfuric acid and caustic soda in refining.
EC 2d	Barrel as the first container and unit of measure for storage and transportation.
EC 2e	Marketing of petroleum products through general purpose channels of trade.
EC 3	Business practices and ethics, emphasizing secrecy and survival of the fittest, from the pioneering laissez faire virtues of individual freedom, self-reliance, privacy, and triumph over any obstacle.
EC 3a	Industrial Revolution—newness of large-scale enterprise.
EC 3b	End of Civil War releases many young men seeking their fortunes.
EC 3c	Commercial merchant's view of business preceding the Industrial Revolution: charge what that market will bear and turn over the inventory as rapidly as possible.
EC 3d	Short run view of business from EC 3a,b,c.
EC 3e	Absence of national markets.
EC 3f	Local market conditions produce violent price fluctuations due to EC3d,e, and EC4.
EC 3g	Lack of restraint in competitive practice due to EC 3.
EC 3h	Secrecy as the only way to protect commercial advantages from competitors.
EC 3i	Practice of testifying only to the legal truth from right to privacy (EC 3) and secrecy (EC 3h).
EC 3j	Practice of spying on competition to counter EC 3h.
EC 4	Rule of capture.
EC 4a	Owner of land also owned rights to minerals beneath the land.
EC 4b	Boundaries of mineral deposits often overlap surface land ownership boundaries giving several owners rights to a deposit.
EC 4c	Since oil is fugacious, the landowner or leaseholder who brings it to the surface owns it (rule of capture).
EC 4d	Practice of drilling as many wells as possible to capture maximum production before one's competitors leads to maximizing the rate of production not total recovery due to EC 4c.
EC 4e	Boom and bust cycles due to EC 4d, 3b,d,g.

Exhibit 2-2
Average Annual Price of
Pennsylvania Crude Oil: 1860-1906*
(Per Barrel of 42 Gallons)

Year	Price	Year	Price
1860	$9.59	1883	$1.00
1861	.49	1884	.84
1862	1.05	1885	.88
1863	3.15	1886	.71
1864	8.06	1887	.67
1865	6.59	1888	.88
1866	3.74	1889	.94
1867	2.41	1890	.87
1868	3.63	1891	.67
1869	5.64	1892	.56
1870	3.86	1893	.64
1871	4.34	1894	.84
1872	3.64	1895	1.36
1873	1.83	1896	1.18
1874	1.17	1897	.79
1875	1.35	1898	.91
1876	2.56	1899	1.29
1877	2.42	1900	1.35
1878	1.19	1901	1.21
1879	.86	1902	1.24
1880	.95	1903	1.59
1881	.86	1904	1.63
1882	.78	1905	1.39
		1906	1.60

* Based on reports of U.S. Geological Survey on Mineral Resources except 1906, for which year prices were computed from the quotations in the Oil City Derrick.

Source: U.S. Bureau of Corporations, Report of the Commission on the Petroleum Industry, Part II, "Prices and Profits" (Washington, Government Printing Office, 1907), 84.

Exhibit 2-3

Environmental Changes and Strategic Consequences

Variable	Year and Event	Consequences
Geology	1. 1885 Discovery of Russian oil	Beginning of competition for Standard and American oil
	2. 1901 Texas Gulf coast discovery	Gulf and Texaco are born as Standard loses out
	3. 1909 Discovery in Iran (1904)	Anglo-Iranian (BP) is born and British discover oil in the Empire (1904)
	4. 1908 and 1924 Discoveries in Mexico and Venezuela	U.S. companies enter foreign production
	5. 1930 Discovery in east Texas	Proration of production
	6. 1943 DeGolyer report on Saudi Arabia	Recognition of the size of Saudi Arabian reserves and the shift of the center of gravity of world production to the Middle East
		Post World War II—Europe converts from coal to oil
Technology *Exploration and Production*	1. Post WWI Application of scientific geology and geophysics using seismograph, gravimeter and magnetometer	Many new discoveries ending post World War I feared shortage. Market turns to glut.
Refining	2. 1920's Catalytic cracking	Gasoline demand is met and product mix can be controlled
	3. 1960's petrochemicals	First industry diversification
Transportation	4. 1886 Marine Tankers	Change in distribution channels from barrel and case bulk. Companies begin wholesale bulk distribution through foreign marketing affiliates, replacing commission agents.

Category	Event	Description
Marketing	5. 1908 The automobile; 1910 The British Navy converts from coal to oil	Primary demand for petroleum products shifts from illumination to fuel. Refinery location shifts from source to market. Specialized distribution channels (the filling station) appear and companies go into retail marketing.
Oliogopolistic Interaction	1. 1947 Socal and Texaco sell 40% of Aramco	Exxon and Mobil enter Saudi Arabia. Shell and the French are shut out. The Middle East realigns.
	2. 1948 (Kuwait); 1949 (Saudi Arabia); 1955 (Iran) Aminoil, Getty, and Iricon gain concessions in the Middle East	The "Independents" (non-Majors) enter the Middle East
Government action	1. 1911 *United States v. Standard Oil*	Standrd is split into 34 companies, giving birth through fission to Socal, Mobil, and Exxon
	2. 1918 Russian revolution	Oil nationalized. Nobels and Rothschilds eliminated from the industry. Russian oil disappears from world markets for 40 years.
	3. 1919 Treaty of San Remo	French (CFP) replace Germans (Deutsche Bank) in Turkish Petroleum (later Iraq Petroleum). French enter world petroleum and Germans leave.
	4. 1928 Red Line Agreement	American companies enter the British controlled Middle East
	5. 1938 Mexico nationalizes oil	The power of foreign governments to expropriate the properties of foreign companies is affirmed
	6. 1948 Venezuela passes 50% income tax on oil	50-50 profit sharing spreads through the industry
	7. 1967 Closure of Suez Canal	Transportation routes and scale of tankers and refineries change: advent of supertankers and huge coastal refineries
	8. 1973 OPEC Embargo	OPEC takes over Middle East production and world pricing

3

The Legacy of Standard Oil

3.1 Effect of Standard on the Majors

Because strategy has its roots so deeply anchored in the past it is quite difficult to explain the strategies of any of the Majors without reference to Standard Oil, the granddaddy of them all. Exxon, Mobil, and Socal are its offspring. Their strategies are dictated by their inheritance from the *United States* v. *Standard Oil,* the 1911 antitrust case which resulted in the dissolution of Standard Oil into thirty-four companies, nine of which had foreign operations.[1] Gulf and Texaco were born with the discovery of oil in Texas in 1901, an area which Standard refused to enter. Their strategies were consciously shaped to compete with Standard Oil: Gulf as an integrated producer to avoid the problems its founder W.L. Mellon had had with Standard in his oil operations in Pennsylvania in the 1890's; and Texaco as a marketer of gasoline when Standard Oil's principal product was kerosene.[2] Shell competed vigorously with Standard in the Far East and Europe from the 1890's on, and Churchill, speaking for the British government in 1914, approved the purchase of equity in BP to avoid dependence on Standard and Shell.[3]

These events in the dimly remembered past, reinforced by years of capital investment and development of expertise, are the foundations upon which the strategies of the Majors and a good many others were built. To understand how these foundations were laid, this chapter will examine briefly the heritage Standard Oil left behind to its successors and competitors.

3.2 Standard's First Decade: 1871-1881

In 1863 John D. Rockefeller and his partners began to manufacture kerosene (PL1,2)* at the Excelsior Works in Cleveland.[4] By 1865 it was the largest refinery in Cleveland with a capacity of 505 barrels per day (PL4). Rockefeller's brother, William, joined the business and was soon dispatched to New York to

---

*Policies (PL #) and Environmental Conditions (EC #) are keyed to Exhibits S-5 and S-6 at the end of the chapter.

handle the export trade (PL3). In 1870 The Standard Oil Company (Ohio) was organized as a million dollar corporation with the largest refining capacity of any firm in the world (PL4). The Rockefellers had already begun to venture into auxiliary businesses—lumber, paint, barrel making, tanks, tank cars, warehouses, and lighters (PL5). But these three elements, (1) refining (PL2); (2) kerosene (PL2); (3) exporting (PL 3), were already firmly fixed in Standard's strategy.[5]

The depression from 1869-1874 (EC1), the second worst in U.S. history,[6] placed a severe strain upon the oil industry, especially at the refining level where capacity was estimated to be three times the average crude oil production in 1870-1871 (EC2).[7] To avert bankruptcy and to stabilize the chaotic conditions which plagued the industry, a number of refiners, marketers, purchasers, gatherers, and others formed an alliance of shareholders with Standard Oil (Ohio). Those segments of the industry were nearly cartelized (EC3, PL6). Agreements were signed with railroads to control costs and allocate the flow of crude. These transportation agreements with their rebates, kickbacks, and secret rates were the largest area of complaint about Standard's abuse of power, but they were at least partly based on real reductions and costs (PL9). Competitors often complained that Rockefeller sold below cost. Indeed he did—below their costs. An example is given in Exhibit S-4. At some time the combination had cornered the market for barrels, staves, and tank cars, had engaged in local price cutting, and had negotiated advantageous freight rates from shippers which included rebates on rivals' shipments as well as their own.[8] By 1877 the combination was a success.

Between 1877 and 1881 it began gobbling up rivals.[9] The record indicates that Standard followed a policy of purchasing rather than fighting rivals, paying good but not exorbitant prices (PL8).[10] By gaining the goodwill (PL9), as well as the properties, of sellers, future acquisition was made easier. Similarly, by refusing to pay exorbitant prices, future acquisitions were made cheaper and entry for the purpose of selling out at a high price to Standard was discouraged. Standard often took great pains to cultivate the goodwill of competitors (PL9).[11] As with other policies, Standard's acquisition policies took a long run view of the business (PL7).

By 1881 Standard was dominant in gathering and refining, but it had only limited influence in marketing, and very little in production, aside from its position as a buyer. Trunk pipelines were just coming into existence.[12] But through acquisition and expansion by 1881 Standard's net book value had grown to $55 million.[13]

In its first decade a number of important domestic policies had emerged. To Rockefeller and his associates, the petroleum industry was a different kind of business than it was to the producers. Standard had achieved control (Exhibit S-1) through its positions in purchasing, gathering, storage, and refining, and it had a strong position in transportation (PL5,6).[14] These were the capital intensive areas of the industry which required large fixed investments and high levels of steady operation to operate efficiently (EC4, PL4,10). There were scale economies to be realized from large size in purchasing, storage, and the logistics of supply as well as in refining. Large fixed investments required a long run view of the business. Rockefeller and his associates were the first to believe in the permanence of the industry. The rise of Rockefeller and Standard Oil, with its strength in supply and refining, and emphasis on steady operation, scale economies, and the long run stands in direct contrast to the vigorous, wasteful, competitive chaos of the producing industry, a group whose organization could have provided a model for the Oklahoma land rush.

One policy consequence was that Standard stayed out of production (PL11). In April 1878, the directors voted not to buy any more producing properties and the same year they decided to dispose of the few existing properties which had come with various acquisitions.[15] The instability, rapid exhaustion of properties, inability to gain control, and uncertainty of discovery were unappealing (PL11).

Another policy consequence was that capital intensity, scale economies, a long run view of the business, a variety of businesses, and many partners required a financial view rather than an operating view of the business (PL12). Standard's financial policies were the key to its success. Since there was no price competition within the combination, the aggressive entrepreneurs who joined Rockefeller competed to reduce costs, improve quality, standardize products and processes, develop new refining methods, and raise profits. Duplicate, inefficient, or poorly located facilities were eliminated. In January 1877, the company began self-insurance against fire up to $100,000. Risks were spread over all the activities of the combination and funds flowed from one place or company or function to where they were needed. The most important financial policy was internal financing of all ventures and maintaining liquidity at all times (PL13). As Rockefeller expressed it, "I think a concern so large as we are should have its own money and be independent of the 'Street'."[16] No later than 1882 Standard had no long-term debt and no later than 1886 lent far more short-term money than it borrowed.[17] Dividends were modest to permit the accumulation of capital. These policies gave Standard financial strengths that were unmatched by any competitor. And Standard used them.

Aside from its financial policies and view of the business, the most notable difference between Standard and its competitors or even other large firms of the day in other industries was its system of management. In its first decade Standard was an alliance of shareholders, most of whom were aggressively individualistic entrepreneurs who had founded their own companies and still ran them within the Standard fold (PL14). For this reason and the geographical and functional spread of operations, managerial authority was necessarily decentralized from the beginning. Standard was not a single organization with geographical and functional subsidiaries. It was a confederation of formerly autonomous and competitive units.

Long-term investments required planning; planning required information (PL15). As the organization grew, the information required to make intelligent decisions grew apace. To gather, process, and analyze this information on costs, supplies, market conditions, and competitors, staffs were assembled in New York, Cleveland, and elsewhere (PL16).[18] Standard operatives and agents gathered information on every phase of operations and every move by competitors, a practice whose extremes caused as much criticism as spying.

With large-scale, long-term investments, planning requirements, the need for experimentation and information, with so many elements interacting, and with so many strong individuals running their own operations, the task of central coordination was necessarily diplomatic and deliberative (PL17). Consultation and study on decisions took time. Standard reached decisions cautiously and carefully, but once they were reached, it used its financial strength and efficient organization to carry them out vigorously on a scale others could not match (PL18,19). In the 1870's critics opined that since Standard had invested so much in refineries in the oil region of Pennsylvania, it could not afford to take advantage of the development of trunk pipelines and establish large refineries on the coast. The feasibility of trunk pipelines was established in 1879 by others. In 1881 Standard formed National Transit Company and by the next year it had built over 1000 miles of trunkline.[19] Standard was often not the leader in innovation, but it was a vigorous follower (PL19). Once an innovation was proved, Standard was flexible enough to take decisions on as large a scale as it operated (PL20).

It is worth a moment to evaluate the 42-year-old John D. Rockefeller in 1881 at the end of Standard's first decade. He is thought to be the greatest of the robber barons of the nineteenth century. Yet this reputation does not capture his genius. Had he been simply a ruthless competitor, which he may have been, Standard Oil would never have been formed. Rockefeller's genius was not just in his decisively different view of the petroleum industry. It lay in his ability to attract to his organization a number of able and aggressive entrepreneurs— founders, owners, men accustomed to running their own businesses as they pleased, the most independent spirits of the most individualistic period of U.S. business history—to lead them, harmoniously, to a common view of the business, and to organize a system of management that accommodated their differences and effectively used their abilities. Rockefeller did not kill off his rivals. He converted them. One need not condone his nineteenth-century ethics and actions to pay tribute to the genius of his leadership, not only in bringing the pieces of Standard Oil together, but also in laying a foundation of policy and a system of management whose excellence in developing managers and producing results has stood the test of time. It would be nearly 50 years before Alfred P. Sloan performed a similar task at General Motors.[20] In the world of business, no one had done such a thing before. Except for Sloan in the 1920's and possibly Thomas Watson of IBM in the 1960's, no one has been able to duplicate his accomplishment since.

3.3 Standard's Second Decade: 1882-1891

In its first decade Standard Oil and many of its policies were formed. In its second decade, it matured. There were fewer acquisitions. Operations were systematically studied. Waste and duplication were eliminated. Costs were reduced. In the area of management, it was a decade of consolidation (PL10).

From 1882 until 1892 Standard was organized as a trust (PL21). This change in formal organization occurred in response to a suit won by the state of Pennsylvania upholding its right to tax all of the capital stock and dividends of a corporation operating within the state, not merely those derived from business within Pennsylvania. It was imperative that Standard create separate corporations to hold its properties in each state, and a central legal entity to hold them.[21] These legal difficulties arose in part because corporations were state chartered (there are no nationally chartered corporations in the U.S.) and partly because of the novelty of a corporation which did business and owned assets in more than one state (PL22).

Standard also recognized the need for other administrative changes, including moving its headquarters from Cleveland to New York, its financial and business center, and obtaining more complete control with the reduction of minority interests and administrative simplification. Minority interests could be reduced by inducing minority shareholders to accept shares in a parent organization. The trust form of organization was selected because it fulfilled these prerequisites and for another reason which was very important to Standard's management, the maintenance of a veil of secrecy over the operations of the combination (PL22).[22] The trust form accomplished the creation of a central office with geographical divisions, the second step on the ladder of organizational advances from proprietorship to multidivisional enterprise, which improved product and information flows and efficiency (PL21).

The Standard Oil Trust held shares in the operating companies and 41 investors held shares in the trust.[23] The legal innovation used by Standard was that the nine trustees were authorized to exercise managerial authority over the operating companies. When one of the nine original trustees chose not to move to New York, the trustees took advantage of a provision of their bylaws to permit the other eight to manage as the Executive Committee.[24] The number required for a quorum was gradually reduced as others were temporarily absent. In practice, as before, the Executive Committee constituted the day to day management. The concept of management by owners had evolved to the policy of mananagement by an active inside board who met daily to determine what policies and decisions were in "the general interest" (PL23). [25]

By controlling all appropriations over $5,000, beginning in 1882, the committee controlled the rate of growth of each segment. In 1886 annual reports were required for all salaries over $600 per year and the committee passed on all recommendations for raises. [26] Although lines of authority at the top were somewhat vague (due to the legal ambiguities of the trust form and the novelty of its use by Standard), with these two tools, control over appropriations and salaries (PL24), the Executive Committee controlled the organization.

To focus attention and bring expertise on particular functional issues which affected parts of several of the statewide operating corporations, each with its own limited group of functions, and to relieve the Executive Committee of some of its burden, a series of coordinating committees with advisory powers emerged (PL25). By 1887 the major committees were Manufacturing, Transportation, Case and Can, Export Trade, Lubricating Oil, Cooperage, and Domestic Trade. [27] The committees were forums to assess information, gather views, and achieve unanimity. Cooperation was obtained through consultation, discussion, and compromise. Decisions of the committees took the form of requests, suggestions, and recommendations rather than orders to the operating managers of the companies. [28] Due to the somewhat undefined legal status of the committees, there may have been legal reasons as well as policy reasons for this deliberative, considerate style of management (PL17). But, in the main, the committee system served admirably to achieve consensus and support for "the general interest" while permitting the delegation of authority to operating managers to manage their businesses, subject to general policy guidance.

The necessary concomitant to a deliberative, decentralized form of management is the ability to act decisively when a decision has been reached (PL14,17,18,20). There were three environmental changes in the 1880's which required such action.

The first was the introduction of trunk pipelines at the end of the 1870's. Since Standard was already dominant in the gathering and storage of oil, in purchasing, and in refining, it is only natural that when crude began to move by pipe rather than rail Standard would move into trunk pipelines to strengthen its hold over oil transportation (PL26). It was familiar with pipeline technology from its gathering activities. Negotiating with the railroads had sometimes been difficult and had occasioned criticism. The opportunity to forge a stronger link in the chain was not to be missed. By 1885 the National Transit Company, Standard's trunk pipeline subsidiary, had four lines in operation which blanketed the industry: one from Pittsburgh to Cleveland which passed through the oil regions; a spur from the oil regions to the Buffalo terminal; a line from the regions to New York; and another to Philadelphia and Baltimore.[29] The strategic entry into pipelines was to remain a Standard strength and characteristic, through its successors, worldwide, permanently.

Whenever possible the pipelines were located on the railroad's right of way, to minimize both lease prices and the cost of hauling building and maintenance materials.[30] To cushion the effect on railroads of loss of oil freight to pipelines Standard even gave a rebate of 26% of the transportation revenue to the seaboard to the Pennsylvania Railroad, whether the crude went by rail or not (PL9).[31] Standard's policy, as with other acquisitions was to pay good prices (PL8), a policy rivals sometimes deemed anticompetitive. Another important pipeline policy, also characteristic of its other operations, was that Standard built for permanence (PL7) where some of its rivals built cheaply and hastily. Like National Transit, the Tidewater Pipeline Company had to bury its pipe, when temperature changes caused exposed pipe to whipsaw.[32]

The completed gathering, storing, and transportation system deserves comment because it demonstrates the systematic rationality of Standard's management and the relation of Standard to the largely independent producers.[33] When a producer requested a run from the tank at his well to the gathering system, Standard's United Pipeline, a company representative called a "gauger" would strap a measuring device to this tank. The data from the run would be wired to district headquarters where it was credited to the balance of the leaseholder. Royalties to the landowner were deducted and credited to the landowner. Three percent of the volume was deducted as an allowance for evaporation, leakage, and B.S.& W. (bottom settlings and water). An oil certificate was issued for each 1000 barrel lot and fractions were carried on the books as a credit balance. The charge for pipage after 1879 was a flat 20¢ per barrel, regardless of distance piped. Storage was free for 30 days. After that the storage charge was 1-1/4¢ per barrel per month. Monthly and annual reports of stocks on hand were publicly issued according to Pennsylvania law. Fire losses were assessed to all owners of credit balances according to the size of their balance. For its accuracy and efficiency in extending gathering systems to new fields, Standard earned praise even from its critics.[34] It applied the same thoroughness and dedication to this field (PL12) as it did to preventing price competition and driving out or absorbing competitors.

Oil certificates were traded on exchanges in the oil regions and in New York. Standard purchased almost all of the crude for its refineries on the exchanges. In 1885 its refineries consumed 16.1 million barrels when production was 21.5 million barrels and stocks in storage were 35.7 million barrels.[35] It was the largest but far from the only purchaser. Its practice was to pay the average between the high and low price for superior quality crudes. On at least one occasion it paid a premium price to encourage drilling and development of production in an area where it had a refinery. It sometimes paid higher prices in local areas to bid oil away from competitors, raising the cost of their raw material.[36]

It may come as a surprise that in so many instances Standard's policy was based on paying high rather than low prices (PL8). Rockefeller argued that Standard should buy certificates in excess of consumption of crude when prices were low. Surplus funds earning a low rate of interest should be put to work at this potentially profitable use. They might help to stabilize falling prices. A falling price of crude was not felt to be in Standard's best interest. It discouraged exploration, antagonized producers, and endangered future supplies. Large purchasers of refined products, Standard's principal customers, postponed purchases in hope of a lower price, adding to inventory problems. As a large owner of crude in tanks, Standard also had to be concerned about maintaining the value of its inventories.[37] Hidy and Hidy concluded that Standard did not speculate on the exchanges, aside from purchasing excess crude when prices were low.[38]

The concern over long-term supplies of crude (PL7) eventually propelled Standard to take the second major step of the decade: investment in Lima (Ohio) crude (PL27). Pipeline men with Standard had long wanted to get into production. Pipeline loads would be stabilized; competition from other tranporters would be less important; public criticism from producers would be minimized; and the opportunities for profit appeared large. Refiners, on the other hand, viewed Standard's business as the manufacture of refined products (PL1).[39] Crude was simply raw material in which investment should be minimized. Crude production was a chaotic, unstable business, a speculation better left to others (EC5).

Two factors brought about a change in Standard's strategy of production. Consumption of Pennsylvania oil had begun to exceed its production in 1885 and the prospects for new reserves as large as the Bradford field appeared unlikely.[40] But it was the challenge of Lima crude to the stability Standard had labored so hard to achieve which proved decisive. To retain that stability Standard decided that it had to conquer Lima crude.

The discovery of Lima crude in 1884 challenged both groups within Standard. It was vile smelling, sulphurous stuff, nicknamed (with reason) "skunk oil." It had a smaller percentage of lighter fractions, similar to the petroleum produced in Ontario, across Lake Erie. It defied refining by existing methods, producing a smaller fraction of inferior kerosene at a cost at least 10¢ per barrel higher than Pennsylvania crude.[41] It was located in northwestern Ohio (near Lima) and northeastern Indiana, an area geographically distinct from the Appalachian region, which would require a new system of gathering and trunk pipelines. Standard executives perceived immediately the importance of Lima crude to Standard's strategy. They could not afford to allow anyone else to handle it; neither could they afford to pay more than it was worth in the form of manufactured products. As Benjamin Brewster, one member of the Executive Committee wrote, "The question is—how to utilize it and where.... We can afford to spend time and money on [the problem] to answer [it] intelligently."[42] It was a response typical of Standard's approach to problems (PL12,15,18).

Their first move was to control the crude. The Buckeye Pipeline Company, organized March 31, 1886, was authorized to construct a gathering and storing system. In June 1886, Standard's purchasing agency began to buy all the oil Buckeye took in, about 85% of Lima Indiana production. Standard then hired Herman Frasch, the outstanding researcher on sulphurous oil, from the Imperial Oil Company, Limited in Ontario to solve the problem of refining "sour" crude.[43] On July 1, 1886 the Solar Refining Company was formed with a plant at Lima to process "sour" crude. Frasch took longer than expected, until October 1888, to conquer Lima crude. In May 1889 Standard then began construction on a huge new refinery at Whiting's Crossing near Chicago (PL4). With a 36,000 barrel per day capacity, the largest in the world, it was designed to supply the entire Mississippi valley. The Whiting refinery was a monument to the victory over Lima crude and to the policy of having ample funds available when needed (PL13).[44]

The victory was completed when in June 1889 Standard made its first major investment in producing property with the purchase of the Ohio Oil Company (PL27). In 1888 Standard's total production was only 200 B/D out of 75,649 B/D in the U.S. By 1891 it was 38,994 B/D out of 148,747 B/D, over 25% of the U.S. total (Exhibit S-1).[45]

By the end of 1891 Standard had invested in Lima crude $9.6 million in pipelines, $7.2 million in inventories, $7.3 million in refineries, and $8.1 million in producing properties, a total of $32.2 million.[46] It had also put $14.5 million into other producing properties.[47] The cost of these massive strategic investments, a total of $46.7 million, compares to net assets of $55 million in 1881 and $102 million at the end of 1892 (PL20).[48]

The third major event which profoundly affected Standard's strategy was the threat to its foreign marketing by Russian oil (EC6). It led to the bulk distribution of products at home and abroad and the establishment of foreign marketing affiliates (PL28,29). The foreign market was very important. In the early 1880's more than half of U.S. refined products were exported. Two-thirds went to Europe. Kerosene accounted for more than 80% by value of refined product exports. Standard handled 90% of the exports.[49] Standard had a virtual monopoly in foreign kerosene, which was shipped abroad in barrels painted blue with white ends bearing Standard's brand names and in cans to the hotter climates. But most sales were through commission agents f.o.b. New York so that Standard's control ended at the edge of New York harbor.[50]

By 1888 Russian kerosene had seized 22% of the world export market.[51] With high quality kerosene, highly productive wells, production costs one-third to one-half those in the U.S., continuous distillation, proximity to markets in Europe and the Eastern Hemisphere, cheaper bulk transport in tank steamers, an internal market for fuel oil which made kerosene a byproduct that could be sold for whatever it would bring, regulations promoting export, and financed by Europeans like the Nobels and Rothschilds, Russian products were formidable competitors to Standard's.

Standard met the Russian competition head on. To hold on to market share it cut prices, to the dismay of refiners. It upgraded its services. It did not market inferior kerosene made from Lima crude abroad. These efforts were to little avail. By 1888 when its volume in the U.K. showed an absolute decline, a more effective response was needed.[52]

It came with the overhaul in the next three years of Standard's entire system of foreign distribution. Bulk marine transport led to the establishment of foreign marketing affiliates to distribute oil under the Standard brand names in new, efficient horse-drawn tank wagons (PL28,29).

Wilhelm A. Riedeman, a leading German importer of Standard products, ordered a new vessel, the *Glückauf*, which was launched June 16, 1886 (EC7).[53] It carried 20,000 barrels of oil contained in its skin rather than in tanks or barrels. The cost was about three-fourths as much as by sailing ships carrying 5000 barrels each. By 1892 three-fourths of the total export of American crude and kerosene, except that in cases, was in bulk.[54]

Transport in bulk required a brand name marketing affiliate at the other end to package and distribute the product. The Anglo-American Oil Company, Limited, Standard's first foreign affiliate, was organized on April 24, 1888 to import and wholesale in the British Isles. The following year it delivered to jobbers and retailers 71% of all American oil imports to Britain.[55] It had also begun selling under its own name to commission merchants throughout the British Empire in the Eastern Hemisphere.

On the Continent the need to compete quickly, the lack of knowledge of the wholesale trade in each country, language barriers, and myriad legal regulations such as restrictions on warehousing and storage permits required alliance with established local merchants, the largest and best managed possible. DAPG, the Deutsche-Amerikanische Petroleum Gesellschaft (German American Petroleum Company), approximately 40% owned by Standard, appeared on February 22, 1890 to market in Germany; the American Petroleum Company (51% owned) (Holland and Belgium) on March 10, 1891; two companies (60% owned) in Italy in 1891; and a 21.45% interest was purchased in a Scandinavian firm late in 1891.[56]

By the end of that year the policies of bulk marine transport and wholesale distribution through partly owned foreign affiliates were well established (PL28,29,33). But, though it had several opportunities, Standard declined to enter either foreign production or refining at this time outside the Western Hemisphere.[57] It is also interesting to note that specialty products such as lubricants, wax, and vaseline maintained their own separate foreign marketing organizations and had established offices abroad in the early 1880's.[58] The older pattern of specialized marketing, a necessity for developing the market for new products (PL30), preceded bulk marketing of kerosene and the establishment of foreign affiliates. It remained at the time of divestiture in 1911, when nine of the thirty-four pieces into which Standard was split had foreign operations, mostly in marketing.[59]

Bulk distribution produced similar changes in domestic marketing between 1889, when Standard had only 252 tank wagons east of the Mississippi, and 1892 when one-third of its kerosene deliveries to retailers were by tank wagon.[60] Control of marketing tightened, as minority interests were bought out; control over brands increased; efficiencies of centralization were sought. Prices were set by the general policy of maximizing market share (PL31), a refining oriented policy that would maintain volume and the efficiencies of scale. As Rockefeller put it in 1888, "We want to continue, in reason, that policy which will give us the largest percentage of the business."[61] Prices were cut where local or foreign competition required and raised to the limit where it did not.

These incidents show Standard Oil at the peak of its power—a Goliath—moving rationally, deliberately, systematically, using its massive financial resources in the attempt to impose its will on the industry (Exhibits S-1, S-2). They show two other things. Even in the 1880's the industry was already growing and changing too rapidly to be dominated by any single competitor. Also, Standard was never immune to competition. In each of these three incidents Standard's strategy was a response to competition, in spite of all it could do to eliminate it.

Its dominance of the industry, however, was awesome (PL31). Even as late as 1899 it produced one-third of U.S. production and refined an astounding 80% of it.[62] If anything, in gathering, storage, and transportation of crude it was even more dominant. In the export market it marketed the products of other refiners as well as its own, bringing its share of U.S. exports to 90% or more on at least some occasions.[63]

However, because of the nature of channels of distribution this may overstate Standard's position in marketing somewhat. Until 1890, when foreign affiliates were established, marketing consisted of little more than wholesaling at the refinery gate or New York harbor in branded barrels. All exports were governed by the rules of the New York Produce Exchange, like other commodities then or later.[64] Commission agents who handled general merchandise were the main channel of distribution. After the 1880's Standard faced substantial competition everywhere abroad from Russian products. Its share of the foreign market was perhaps only a third of petroleum product sales. It had significant competition from other products and technologies at home and abroad. Not only fats, oil, wood, and coal competed with petroleum products, but also natural and manufactured gas, refined products from shale and, most importantly, electricity.[65] Even after the establishment of foreign affiliates, Standard was still integrated forward into marketing only to the bulk distribution wholesaling level both in the U.S. and abroad (PL28). Retailing was still done through the general store. Specialized retailing outlets for petroleum products did not appear until the invention of the gas station about the time of dissolution in 1911.

In production its influence was far more dominant than its market share indicates. As W.L. Mellon, the founder of Gulf and a man who had considerable experience with Standard, recalled:

> The Standard Oil Company at that time was the only oil company of major importance in existence, but it was not primarily a producing company. In fact, except in California, it had relatively little production of its own and had never made it a general policy to own or control production in a large way because it was practically the sole buyer of crude oil (PL11).
>
> Consequently, every wild-catter, every lease buyer, every owner of a producing well, was, in effect, a part of the Standard Oil Company's producing department. Whatever price was posted by the Standard Oil Company at the wellhead was the price the producer got for his crude. Standard made the price.
>
> In saying this I do not mean to imply that the company took unfair advantage of the producers whose oil it bought. As a matter of fact, the producers who traded with the Standard Oil Company did very well, though some of them grumbled. Nevertheless, these producers were at the mercy of this company, which had a complete pipe line system. Normally, this was extended to any new field brought into production. Texas had been a significant exception. So it was to be expected that, as other fields were discovered and developed, Standard's pipelines would be extended to these new fields."[66]

Mellon's response to Standard's strategy is also worth noting:

> This prospect did not leave much excuse for anybody to think that he could operate an oil business on a big scale in competition with Standard if he tried to operate in the same way as Standard. I concluded that the way to compete was to develop an integrated business which would first of all produce oil. Production, I saw, *had* to be the foundation of such a business. That was clearly the only way for a company which proposed to operate without saying "by your leave" to anybody."[67]

Standard's strategy embraced both horizontal and vertical integration as cornerstone policies from its beginning. It sought and achieved, directly or indirectly, control of each segment of the U.S. industry: production, gathering, storage, transportation, refining, and wholesale marketing. No organization ever fell more clearly into Section 2 of the Sherman Antitrust Act than did Standard (PL31).[68]

3.4 Twilight of the Combination: 1891-1911

There were fewer strategic changes in the final period of Standard's history. Much of the time Standard was on the defensive legally. Five changes bear mention: the retirement of Rockefeller, the difficulties in Texas, foreign refining and producing, and the rise of Jersey Standard. Finally, there was the dissolution.

The retirement of John D. Rockefeller from day to day affairs, gradually after 1891 and completely after 1896, was a non-event.[69] Though Rockefeller remained nominally president until 1911, the reins of power passed smoothly from the founder to his successor, John D. Archbold, a brilliant, capable, conservative man, testifying to the success of Standard's system of management in developing and retaining extremely able men (PL14,32).

The next was Standard's difficulties in Texas which led to a minimal position there, allowing the rise of Gulf, Texaco, and other companies. The Waters-Pierce Oil Company was majority owned by Standard but managed by one of its founders, a sharp operator named Henry Clay Pierce (PL14,33).[70] Pierce had been a thorn in the side of Standard's management since it purchased half of his company in 1878. Pierce's aggressive tactics, excessive profits, and independence were always a source of worry to the Domestic Trade Committee. In 1900, as the result of a state antitrust suit, the company was expelled from Texas. Partly from the hostility toward Standard in Texas and partly from the impossibility of controlling Pierce, when the Mellon family, founders of Gulf, offered to sell its producing properties in Texas to Standard in 1902, they received the reply, in an exception to his retirement, that "Mr. Rockefeller would never put another dime into Texas."[71]

The entries into foreign producing and refining were undertaken gradually as competition required and opportunities permitted (PL34). The extent of this activity was rather small until after World War I. Standard had financed a refinery in Galicia (Poland) in the 1870's for a time. In the 1880's Waters-Pierce had established two small refineries in Mexico. Standard also owned part of a refinery in Havana, built to avoid tariffs on product imports. The Imperial Oil Company, an integrated Canadian company, was acquired in 1898.[72] DAPG and American Petroleum invested in a refinery in Germany and later in naphtha re-run plants. In 1904 an integrated operation, Romano-Americana, was established in Romania.[73]

The organizational changes after 1892 are more useful as a historical determinant of the industry after 1911 than as evidence of strategic change. They show why Jersey Standard (now Exxon) rather than New York Standard (now Mobil) became the principal heir to Standard Oil after 1911.

Until 1899 New York Standard was the dominant financial unit of the combination.[74] It had been the headquarters of the combination since the move from Cleveland in 1881. It was the banking and financial center as well as the chief exporter.[75] Internally it made loans and investments in subsidiaries. Externally, it settled foreign trade accounts and invested the combinations' short-term funds.

As the result of a suit won against Ohio Standard in 1892, the trust was dissolved.[76] Its 92 operating units were consolidated into 20 companies. Taking advantage of a change in 1889 of the state law in New Jersey, which now permitted ownership in the stock of other corporations, Jersey Standard ended up with 21 of the 92 units and became a major part of Standard.[77] Shares in the 20 companies were then distributed to trust certificate holders in proportion to the ownership in the trust. About half of the holders refused the exchange. The trust then exchanged all its remaining investments from the uncovered certificates for shares of Jersey Standard. This left each of the 19 companies owned about half by the ten families of major shareholders and half by Jersey Standard, whose shares in turn were owned half by the trust and half by the families of major shareholders. Since the trust was prohibited from voting its ownership by the suit dissolving it, the Standard Oil Interests, as they were called from 1892-1899, were held together by the major shareholders who were steadily growing older.[78] This confusing arrangement, which would eventually have dissolved the combination as relatives and heirs sold off their holdings, was changed in 1899 when Jersey Standard was recapitalized as a holding company with $196 million[79] in net assets and issued new shares for those outstanding in the other 19 companies, both those from the major shareholders and the remaining holders of trust certificates who now accepted the exchange. Thus Jersey Standard became the parent corporation and the combination was preserved.

In the 1890's Standard was a mature organization and had already begun to reap the consequences of its policies. As an economic organization, it was splendid. Its public image, however, was a disaster. Two decades of secrecy, sharp practices, and size finally brought the giant down.

The American attitude toward Standard Oil was curiously ambivalent, gladly embracing its products, quality, and service and its contributions to their lives, while detesting with equal enthusiasm its abuses. They desired the golden eggs, while wishing to be rid of the golden goose. It is an attitude which executives today must ponder.

Partly it was an expression of American character. Whatever the organization called the Standard Oil Trust or "interests" or combination was, no one, even its executives, was quite sure, but it clearly was not an entrepreneurial organization. It was large and powerful, and often seemed sinister. Americans had won their freedom from political kings and they were not about to subjugate themselves to economic kings in the form of what a generation hence would be called modern corporate capitalism. Large business organizations were going through the throes of being invented, as witnessed by

Standard's numerous reorganizations, and many Americans wanted no part of them. Any unit of power larger than the individual was bound to stir resentment. And Standard did—like no other. The increasingly sovereign American consumer wanted the products of material prosperity set before him in glittering array by numerous, anonymous, silent servants of capitalism. He identified with the products, but never with the producer.

The end came in 1911 when the holding company was struck down and dismembered into 34 companies. At the time of the dissolution the net value of Standard's assets was $660 million. The largest piece by far was Standard Oil (New Jersey) with $285.5 million in net assets, followed by Standard Oil (New York) with $60 million, National Transit with $52 million, the Ohio Oil Company with $44 million, and Standard Oil (California) with $39.2 million.[80] The course of events for the 34 pieces since 1911 is presented in Exhibits S-3.

3.5 The Strategy of Standard Oil

It is appropriate to close this chapter with a summary of what is meant by a "Standard Oil Strategy" (Exhibit S-5).

Standard's strengths lay in refining, gathering, storage, pipelines, rail transportation, and finance (PL1,5). It was deeply committed to kerosene manufacture for export as well as domestic trade (PL2,3). It sought worldwide stabilization of the industry under its control, but never quite achieved it (PL6). Its weak areas were new field exploration, production, and fuel oil (PL11). Most of its crude was purchased (even after 1888) at daily average prices on the oil exchanges (PL8). After 1888 it purchased and developed producing properties but did little wildcatting in nonproducing areas (PL27). Until the late 1880's it sold its refined products for export through commission agents in New York. After that it sold wholesale through foreign affiliates in Europe and Canada. It practiced limit pricing but not monopoly pricing. Its policy was to maximize market share and cut price when it had to (PL6,12,31). It believed in secrecy, narrow legal interpretations, and used its power freely (PL22).

Standard Oil was a financially oriented organization with a large-scale, long-term perspective (PL,7,12). New York Standard and later Jersey Standard functioned like banks, investing loans and equity in subsidiaries, receiving much of their income from interest and dividends rather than from operations (PL13). Standard sought to increase profits by reducing costs and expanding volume (PL12). Its policy was to buy out trouble rather than fight, paying good but not exorbitant prices (PL8). It bought into new areas rather than innovating or starting from scratch (PL18). It was often criticized for paying high rather than low prices for assets (PL8). While often not the first to innovate, it was a vigorous follower (PL19), whose expertise was implementing proven ideas on a large scale (PL4,19).

Standard was managed by an active board of inside directors (no outside directors) through a system of specialized committees (PL23,25). Its management style was deliberative and consultative (PL17,18). Unanimity was sought and usually achieved. Standard kept track of everything that occurred in the industry through a network of its employees, agents, and staff so that decisions might be intelligently made (PL16,18). It was slow to make decisions, but once made, it moved massively to implement them (PL19,20). It acquired and developed able management and decentralized authority to manage operations subject to guidelines in "the general interest" (PL14,32).

Altogether these policies (Exhibit S-4) comprise the legacy of Standard Oil.

Exhibit S-1
Standard Oil Production and Refining

	(a)[1] World Production	(b)[1] US Production	(c)[2] Standard's Production	(d) US % of World Products (b)÷(a)	(e)[4] Standard's % of US Production (c)÷(b)	(f) Standard's Production ÷ Refinery Runs (c)÷(b)	(g) US Production Controlled by Standard (h)÷(b)	(h)[3] Standard's US Refinery Runs	(i)[5] Standard's Refinery Capacity	(j)[7] Standard's Foreign Refinery Capacity
	TBD	TBD	TBD	%	%	%	%	TBD	TBD	TBD
1857-60	1.43	1.37	0	96						
1870	15.9	14.4	0	91					10+[6]	1872
1880	82	72		88						
1881	88	76		86						
1882	98	83		85						
1883	83	64		78						
1884	99	66		67						
1885	101	60		59		0	80	48		
1886	129	77		59		0	68	52		
1887	131	77		59		0	70	54		
1888	143	76	0.5	53	0.7	1	71	54		
1889	169	96	15	57	16	24	66	63		
1890	210	126	30	60	24	44	54	68		
1891	250	149	39	60	26					
1892	243	138	33	57	24	31	77	106		
1893	252	133	38	53	29	34	85	113		
1894	245	135	38	55	28	33	86	116		
1895	284	145	44	51	30	39	77	112		
1896	167	167	48	53	29	40	71	119		
1897	334	166	48	50	29	36	80	132		
1898	342	152	51	44	34	38	88	133		
1899	359	156	50	44	32	40	81	126		
1900	409	174	55	43	32	45	70	122		

Exhibit S-1 (Cont.)
Standard Oil Production and Refining

	(a)[1]	(b)[1]	(c)[2]	(d)	(e)[4]	(f)	(g)	(h)[3]	(i)[5]	(j)[7]
	World Production	US Production	Standard's Production	US % of World Products (b) ÷ (a)	Standard's % of US Production (c) ÷ (b)	Standard's Production ÷ Refinery Runs (c) ÷ (b)	US Production Controlled by Standard (h) ÷ (b)	Standard's US Refinery Runs	Standard's Refinery Capacity	Standard's Foreign Refinery Capacity
	TBD	TBD	TBD	%	%	%	%	TBD	TBD	TBD
1901	377	190	52	41	27	39	70	133		
1902	498	243	49	49	20	35	58	142		
1903	534	275	48	52	17	34	52	143		
1904	597	321	47	54	15	31	48	153		
1905	589	369	42	63	11	26	44	164		
1906	584	347	40	59	11	23	50	172	217	7.8
1907	723	455	53	63	11	27	43	197		
1908	782	489	58	63	11	28	42	205		
1909	818	502	64	61	12	29	44	222		
1910	898	574	82	64	13	32	45	258		
1911	943	604	90	64	14	33	45	273	303	16.2

1. API, *Petroleum Facts and Figures*, 1930 3rd ed. pp. 8-9.
2. Hidy and Hidy, 187, 271, 374.
3. Hidy and Hidy, 187. Data for 1899-1905 include amounts purchased and resold. This amounted to 6 TBD in 1906.
4. Hidy and Hidy, 407.
5. Hidy and Hidy, 414.
6. Hidy and Hidy, 413n. Nevins, *Rockefeller*, I, 366.
7. Hidy and Hidy, 514.

Exhibit S-2
Standard Oil Product Mix and Sales

	(a) Crude Oil Exports	(b) Kerosene U.S.	(c) Kerosene Exports	(d) Naphtha & Gasoline U.S.	(e) Naphtha & Gasoline Exports	(f) Gas Oil U.S.	(g) Gas Oil Exports	(h) Fuel Oil Sales—U.S	(i) Fuel Oil Exports	(j) Lube Oil U.S.	(k) Lube Oil Exports	(l) Wax
SALES												
1910												
Volume (TBD)	2.65	23.84	50.84	25.23	4.86	?	3.02	?	1.18	?	?	?
Mkt. Share (%)	?	75.3	?	74.3	?	?	?	?	?	?	?	?
1906												
Volume (TBD)	?	31	?	16	?	?	?	?	?	8.97	6.89	2.86
Mkt. Share (%)	?	82.1	?	84.4	?	?	60.2	?	?	?	72.8	88.3
1892												
Volume (TBD)	?	?	?	?	?	?	?	?	?	?	?	.73

Exhibit S-2 (Cont.)
Standard Oil Product Mix and Sales

	Kerosene	Naphtha and Gasoline	Gas Oil	Fuel Oil	Lube Oil	Wax
REFINERY RUNS						
1897						
Volume (TBD) 112.53	72.15d	16.19d	?	18.08d	4.76d	1.35d
% Total Volume 100	64.1	14.4	?	16.1	4.2	1.2
Value ($10^6) 59.3	39.2	7.8		3.9	5.8	2.7
% Value 100	66.1	13.2		6.6	9.9	4.5
1890						
Volume (TBD) 62.58	47.22d	9.88d	?	2.14d	2.61d	.73d
% Volume 100	75.5	15.8	?	3.4	4.2	1.2
Value ($10^6) 48.2	37.4	5.8		.37	3.8	.9
% Value 100	77.6	12.0		.8	7.9	1.9

Source: Hidy and Hidy, 473-74, 528, 289, 492-93, 572-73

Exhibit S-3
Fates of the 34 Standard Affiliates after 1911

Name in 1911	Fate
A. *Still in independent existence today*	
1. Standard Oil Company (New Jersey)	Exxon Corporation
2. Standard Oil Company (California)	Standard Oil Company of California (now Chevron)
3. Standard Oil Company (Indiana)	same
4. The Ohio Oil Company	Marathon Oil Company
5. Continental Oil Company	same
6. Borne, Scrymser Company	Borne Chemical Company
7. South Penn Oil Company	Pennzoil Company
8. Washington Oil Company	same
B. *Merged with or acquired by Standard affiliates*	
9. Standard Oil Company (New York)	Merged with Vacuum (1931); Mobil today
10. Vacuum Oil Company	Merged with Socony (1931); Mobil today
11. Anglo-American Oil Company	Acquired by Jersey Standard in 1930
12. South-West Pennsylvania Pipe Lines	Acquired by National Transit in 1952
13. National Transit Company	Acquired by Pennzoil in 1965
14. Eureka Pipe Line Company	Acquired by Pennzoil in 1947
15. Standard Oil Company (Kentucky)	Acquired by Socal in 1961
16. Standard Oil Company (Kansas)	Acquired by Standard Oil Company (Indiana) in 1948
17. Standard Oil Company (Nebraska)	Acquired by Standard Oil Company (Indiana) in 1939
18. Solar Refining Company	Acquired by Sohio in 1931.
19. Indiana Pipe Line Company	Acquired by Buckeye in 1942
20. Northern Pipe Line Company	Acquired by Buckeye in 1964
21. New York Transit Company	Acquired by Buckeye in 1964
C. *Merged with or acquired by non-affiliates*	
22. Chesebrough Mfg. Co. Consolidated	Chesebrough-Pond's Inc.
23. Atlantic Refining Company	Merged with Richfield in 1960 to become Arco
24. Cumberland Pipe Line Co., Inc.	Acquired by Ashland Oil Co. in 1931
25. Southern Pipe Line Company	Acquired by Ashland Oil Co. in 1949
26. Prairie Oil and Gas Company	Acquired by Sinclair in 1932 which was acquired by Arco in 1969; some properties sold to BP in 1969 and Pasco in 1973
27. The Standard Oil Company	The Standard Oil Company (Ohio) (Sohio); merged into BP in 1970

28. Buckeye Pipe Line Company	Became division of Pennsylvania Company (Penn Central) in 1964
29. Union Tank Car Company	Union Tank Car Compay, division of Trans Union Corporation
D. *Liquidated or fate unknown*	
30. Swan and Finch Company	In liquidation mid-1960's
31. The Crescent Pipe Line Company	Liquidated in 1920's
32. Galena-Signal Oil Company	Liquidated in 1920's
33. Colonial Oil Company	Fate unknown
34. Waters-Pierce Oil Company	Fate unknown

Source: Standard Oil Company of California.

Exhibit S-4
Standard Oil Scale Economies

The Conquest of Cleveland

Once their enterprise had been incorporated as the Standard Oil Company, Rockefeller, Flagler and Harkness began a concerted effort to dominate oil refining in Cleveland. Their strategy was to obtain still more favorable rates from the railroads and to do so by taking advantage of the large output of their two refineries. Sometime in 1870 Flagler offered James H. Devereux, the General Manager of the Lake Shore Railroad (a subsidiary of the New York Central) to ship 60 carloads of oil a day every day if the Lake Shore and the Central would give his company a rate of $1.30 a barrel from Cleveland to New York (the published rate was $2.00 a barrel) and 35¢ a barrel for crude from the Regions to Cleveland.[81] Devereux quickly accepted. As he later testified the assured regular flow permitted him to run a single train daily made up wholly of oil cars instead of putting oil cars on trains mixed in with other types of freight cars. The improvement in scheduling meant that fewer cars were needed. The resulting lower investment and maintenance costs and the small cost of providing a locomotive and crew for the daily trip, meant that the railroad could still make a good profit on the rate. Of course, Devereux knew that if he refused, Flagler could certainly get the same deal from the Erie. When other Cleveland refiners protested after hearing of the rate cut, Devereux told them "that this arrangement was at all times open to any and all parties who would secure or guarantee the like amount of traffic or an amount to be treated and handled in the same speedy and economical manner. . . . " But no other Cleveland refiner could guarantee 60 carloads a day.

Armed with this contract Flagler and Rockefeller first invited their two major competitors (Clark, Payne and Company and Westlake, Hutchins and Company) to join forces—an invitation the two readily accepted. Then the Standard Oil approached the other Cleveland refiners. By the end of 1871 five large and seven small firms had sold their properties to Standard Oil and the conquest of Cleveland was practically complete. At the same time to strengthen the position in the all-important foreign market the Standard partners bought Jabez A. Bostwick and Company of New York, a leading exporting firm with a refinery on Lond Island. In Cleveland the larger and more efficient works continued to operate under Standard's control although the change in command was kept secret. Competent refiners like Payne and Bostwick became senior executives of Standard Oil. On the other hand, smaller refineries not producing specialized products were closed down and their owners had to find a new business.

Source: Affidavit of James H. Devereux in the case of *Standard Oil Company* v *William C. Schofield, et al.* quoted in Tarbell, *History of the Standard Oil Company*, I, 277-79 and Case BH120 "Standard Oil and the Early Development of the American Oil Industry," Harvard Graduate School of Business Administration, 1961.

Exhibit S-5
Standard Oil Policies

PL 1	Refining as principal activity: (a) V; (b) 1863-1911.
PL 2	Kerosene as principal product: (a) P; (b) 1863-1911.
PL 3	Sale to export as well as domestic trade: (a) G; (b) 1865-1911.
PL 4	Large scale: (a) S; (b) 1870-1911.
PL 5	Vertical integration into supplier businesses, gathering and storage, and rail transport: (a) V; (b) 1870-1911.
PL 6	Horizontal and vertical combination of competitors—refiners, marketers, gatherers, and purchasers under the leadership of Standard Oil: (a) H,V; (b) 1870's-1911.
PL 7	Long run view of the business: (a) A; (b) 1870-1911.
PL 8	Pay good but not exhorbitant prices for supplies and acquisitions: (a) A; (b) 1870's-1911.
PL 9	Cultivate goodwill of competitors: (a) A; (b) 1870's-1911.
PL 10	Reduce costs through efficient steady operation, large scale, heavy fixed investments, and long run view: (a) A; (b) 1882-1911.
PL 11	Stay out of production and exploration: (a) V; (b) 1878-1888 (production), 1865-1911 (exploration).
PL 12	Rational, systematic, thorough financial view of the business —no price competition within the combination, reduce costs, improve quality, rationalize facilities: (a) A; (b) 1878-1911.
PL 13	Financial strength—internal financing of all investments, no long-term and no net short-term debt, modest dividends, maintain large cash balances: (a) A; (b) late 1870's-1911.
PL 14	Decentralized authority to manage PL 6: (a) A; (b) late 1870's-1911.
PL 15	Long-term planning: (a) A; (b) 1882-1911.
PL 16	Systematic accumulation of competitive information: (a) A; (b) 1870's-1911.
PL 17	Diplomatic, consultative, advisory rather than autocratic, authoritarian management style: (a) A; (b) 1880's-1911.
PL 18	Cautious, careful, deliberate, often slow, decision making. Consensus "in the general interest" sought and achieved: (a) A; (b) 1880's-1911.
PL 19	Vigorous implementation with financial strength and effective organization—not a leader in innovation, but a vigorous follower: (a) A; (b) 1880-1911.
PL 20	Flexibility to take large-scale strategic decisions and not become hidebound by past—ability to act decisively: (a) A; (b) 1880-1911.
PL 21	Trust organization: creation of a central office: (a) A; (b) 1882-1892.
PL 22	Secrecy: (a) A; (b) 1870's-1911.
PL 23	Day to day management by an active, inside board of directors: (a) A; (b) 1882-1911.
PL 24	Control maintained by control over appropriations, salaries, and assignments: (a) A; (b) 1882-1911.
PL 25	Committee system of management: (a) A; (b) 1882-1911.
PL 26	Integration into pipelines: (a) V; (b) 1881-1911.

PL 27	Purchase of producing properties—integration into production: (a) V; (b) 1889-1911.
PL 28	Integration forward into bulk marine transport and wholesale marketing: (a) V; (b) 1888-1911.
PL 29	Establishment of foreign marketing affiliates: (a) G,V; (b) 1888-1911.
PL 30	Separate marketing networks for different products: (a) P,A; (b) 1880's-1911.
PL 31	Prices set to maximize market share: (a) H,A; (b) 1880's-1911.
PL 32	Strong management development: (a) A; (b) 1880's-1911.
PL 33	Partial ownership of some marketing affiliates: (a) A; (b) 1870's-1911.
PL 34	Gradual entry into foreign producing and refining: (a) G; (b) 1890's-1911.

Notes:

(a) Type of strategic choice (see Part 1, Ch. 1, Fig. 2)

S—Scale	PL 4
G—Geography	PL 3,29,34
H—Horizontal concentration	PL 6,31
V—Vertical integration	PL 1,5,11,26-29
P—Product diversification	PL 2,30
A—Administration	PL 7-10,12-25,30,32,33

(b) dates policy was in effect

Exhibit S-6
Standard Oil Environmental Conditions

EC 1	Depression 1869-74.
EC 2	Overcapacity in refining industry—1870's.
EC 3	Horizontal combination in refining—1870's.
EC 4	Heavy fixed investments required in refining, gathering, storage, and transportation.
EC 5	Vigorous, wasteful, competitive, short run chaos in the producing industry—1860's-1930's.
EC 6	Foreign marketing of Russian kerosene—1880's.
EC 7	Invention of tanker—bulk marine transport.
EC 8	Standard barred from Texas—1900.
EC 9	Dissolution of the trust as a result of a suit won against Ohio Standard in 1892.

4

The Strategy of Exxon

4.1 Overview

The Standard Oil Company (New Jersey), renamed Exxon in 1972, is the most direct descendant of Standard Oil. It inherited the policies, the lion's share of the assets, and the people who ran Rockefeller's combination from the 1911 divestiture (EC1).* Many of those policies are just as characteristic of Exxon today as they were of Standard at the turn of the century. There have been few changes and no major shifts inconsistent with the strategy of Standard Oil. Jersey's story is the continuity of a very successful strategy and system of management, both tributes to Rockefeller's extraordinary abilities. It also demonstrates the persistence of strategy despite great environmental change.

Jersey's size and financial strength in 1912 (R11a,b; PL1,17), though reduced from that of Standard, were still awesome. Jersey received $285.4 million in net assets, 43% of the combination's $660.4 million total in 1911. The rest, $375 million in net assets with $86,774,000 (91%) in 1911 earnings, was dispersed among the 33 divested affiliates. It was still the largest oil company in the world. In the U.S. only U.S. Steel was larger.[1] In 1911 Standard produced 90 TBD (= thousand barrels per day), refined 273 TBD, and marketed at least 165 TBD (perhaps as much as 200 TBD). In 1912 Jersey produced 11 TBD, refined 101 TBD, and marketed 43 TBD of refined products (R11c,d).[2]

In 1976 Jersey was still the world's largest oil company and, except for Shell, is two to three times the size of the other Majors (PL1,3). From the strategic comparisons in Table 1-2 (Chapter 1), Jersey is still stronger in refining and marketing than production (PL7,8). It has rebuilt Standard's worldwide organization, being almost evenly balanced between Eastern and Western Hemispheres (PL2). It is still a large volume wholesale marketer to a greater degree (PL8) than any other Major. Its affiliates are regionally and

---

*Policies (PL#), Resources (R3), Handicaps (H#), Environmental Conditions (EC#), Personal Values (PV#), and Social Responsibilities (SR#) are keyed to the exhibits at the end of the text. These categories are the strategy components of the business policy model (see Appendix A) and will be used for analysis in Chapter 11.

functionally (PL13,26) decentralized. Policies not shown in Table 1-2 that remain important are the committee system of management, an active board of inside directors (until 1966), little debt and large liquid assets (financial strength), and that most crude supplies have been acquired by purchase of crude or producing property rather than exploration (PL4,5,12,15,17). In diversification Exxon falls about in the middle of the Majors with a little bit of everything but approximately 98% of assets and income from oil, gas and chemicals. Exxon still has the strengths and the weaknesses of Standard Oil (H1-4, R11).

4.2 1912-1918: Dependence on Purchases

Though its size was greatly reduced from that of Standard, Jersey's strategy was virtually identical (PL1-21). From 1912 to 1918 Jersey stabilized its operations and attempted to rebuild what it had lost through divestiture. Though Jersey was not all of Standard Oil it is probably more accurate to think of it as Standard minus what was lost by divestiture rather than as a separate company. Its assets were diminished by divestiture but it inherited the Standard strategy intact.

Crude supply was the most critical strategic issue for Jersey for at least 40 years after 1911. There are four crude acquisition policies differing in the degree of vertical integration:

(1) spot purchase
(2) long-term purchase
(3) purchase and development of producing properties after discovery
(4) wildcat (new field) exploration—purchasing concessions in unproven areas and drilling discovery wells on them.

Jersey replied on the first three but rarely the fourth until the mid-1950's (PL4,5).

Jersey in 1912 was left with production in Appalachia, Louisiana, and Romania (R11c). Producing companies in Ohio (Ohio Oil), Kansas (Prairie Oil and Gas), Appalachia (South Penn Oil Company), and California (Socal) were divested.[3] Fortunately the dissolution decree did not prohibit doing business with former affiliates, so Jersey was able to obtain most of its requirements by long-term and spot purchase contracts with the divested affiliates. Beginning in 1911 Jersey had also made large purchases of Mexican crude at the wellhead for 10¢ per bbl. In 1912 Jersey produced only 11% of its U.S. refinery requirements (H1) and purchased 90% of the remainder from former affiliates (PL5,22).[4]

By 1918 it produced 27 TBD (19 TBD U.S.), 17% of its refinery requirements. Former affiliates, whose production was declining as they did not take the risks to expand production, supplied only 57% (78 TBD) of purchases and Mexican sources 28% (40 TBD).[5] The largest part of the increase in its own production came after 1915 when the Carter Oil Company, Jersey's Appalachian producing subsidiary, spurred by the discoveries at Cushing and Bartlesville, Oklahoma, established a western division in Tulsa under J. Edgar Pew, a member of the family that founded the Sun Oil Company in Pennsylvania. Using Jersey's great financial strength (PL15) Pew vigorously pursued the conservative and expensive policy of buying leases in proved producing areas, checkerboarding western Oklahoma with leases for Jersey from 1915 to 1917 (R15), when he quit, as well as buying up large amounts of flush production.[6] Carter drilled the discovery well in only one of the 31 fields in which it operated leases between 1912 and 1918 (PL4,5). Louisiana Standard, which had purchased its first producing leases in 1910, relied even more heavily on purchases (27.4 TBD purchased vs 3.2 TBD produced in 1918).[7]

Jersey Standard followed the same policies abroad (PL4,5) between 1912 and 1918. It purchased producing properties in Peru in 1913 (R13) through a Canadian subsidiary and in Mexico in 1918 (R18). Its two foreign explorations came as a counterattack on Shell (EC2) in 1912 when it formed NKPM, an organization to explore and produce in the Dutch East Indies in Shell's backyard and in Canada in 1917, also in response to Shell's interest there (EC2).[8]

In refining and marketing between 1912 and 1918 Jersey held its own. Its share of world refinery runs increased from 12-1/2% to 14% but declined in the U.S. from 18% to 15%. It expanded its marketing volumes within existing U.S. territories but entered no new ones and it continued to market through former affiliates (PL22).

Refining continued to be the bulwark of Jersey's strength (PL7). Refinery runs worldwide increased to 163 TBD (132 TBD U.S.) compared to net production of 27 TBD (19 TBD U.S.) and refined product deliveries of 62 TBD (28 TBD U.S.). The percentage of income before overhead charges from refining activities increased to 52% in 1917 from 22% in 1912 while that from both marketing and production fell.[9] In the divestiture Jersey kept six U.S. export refineries (R11d), including the world's largest at Bayonne, New Jersey. Four were on the Atlantic coast (three had a full product range; one produced lubricants only), one in Appalachia (lubricants), and one in Louisiana (Baton Rouge). All that was left of its domestic marketing affiliates was in eight states (New Jersey, Maryland, Virginia, West Virginia, North and South Carolina, Louisiana, Tennessee), natural gas sales to industrial and government units, and the wholesale trade (R11e). Abroad it retained marketing affiliates in Canada, the Caribbean and continental Europe (R11e). It lost marketing (H3) affiliates in all other parts of the U.S., Mexico, Great Britain, Africa, the Far East, and all manufacturers and marketers of specialty products (Vacuum—

premium lubricants; Chesebrough Mfg.—vaseline; Galena Signal—railroad oils).[10]

Jersey inherited the top management of Standard Oil (R11g). At the core of that management were John D. Archbold, A.C. Bedford, and Walter C. Teagle, the chief executives of Jersey for the next 25 years, and Heinrich Riedeman, the chief European voice within the organization, who combined Old World diplomacy with Prussian efficiency as the head of the German affiliate DAPG. The reputation of its management invited hyperbole (R11g): "We say here in Germany that the three greatest organizations in the world are our [the Prussian] army, the Roman Catholic Church, and the Standard Oil Company" (PL18).[11]

In 1914 Teagle went to London to set up foreign operations away from U.S. government scrutiny (PL15) and political hostility (EC3). The hostility toward any company which had been associated with Standard, and especially Jersey, was compounded by the fact that with Standard affiliates still sharing the same headquarters (PL22) at 26 Broadway in New York and doing business with each other as usual, the public doubted the efficacy of the divestiture. When the war broke out Teagle moved to Toronto, Canada, as head of Imperial Oil Company (Standard's Canadian subsidiary), from which place he set up International Petroleum Company to acquire South American producing properties. In 1915-1916 he built up Imperial, adding refineries in Vancouver, Alberta, Montreal, and Halifax, expanding marketing, and establishing an exploration unit. In 1914 he also acquired from Standard Oil of Indiana sole rights abroad for the Burton-Humphreys thermal cracking process for Jersey. Jersey also acquired (1913) and expanded a refinery (4.6 TBD) at Talara, Peru (with its producing properties) and built one at Tampico, Mexico (8.6 TBD) (1914) to produce fuel oil (R13-14).

Hostility toward Standard (EC3) and therefore Jersey existed outside the United States as well. Between 1911 and 1914 Jersey fought off a movement to create a German petroleum monopoly by law, a move instigated by Deutsche Bank and Shell, Jersey's two chief competitors, who wanted to take over its 75% share of the German market (EC2).[12] More broadly, the nineteenth century was over and with it the laissez faire attitude toward foreign investment, exploration, exploitation, colonialism, and power. From the first nationalization in Russia in 1920 to the present the pendulum of government control over the oil industry has swung steadily toward limiting the ability of private enterprise to conduct business freely (EC8).

Jersey lost control of nearly all the pipelines which had been such an important tool in Rockefeller's strategy to dominate production (PL3,6) but it continued to use the pipelines of divested affiliates. In 1919 67% (75 TBD) of Jersey's East coast refinery runs moved this way to Bayonnne and Bayway (PL22).[13]

In 1912 even with the 31 tankers (23 German, 8 Dutch) retained by European affiliates (R11f) for carrying refined products to Europe, the world's second largest petroleum fleet, Jersey faced an acute shortage of controlled tonnage (H2) after divestiture (EC1) to haul products and Mexican crude (EC4). The largest fleet, Anglo-American's 34 tankers and sailing vessels, had been divested as had Standard Oil Company (New York's) five ships. Jersey expanded its European built fleet before the war (EC5), transferred most of them (26 of 35) to American registry in 1914-1915, and began building ships in the U.S. well ahead of competitors to increase control over its tanker transportation (PL6,22). By 1919 the fleets of Jersey and affiliates numbered 79 ships, second only to Shell's 100, and far ahead of New York Standard (33), Anglo-American (29), Anglo-Persian (24), Gulf (22), Pan-American (20), Texaco (15), Burmah (10), and Socal (10).[14]

The second major event in the petroleum industry between 1912 and 1918 was the shift in product mix away from kerosene (especially in the U.S.) with the rise of the automobile and electricity and toward gasoline as motor fuel and fuel oil as marine, rail, and utility boiler fuel (EC6). Jersey, for example, built the Tampico refinery (R14) to refine fuel oil from Mexican crude. Jersey's product mix shifted toward gasoline and fuel oil (PL23) from 1912 to 1927 as it sought to exploit its strength in refining (PL7) and the wholesale trade (PL8) and minimize its vertical and geographical weakness in retail marketing (H3). It sold these products in bulk to former marketing affiliates (PL8,22).

In the first years after divestiture (EC1) (1912-1918) Jersey tried to minimize its impact by maintaining its strategy (PL1-21) and relationships with former affiliates (PL20), and by rebuilding where it could not (example: tankers discussed above) (H2). Its greatest weakness was dependence on purchased crude (H1), a consequence of its strategy (PL4,5). In the U.S. it did not expand geographically except into production in Oklahoma, choosing instead to expand volume within existing territories (PL19) as it tried to cope with the changes in its product mix (EC6,PL21). It expanded mainly abroad (PL23) in production (Mexico, Peru, the Dutch East Indies) and refining (Canada, Mexico, Peru), building export refineries abroad for the first time (PL24) and beginning to change its foreign supply by U.S. export policy (PL9). Jersey's concern in this period with expanding its foreign business (PL24), which accounted for over half its marketing volume (R11h), was the result of its worldwide identity (PL2), of hostility and antitrust restrictions at home (EC1,3), compared to the freedom to expand and integrate abroad, World War I (EC5), which confined expansion to the Western Hemisphere, the greater importance of its principal product, kerosene (PL10), in foreign markets compared to its decline in the U.S., and the global interests and leadership of Walter C. Teagle, Everit J. Sadler, and Heinrich Riedeman (PV1-6).

4.3 1918-1927: Expansion of Foreign Production

In the decade after World War I an important shift took place in Jersey's crude supply strategy from dependence on purchases (PL5) to expansion of controlled production, mainly by purchase of proved reserves and producing properties (PL23). It also continued to emphasize foreign expansion (PL22) due to antitrust aftermath at home (EC1,3).

By 1917 the wisdom of depending on purchased crude was already being questioned. Chief critic of dependence on purchases was Everit J. Sadler, who had built up Standard's production in Romania from 1909 to 1916 and who was to be a strong voice within Jersey for production and foreign operations during his tenure as a director from 1920 to 1942.[15] Teagle, who became president in 1917, agreed.[16] Not only was it an expensive policy,[17] but the company had no control over its sources. The former producing affiliates, having no refinery investments to protect, had not been obliged to take risks to expand production and had not kept pace with the expansion of American production, much less Jersey's refining capacity. After World War I crude prices were rising, there were fears of a shortage (EC7), and Shell was expanding rapidly under Deterding. In 1918 Shell produced 83 TBD to Jersey's 26 TBD and had virtually locked up reserves in many places abroad.[18]

A change in procurement policy took shape. Jersey would acquire (PL25) production worldwide, especially foreign production. Sadler wrote in 1919:

> It appears to me that the future of the Standard Oil Company, particularly the New Jersey Company, lies outside of the United States, rather than in it. This is due primarily to the fact that the New Jersey Company's business is largely outside of the United States, its principal refineries are on tidewater, and it seems naturally designated for expansion in foreign fields. It is also true that the trust laws of the United States and their present trend seem to preclude continued expansion in this country... (PL24,R11d,h,EC3).[19]

Jersey spent millions of dollars wildcatting unsuccessfully (H4) in South America in the 1920's in Argentina, Bolivia, Ecuador, and Venezuela (PL1,16,17).[20] It purchased the DeMares concession in Colombia (R20a) in August 1920 with three successful wells. By 1927 its net production of 36 TBD was the second largest of all Jersey's producing subsidiaries. It also acquired Venezuelan concessions in 1920 (R20b) (the second Major after Shell [1912] to do so) but its wildcatting efforts from 1921 to 1928 were so unsuccessful (H4) that the director in charge of the venture began to be known at headquarters as the "non producing production director."[21] By 1926 Jersey was exploring the possibility of purchasing a large share in a Shell producing subsidiary in Venezuela, the Colon Development Company (PL5,24,25).[22]

In 1922 Jersey established a Foreign Producing Department (PL22,23) in Paris to keep tabs on postwar oil deals. From 1922 to 1924 possibilities were investigated in France, Italy, Czechoslovakia, Spain, Tunis, Egypt, Arabia, Abyssinia, and the north, east, and west coasts of Africa. Jersey was interested only in large fields (PL1) and these efforts were without result. By 1926 a decision was made to limit European efforts to Romania, Poland, and Russia. Efforts to revive Romanian production were hampered by inefficient management, war damage claims, and a 1924 law requiring foreign concerns to turn over 55% of their stock over to bona fide Romanian citizens (EC8). In 1919 Jersey bought a 25% interest in the Nobel brothers' vertically integrated company in Poland, but although its sales were the largest in Poland, its production was never sufficient to supply the market with the financial result of consistent losses until it was sold to Socony-Vacuum in 1937. In 1920 the Nobels offered Jersey a 50% interest in their properties in Russia, where they were the largest producers. The properties fell into the hands of the Bolsheviks in July 1920, but, anticipating their defeat, Jersey actually paid the Nobels $6.5 million in a gamble to acquire half or more of the Russian petroleum industry for $14 million (PL1).[23]

Jersey also ran up against politics elsewhere (EC3,8). In the Sykes-Picot Agreement of 1916 and the San Remo Oil Agreement of 1920, Britain and France set up a petroleum alliance in the Middle East whereby Britain would be assigned a mandate over Palestine and Mesopotamia while France would get Syria, Lebanon, and the Germans' 25% interest in the Turkish Petroleum Company with the Americans, and specifically Standard, excluded. Jersey had faced similar efforts to exclude them and favor nationals by the Dutch, Germans, Mexicans, and several South American countries (EC3,8). Establishment of foreign production was one way of politically bolstering the company's right to compete abroad in the more important refining and marketing businesses as well.[24]

From 1919 until 1928 Jersey led a long and ultimately successful struggle, with frequent proddings by the U.S. State Department for Britain and France to honor the Open Door policy, to be admitted to the Middle East. On July 31, 1928, in the famous Red Line agreement (EC9) a group of American companies led by Jersey obtained a 23.75% interest from Anglo-Persian in the Turkish Petroleum Company (R28), which had discovered oil in Iraq on October 15, 1927. They were limited to conduct joint operations only by the agreement in most of the Middle East, precluding independent expansion. Jersey also negotiated extensively between 1921 and 1923 without result for the Khostaria concession in the five most northern states in Persia.[25]

Among Jersey's foreign producing efforts begun previously the results were mixed. In Canada there were dry holes. Romania struggled along at less than 4 TBD, which was below prewar levels. In Peru production expanded from 5 TBD to 20 TBD from 1918 to 1927, but the production of Transcontinental, the Mexican subsidiary acquired in 1918 (R18), rose sharply to 59 TBD in 1923 then fell below 10 TBD by 1927. NKPM, in the Dutch East Indies, made a small discovery in 1922. Sadler and Seth B. Hunt, a refiner who became Jersey's producing director, favored selling NKPM to Socony. But Teagle argued successfully against giving up a property in Shell's home territory (EC2), paving the way for the formation of Stanvac a decade hence.[26]

Despite its desires (PL24-5), Jersey's best producing investment from 1918 to 1927 was at home. In 1917-1918 it was considering re-entry into Texas, having been expelled in 1909 by a state antitrust decision.[27] In 1918 a purchase of the Texas Pacific Coal and Oil Company, which had just brought in the important Ranger field in north Texas, fell through. On January 19, 1919 Jersey acquired 50% of the Humble Oil and Refining Company, a group of independent Texas producers led by W.S. Farish, who later succeeded Teagle as Jersey's chief executive. For $17 million it got an interest in Humble's 16.5 TBD production and 7.5 TBD wellhead purchases and, most importantly, an able producing organization. In 1919 under Wallace E. Pratt Humble began to use subsurface geologists and by 1924 geophysics in exploration (R19,H1).[28]

But it did not get control, in spite of the fact that Jersey's legal department instructed its Houston attorney to buy five shares of Humble's stock on the open market just after the acquisition to assure majority ownership. The fear of future antitrust actions against Standard Oil interests (EC1,3), Jersey's tradition of local autonomy (PL13), plus the native independence of the Texans gave Humble Jersey's financial backing (PL17,R11b) without sacrificing its autonomy, beginning with the circumstances of its acquisition a long tradition of autonomy within Humble from Jersey. By 1921 Humble had risen from fifth to first in production among Texas oil companies.[29]

Both Louisiana Standard and Carter found success with protective leasing—leasing in promising but unproved areas, frequently paying high lease bonuses—as others made discovery after discovery from 1921 to 1924. Neither did much wildcatting and their efforts became increasingly conservative. Carter, it was said, never adopted any new technique until it had been declared obsolete by its rivals (PL4,14). Carter concentrated its efforts on one field at a time. Louisiana Standard continued to prefer purchase to production, especially after the decline of crude prices in 1921. When production tripled between January 1922 and September 1923 in California, Jersey purchased large amounts (26% of 1923 refinery runs) from Socal and others (PL1,5,22).[30] The variety of approaches of the U.S. producing subsidiaries is evidence of the considerable degree of local autonomy Jersey permitted (PL13).

By 1927, though Jersey remained a very large purchaser, it had increased its net production since 1918 from 14% to 32% of refinery runs in the U.S., from 25% to 91% abroad, and from 17% to 44% worldwide.[31] It had tried to establish itself as a producer in every place where significant production existed, a large-scale worldwide strategy which demonstrates its massive ability to respond using its financial strength (PL1,2,16,17). This record also shows its lack of success in exploration (H4,PL4).

By 1927 Jersey's refining volume had risen to 434 TBD (349 TBD U.S.), 14.3% of worldwide refinery runs (15.4% of U.S.). Its U.S. share was more than twice that of the next largest refiner, Socal, and 2.5 times the combined total of Texaco, Gulf, and Shell in the U.S. It was still the world's dominant refiner, with a market share little changed since 1918, and refining was still its central business (PL3,7). It marketed 247 TBD (110 TBD U.S.), 57% of its refinery runs compared to 38% in 1918 (PL7). The rest was sold wholesale (PL8). It held 10% of the U.S. market and 23% of the foreign market for refined products. Abroad Shell held 16%, Anglo-Persian 11.5% and Russian suppliers 6.5%, making Jersey still by far the largest foreign marketer (PL1,3).[32]

Exports were still large (51 TBD) (PL9) but supplies from foreign refineries had risen to 38% of total foreign deliveries from 34% in 1918, 12% in 1912 (PL24). Local refineries were built abroad at Barranca, Colombia (1922), Calgary, Canada (1923) (4 TBD), and acquired by European affiliates at Libruza, Poland (1919), Trieste, Italy, Alicante, Spain, and Marseilles, France between 1918 and 1927. The European additions, like existing refineries and specialty plants in Valloy, Norway, Germany, Argentina, and Cuba, were a response to tariff barriers on refined products (EC10), necessary to maintain a competitive marketing organization. An export refinery was built at Palembang, Sumatra in 1925.[33]

In the U.S. a fuel oil and asphalt refinery was opened in Charleston, South Carolina in 1920. But the major downstream shift from 1918 to 1927 was the rise of the Gulf coast refineries, following the shift of production from Pennsylvania and Ohio to the mid-continent and Gulf coast (EC11). Humble built a coastal refinery at Baytown, Texas in 1920. Jersey's Gulf coast refinery runs, half as large as its Atlantic coast runs in 1918, exceeded them in 1926. Despite the addition of seven local or interior refineries compared to two export refineries, 82% of Jersey's refinery runs (86% in 1918) came from its six U.S. and four foreign export refineries (PL1,9).[34]

The cost and obsolescence of the Burton-Humphreys thermal cracking batch process (EC13) convinced Teagle by 1918 that Jersey should develop its own gasoline cracking process. He hired E.M. Clark away from Indiana Standard to set up a research organization at Jersey (PL27). A gifted administrator as well as scientist, Clark was for many years a leading voice in Jersey, persuading its reluctant refiners of the value of new processes (PL13,18).[35] For example, Louisiana Standard did not adopt Clark's improvements until 1931. Reversing Jersey's lag in innovation (PL16) by 1921 Clark's organization had developed a continuous tube and tank (called the "Ellis") cracking process (R21). New patents led to large lawsuits for infringement by competing processes that threatened to retard development. In

1923 the three leaders in refining technology, Indiana Standard, Texaco, and now Jersey, signed a patent pooling agreement ending the lawsuits and allowing each company to pursue future developments without infringement.[36]

Evidence of the change in Jersey's policy toward innovation (PL16) from follower to leader (PL27) were the patents which led to its 50-50 joint venture with General Motors (called Ethyl Corp.) to manufacture and market tetraethyl lead, a key component of the new higher octane ethyl gasoline. Jersey's first industry diversification moves, purchase of the Ellis patents for making isopropyl alcohol from lighter petroleum fractions in 1918 and negotiations beginning in 1920 with I.G. Farben interests which led to a 1927 partnership to develop the Bergius coal liquefaction patents, came from its new interest in research.[37]

In spite of its strength in refining gasoline, Jersey fell (H3) further behind in retail marketing as competitors, especially Gulf and Texaco, entered its territories and rapidly expanded their retail organizations by leasing and licensing retail stations to independent jobbers and operators. Jersey confined its marketing efforts to 11 states, continuing its policy of expanding volume within existing territories (PL22). By contrast in 1926 Gulf marketed in 26 states and Texaco in 44. From the refinery gate in 1919 50.9% and in 1924 47.8% of its domestic gasoline was sold wholesale to Standard Oil of New York. In 1927 Teagle testified that over 60% of Jersey's sales of all products were made in tank-car and barge lots (PL8).[38]

In addition to its historic policy of wholesaling (PL8) there were several reasons why it did not keep up in retail marketing (H3). Until 1924 the Marketing Committee (PL14) was headed by a succession of experienced executives nearing retirement. It had no brand name under which it could enter the territories of former Standard affiliates. On the advice of its legal department to let others test its legality Jersey did not adopt the lease and license policy which enabled its competitors to expand retail outlets so rapidly. It chose to own and operate its gasoline stations. Even after J.A. Moffett, Jr., a younger and more energetic director, took over marketing in 1924 and began to build and acquire stations more rapidly, each acquisiton required advance approval of the Legal Department so as to always leave a reasonable amount of business to competitors, an inhibition to competition brought about by the divestiture (EC1). Another restraint on retail marketing by the Legal Department affected price policy. Fearing prosecution under price discrimination statutes, Jersey could not respond to local price cutting. In July 1925 it finally reversed its tankwagon price policy to reflect statewide averages.[39] This experience is indicative of the difficulty of achieving (H5) consensus (PL18) and performance in a large, rambling (PL13) organization. Jersey often used its size and financial strength (R11a,b,PL1,2,17) to great advantage. But in other areas like retail marketing it had inherent disadvantages and could be easily outmaneuvered by smaller competitors.

Another notable event in the 1918-1927 decade at Jersey was the establishment of the Medical Department (SR1). It was established in 1918 as part of an extensive employee benefits program, coming a year after bloody strikes in 1915 and 1916 at the Bayonne refinery. Its most important work, directed by Dr. Alvin W. Schoenleber, lay with improving sanitation and medical treatment with the producing companies in the tropics. Progress did not come easily as conflicts arose between doctors and producing managers. In one incident in Colombia typical of Jersey's careful personnel management and emphasis (PL16) on cooperation, the local manager obstructed medical work and the chief resident physician was aggressively undiplomatic. Both were replaced. Physicians at Jersey were pioneers in industrial medicine and made a lasting contribution in the countries they served.[40]

From 1919 to 1927 Jersey returned to the use of foreign flag tankers in all but the American coastal trade where they were prohibited by the Jones Act of 1917 (PL6,EC17). It added 27 ships to its fleets, fewer than the 48 and 59 added by Shell and Anglo-Persian from 1920 to 1927. When Mexican production fell off its ships were used to bring crude and fuel oil from California in 1922-1923. Jersey still relied heavily on charters at times, 36% of all cargo movements in 1927, for example. After the post-divestiture shortage was alleviated Jersey's fleet grew with its business and helped maintain a low cost position (PL6) for its East coast refineries.[41]

As production shifted to the mid-continent and Gulf, where it could be transported more cheaply by tanker to Jersey's East coast refineries, and as foreign production expanded, pipelines in the U.S. became a less important component of strategy. In 1921 when Prairie, a former affiliate, refused to cut its pipeline charges Jersey changed policy to regain control over transportation (PL6) and expand within its own pipeline network, supplemented by its strong domestic fleet. By late 1923 Jersey's pipelines had sufficient capacity to move all its requirements to the Gulf coast.[42]

The postwar decade 1918-1927 was particularly important for Jersey strategically as it recovered from divestiture (EC1) and the war (EC5) and began to expand again. What is noteworthy is that it did so without major changes in its strategy. Where it changed policy (PL22,23) the direction was that of return to the strategy of Standard Oil (PL4-6). The challenge of divestiture and events to its strategy had been met. Gibb and Knowlton in their title called this period "The Resurgent Years."[43] The strategy had survived intact and its strengths (R11-21) and weaknesses (H1-5) would guide Jersey's actions for at least the next half century.

4.4 1927-1955: Reorganization, Crude Acquisition, and Research

The most important strategic event of this long period for Jersey was its massive acquisition of crude reserves and production, beginning with purchases in the U.S., Venezuela, and Iraq from 1928 on and concluding with participation in the Iranian consortium in 1954 (PL5,H1). It caught up with and then surpassed Shell as the leading producer and satisfied its need for vertical balance with remarkably little exploration by using the opportunities created by the crude glut (EC29) and Great Depression (EC39), its financial strength (PL17), and a willingness to make massive commitments (PL16-17). The ad hoc reorganizations, as Chandler calls them, led Jersey to address its coordinative problems earlier than any other major and demonstrated the ability of Rockefeller's system of management (PL13,14) to meet new challenges without changing strategy—to face and solve problems—to manage rather than ossifying to the point of merely responding to events. A third set of events began with a massive push into research (PL27), entanglement with I.G. Farben and Hitler, and left Jersey unable to respond to charges of treason in 1942, resulting in tragic career ends for two of its chief executives and realization of the pubic accountability of large private companies (PL15).

As Chandler compares it to DuPont and General Motors, Jersey Standard did not devote the same degree of attention to organizational problems nor was it so consciously creating a new corporate structure of autonomous multifunctional divisions to replace functional departments.[44] But the problems Jersey addressed in the 1920's were quite similar to those Shell and Mobil addressed in the late 1950's and others (Gulf, Texaco) later still. The broad nature of the main problem was how to organize to maintain efficiency and low cost in a large organization as volume and the number of transactions grow. In 1927 Jersey was roughly four times as large in volume and assets as it had been in 1911, a compounded growth rate of 9% per year which placed increasing strain especially at the top levels of the Jersey organization. Reorganizations in 1925, 1927, 1933, and just after World War II were designed to reduce these strains.

Since 1911 the functional committees (Manufacturing, Export Trade, Production, Natural Gas, Lubricating Oil, etc.) (PL12) had changed little and their members had grown old. The directors Teagle appointed were the heads of functional departments and tended to stick to their specialties. These specialist directors had little time for, information about, or interest in the administration of the corporation as a whole or in other functions. An inventory crisis in 1925 in which refineries were continuing to manufacture products that marketers could not market triggered the formation of a Coordination Committee and department to manage product flows and a finance or Budget Committee and department to manage financial flows (PL14). Heading the Coordination Department was the first senior job for one of Jersey's ablest executives, Orville Harden. Though its powers were only advisory (PL16), it soon became essential and its recommendations were followed because it provided important information in the form of short and long-term supply and demand forecasts for each product. Soon it was approving product specifications and reviewing capital expenditures. The Finance Department sought to develop an accounting system that produced financial data that could be used to measure return on investment and managerial performance.

Another problem was that Jersey was both an operating company with refining activities and a holding company with interests in 90 subsidiaries. Board members found themselves constantly preoccupied with operating problems rather than long range plans. Most subsidiaries were active only in a single function and their coordinative problems with other subsidiaries forced their way up the organization to surface at meetings of the directors. As refining was the strongest function it had multiple representation on the board. It was difficult to get specialist directors to focus on broader issues with the result that some major strategic moves like the expenditure of $8.8 million on Russian properties *after* the 1917-1920 revolution were not well thought through. Also the multifunctional subsidiaries like Imperial and Standard Oil (Louisiana) had no board representation and were less closely managed.

In 1927 Standard Oil Company (New Jersey) became a holding company only and its operating subsidiaries were placed in a new company, Standard Oil Company of New Jersey, known within the company as the "Delaware" company after its place of incorporation to distinguish it from the parent. Other subsidiaries were placed in five autonomous multifunction operating divisons (Louisiana, Europe, Latin America, Imperial [Canada], and Humble [Texas]) divided along regional (PL26) lines. Directors were added to the board to represent each major divison, but since most regional directors came to New York only occasionally day to day management continued to fall on the New York officers and an active inside board of directors (PL19) continued. They were, however, relieved of operating problems which had been pushed out to the divisons (PL13). These changes were continued in 1933 with the wholesale retirement of top management personnel as it was decided that board members should not have operating responsibilities but should be policy making general executives only. To separate itself physically from operations the parent company moved from 26 Broadway to 30 Rockefeller Plaza. The Delaware and Louisiana companies were reorganized during the depression to concentrate on a single function as hard times (EC30) caused the divisions to concentrate on their dominant functions. After the Second World War they again became regional multifunction units (PL26) and the Louisiana and Delaware companies were merged to become Esso Standard.

Jersey's acquisition of crude reserves is one of the remarkable stories in the history of the oil industry because (1) it was done almost entirely by purchase of proven or promising properties (PL5) and very little by exploration (PL4) and (2) the motivations were clearly its long run strategic needs at a time when economic calculations indicated the need for stress on holding market share rather than increasing production or reserves. No other Major pursued such a course to any similar degree. Its strategic needs were that the divestiture (EC11) and rapid growth (EC6) had left it crude short (H1) and it wanted to compete with Shell (EC2,PL3) worldwide. These massive purchases are responsible for its strength today and would not have been possible without its strong financial position (PL16,17).

These purchases had begun in 1919 with the acquisition of Humble, the entry into Near East Development (Iraq) in 1928 (R28a), and the failure of exploration in the 1920's to produce significant results, especially in Venezuela.[45] Perhaps this is why Sadler early in 1927 turned down Major Frank Holmes' offer of concessions in the Middle East before it signed the Red Line agreement in July 1928 (EC28). In June 1928 it purchased The Creole Syndicate (R28b) with production from fields on the edge of Lake Maracaibo and the Trinidad Oil Fields (R28c) Operating Company, Limited for its production and advantageous location for supplying marine fuel. The prices were $19 million and $2 million, respectively. From 1929 to 1933 it held merger talks with crude rich, Western located, and Pacific oriented Socal which would have added production (PL5), geographical expansion in the U.S. (PL2), and a competitive threat to Shell in the Pacific and on the U.S. West coast where Shell was strongest in the U.S. But the talks eventually broke down.[46] In 1932 Jersey acquired the 96% interest of Standard Oil Company (Indiana) in the foreign properties of Pan-American Petroleum and Transport Company (R32): a large fleet, 16 TBD production in Mexico (1931), 88 TBD production and 550 million barrels of reserves in Venezuela, a large refinery on the Dutch island of Aruba, and other properties in Latin America and Europe. The price was approximately $140 million in cash (1/3) and stock (2/3) which was only 87.15% of the net book value of the properties. In 1937 Jersey's International Petroleum subsidiary in Canada bought 50% of Gulf's Mene Grande Oil Company in Venezuela for $100 million. When it appeared that large expenditures would be required to develop the property it resold in 1938 25% of Mene Grande to Shell for $50 million. In 1927 after 15 years of fruitless wildcatting its NKPM subsidiary in the Dutch East Indies obtained commercial production on purchased lands and a year later a large new concession from the Dutch.[47]

But all this foreign expansion was merely a prelude compared to the activities of Humble in Texas.[48] Under the leadership of Wallace Pratt Humble became expert at geophysical surveying and reservoir management (PL25). It used these skills to buy proved reserves and to lease large blocks of promising acreage (PL1,5,R33). Between 1927 and 1938 it spent $85 million in cash for leases in Texas, Louisiana, and New Mexico, doubling its leased acreage to 8 million acres even after the acreage it surrendered. But its reserves (R39) rose from 55 million barrels in 1927 to 2.4 billion barrels in 1939 while U.S. reserves doubled to 18.5 billion barrels. Jersey's other U.S. producing subsidiaries, Carter in Oklahoma and Standard of Louisiana, achieved smaller successes.

One of Humble's largest block leases was the 1 million plus acre King ranch in Texas. Humble's practice of large block leasing coincided perfectly with Jersey's large-scale activities (PL1), desire for large market shares (PL3), financial strength (PL15), capacity for massive action (PL14), and the economic efficiency of unitized field management. Though Humble did make discoveries in the 1930's it was more interested in banking reserves for the future (PL5,21) than in adding to the glut (EC29) of production in the 1930's by exploration (PL4). By 1938 Jersey produced 562 TBD (166 TBD U.S.) equal to 69% of its refinery runs of 818 TBD (373 TBD U.S.) while it marketed 70% (572 TBD) (187 TBD U.S.) of its refined products itself.[49] Before World War II it had caught up to Shell in worldwide production.[50]

After others made large discoveries in the Middle East in the 1930's, Jersey desired a larger position there than its participation in Iraq Petroleum provided, and one in which it was not hemmed in by BP, Shell, and the Red Line (H6). It wanted to strengthen its position in the Eastern Hemisphere to at least match that of Shell (PL3, EC2). It had led the American companies into the Middle East and did not want to see others reap the benefits. Throughout the 1930's it held discussions with Socony, Socal, and others to participate in some way in Saudi Arabia and Bahrain. In events related in the Socal chapter in 1947 it was successful in purchasing 30% of Aramco (PL5, R47) for $77 million in cash and $103 million in preferential dividends from future profits, a cheap price indeed. Its principal contributions were really markets and capital to develop the concession. With the Aramco participation it achieved the position it desired. In 1954 it made its last major purchase by participating in 7% of the Iranian consortium (R54). Unlike previous crude acquistions this time its decision was "political." "We would have made more money if we had done added drilling in Saudi Arabia; we had plenty of oil there. We were pushed into the consortium by the United States government," a director later explained. "We recognized the dangers if Russia got in; we had a real interest in seeing to it that the problem was solved. We didn't balk at participation. Had the Russians gotten Iranian oil, and dumped it on the world markets, that would have been serious."[51] Whatever the reason, it did not abandon its policy of purchasing reserves (PL5,25).

Walter Teagle had pushed Jersey heavily into research (PL27) after World War I to keep up with scientific advances being made in refining and avoid costly licensing fees like those Jersey had paid for the Burton-Humphreys process. The first results of this effort were advances in gasoline cracking and the manufacture of isopropyl alcohol from the purchase of the Ellis patents.[52] In 1924 the Ethyl Gasoline Corporation was organized as a joint venture with General Motors to manufacture tetraethyl lead as a gasoline additive. In 1925 it became interested in Colorado shale as an oil source and in the patents of the German firm B.A.S.F. for converting coal to oil.[53] This led to another joint venture in 1927 with I.G. Farben, which had absorbed B.A.S.F. in 1926, to develop the process and split any royalties. In 1929 a further agreement followed to pool patent information on oil and chemicals which had collusive implications, as Sampson quotes, "the I.G. are going to stay out of the oil business and we are going to stay out of the chemical business." (Petrochemicals did not become a significant industry until after World War II.) The Germans got tetraethyl lead, crucial to aviation fuel, while Jersey awaited the Germans' development of synthetic rubber, results which were delayed by Hitler, and held back its own efforts. In 1939 after the outbreak of war in Europe they continued to exchange information and dissolved the venture subject to a profit sharing agreement.[54]

There was criticism of patent arrangements like these in the 1930's as contributing to monopoly over technology and retarding progress, and Jersey, the close-mouthed heir to Standard Oil (PL13) was an obvious target. In 1941, after searching company files, the U.S. government brought two antitrust suits against Jersey: one alleging conspiracy to control oil transportation through pipelines, and a second attacking the patent pooling arrangements with I.G. Farben. The first suit, against the American Petroleum Institute and a score of companies including Jersey (the largest pipeline owner, however) (PL6), was settled by a consent decree limiting dividends to pipeline owners to 7% of property valuation. The second suit included criminal charges against Walter Teagle, then chairman but semi-retired, and W.S. Farish, who had succeeded Teagle as president and chief executive in 1937. A consent decree was offered early in 1942 so as not to hinder the war effort. There was bitter debate within Jersey's board as to whether to accept it, with Teagle being against, but in the end they did. The Standard-I.G. patents were released royalty free and small fines were paid in the belief it would be poor judgment to contest the suit with public opinion so hostile.[55]

Then the antitrust chief, Thurman W. Arnold, testified before Senator Harry S. Truman's Committee on National Defense, alleging Jersey's "cartel arrangements with Germany are the principal cause of the shortage of synthetic rubber." He accused Jersey of having sought a protected market, of limiting competition, of not developing manufacture itself, allowing others to do so, and not sharing its knowledge with the U.S. government while sharing it with I.G. Farben. He specifically disclaimed unpatriotic motives but when a reporter asked Sen. Truman if such action was treasonable, he replied, "Why yes, what else is it?"[56] Having entered a consent decree and paid fines, Jersey and its two top executives stood accused of treason (EC42).

The company, and especially Walter Teagle, were made scapegoats for the wartime shortage of a crucial material. Both men were broken by the ordeal.[57] Teagle resigned; Farish died. Whatever the merits of the charges, Jersey had been manipulated by Hitler. More importantly, while it could defend itself in court, it had prepared no defense against public opinion. For years its legal defenses in antitrust matters rested upon the opinions and distinguished reputation of John W. Davis, former Solicitor General of the U.S. and Democratic presidential candidate in 1924. Since 1911, with the Standard Oil albatross around its neck, it had gone about its business quietly, often secretly, strictly observing the law and honoring its contracts, continuing the old Standard Oil habits (PL15). Teagle had employed agents who reported solely to him. Even within Jersey board members were not acquainted with all of Jersey's activities. In the ordeal of 1942 it found it could not defend itself. It was politically naive and did not realize that is actions had to be justifiable to the public as well as its shareholders. In response it upgraded its public relations efforts, but it has never approached the level of effort Shell, notably free of such problems, has devoted to politics, openness, and public relations.

These tragic events tended to obscure Jersey's real accomplishments in research during the interwar period. It did develop butyl rubber (R42), a 100% Jersey accomplishment, and shared the technology during the war.[58] Two other products, isopropyl alcohol, first produced at the Bayway refinery in 1920, and Paraflow, the first in a long line of lubricant additives from the Baton Rouge refinery after 1927 (R20,27), plus synthetic rubber were the seeds for what would become a huge petrochemical business, second in size only to Shell in the oil industry.[59] The foundation was laid by the same commitment to research (PL27) that produced the Farben debacle.

Another strategic issue Jersey addressed in the late 1920's was the geographic expansion of its retail marketing territory (PL2). In 1926 it had only 5+% of the U.S. gasoline market and marketed (H3) in only 11 states. Five oil companies had larger market shares and several marketed over a wider area (H3).[60] In the U.S. Jersey sought to expand its territory by discussing a merger with Standard of California.[61] It formed an affiliate in Pennsylvania. In 1929 it acquired the Beacon Oil Company with markets in New England and New York and a refinery at Everett, Massachusetts.[62] All its affiliates built or acquired additional retail outlets. Abroad in 1930 it acquired (R30) Anglo-American to hold onto its largest customer for U.S. refined product exports (PL9).[63] A dispute with the Dutch government resulted (R33b) in the formation of Stanvac in 1933, combining Socony-Vacuum's Far Eastern markets with Jersey's production.[64] With these measures its U.S. market share in gasoline edged up to 6% in the 1930's.[65]

4.5 1950-1973: Growth and Stability

The period since World War II has been one of remarkable growth and remarkable stability for Jersey Standard. From 1946 to 1973 its net income, assets, and equity rose approximately tenfold while volumes produced, refined, and marketed rose almost fivefold. Expressed as compounded growth rates, the financial aggregates grew at 9% on average over this time while volume grew at 5%. Jersey's figures were typical of the Majors as a whole. With high demand, cheap crude, and new and better products like higher octane gasoline, additives, and petrochemicals Jersey was successful in adding more value as well as volume to its products. Despite this growth Jersey, like the other Majors, lost market share to new competitors. Its share of world production fell from 11.8% in 1950 to 8.3% in 1973; in refining from 15.6% to 10.4%; and in sales from 15.0% to 8.9%.[66] With rapid growth, increasing profitability, new products, and heavy competition, it is not surprising that new strategic initiatives were minimal.

Another reason that stability rather than change was the keynote of the period for Jersey was that, taking the long view (PL19) during the 1930's, it had built for future prosperity that arrived during the postwar period. In a still longer view (PL21) the rebuilding job begun after 1911, while not finished, was well on its way. It was again represented in every part of the world (PL2), though not in every country or in most of the Midwestern and Western states (PL20) in the U.S. Since World War I it had rebuilt its crude reserves worldwide, not by exploration, but primarily by purchase (PL2,4,5). It now had foreign refineries and had begun to integrate vertically abroad, diminishing supply by export (PL9,24). Its fleet and pipelines had been rebuilt (PL6). It had invested heavily in research and no longer lagged technologically as it had before World War I (PL16,27). Its leaders had demonstrated the ability to adjust the structure of the organization periodically to the demands of change in their business (PL26). Taken together these were the accomplishments of the general of Jersey executives led by Heinrich Riedeman, Everit J. Sadler, William S. Farish, Orville Harden, and, most of all, Walter C. Teagle (PV1-7).

Also lending stability was the fact that all the rebuilding had been accomplished without any radical change in Jersey's strategy. Its product mix had changed from kerosene to gasoline and fuel oil as principal products (PL10,23) as the environment changed but its methods had not. It was still the same large-scale, long-term worldwide organization interested in as large a share of the business as possible (PL1-3,21). Refining was still its largest function (PL7). It did not export products from the U.S. as it once had (PL9) but in 1950 there was still a substantial product flow (about 200 TBD) from the Western to the Eastern Hemisphere from Caribbean refineries. It was not the leader in new field exporation or retail marketing (H3) that it was in refining and wholesale marketing (PL4,5,8,24).

Its management style was the same—deliberate, decentralized, consultative, carefully studied committee decisions (PL13,14,16,18) made by an active inside board of directors (PL19)—but with perhaps more openness (PL15) as it accepted public responsibilities in addition to those owed its employees and stockholders. After the debacle in 1942 each decision was studied by one junior committee, one senior committee, and one board committee before action, a process that did not build a reputation for precipitous action.[67] The company had always taken great care with its personnel decisions (PL20, SR2); they took up more time of the board of directors than all other matters together.[68] Policies of executive development, a critical issue even with John D. Rockefeller, had been polished to a fine art: recruiting for anticipated needs, stockpiling managers like crude oil reserves (PL21) for the long term, moving managers around to add to their experience, and filling each vacancy with the best man regardless of seniority.[69] It continued to take its legal and contractual responsibilities without question (PL15). It had maintained its financial strength and the will to use it (PL17). In the 1960's a director recalled reading out the company's cash reserves daily at each board meeting.[70] While it was still subject to the inertia and habitual weaknesses (H5) of any giant organization its ability to take massive action (PL16) was unmatched. While most if not all of the other Majors were struggling or would struggle with the problem of how to maintain an able management, there was no question inside or outside Jersey that both Rockefeller's strategy and structure of management (PL20) still worked to produce able managers and sound decisions. Of the seven Majors it is at Jersey that one has the most difficulty finding at least one major strategic error which changed its business or viability in a product, vertical or geographic area.

Thus with both its assets and strategy on a sound footing Jersey was the Major best prepared for the postwar era. The measure of its confidence that both the war (EC40) and the depression (EC30) were over was a huge capital spending program aimed at developing the Middle East, rebuilding its Eastern Hemisphere operations, and its now obsolete U.S. pipelines. In each functional area it prospered by continuing to do what it did best.

In exploration and production it continued to participate in many discoveries in the U.S. and abroad by holding good prospective acreage under lease. But it was not the worldwide leader in new field exploration. Often the discovery strikes were made by others. Jersey excelled as a producer and developer of fields rather than as a discoverer (PL4,5). Two exceptions were the discoveries at Le Duc near Edmonton, Alberta, Canada by Imperial in 1947 (R46b) which triggered the entry of Canada as a major producer, and its discovery (R59) of the Zelten field in Libya in 1959.[71] The Zelten field was the largest wildcat discovery ever made by the company, and it was a source of some amusement in the industry that a picture of the Zelten No. 1 well appeared on the cover of the 1959 Annual Report. It shared the discovery of the Groningen gas field in the Netherlands (R62), the world's largest, with Shell in 1962. Spurred by these discoveries its exploration program expanded in the early 1960's when others were trying to cope with a growing crude glut (EC57) from the enormous production in the Middle East. After its DeMares concession in Colombia expired in 1951 and was not renewed (EC51) it made a discovery there later in 1960 (R60). It was also the most successful of the Majors in developing production in Europe where its production came to rival the small amount it got from Stanvac in Indonesia. It was not a leader in offshore exploration in the Gulf of Mexico or in Indonesia where the volume of its production stagnated. Overall, if it had had to depend on its discoveries for supply, rather than on the advantageous acquisitions in the U.S. and Middle East, it would not have done so well (PL4,5).

In refining volume and capacity it continued to lead the industry (PL7). New refineries came on stream nearly every year (R60). It added capacity worldwide (PL2) but especially in the U.S., Canada, the Caribbean and Venezuela, and most of all in Europe. Its markets were large enough to support efficient scale refineries in most places so that it did not have to resort to suboptimal scale or joint ventures as the smaller Majors did. As it had in the U.S. (Baytown, Bayonne, and Baton Rouge) it concentrated its efforts on a few large-scale refineries (PL1). Even in the 1950's several of its refineries processed well over 100 TBD each. Like Shell it took full advantage of the favorable economics of large scale in refining.

The European market, with a heavier industrial demand component of (EC50) product consumption, favored Jersey's specialization in wholesale marketing (PL8) and fuel oil in product mix (PL23). Most oil companies wanted to produce gasoline and dispose of fuel oil. Countries were often more concerned with and placed more controls on fuel oil than gasoline. But Jersey, with vast Middle Eastern reserves suitable for fuel oil (R47,54), became the premier fuel oil seller. Rapid economic growth and rebuilding after the war (EC48) greatly enhanced demand. Exports (PL9) from Venezuela tapered off as North American consumption rose and Middle Eastern sources replaced them. As the glut from 1957 to 1969 (EC57) turned to shortage in the 1970's fuel oil prices rose most as dumping in Rotterdam ended. In 1977 European price indices for fuel oil were 7.7 times the 1965 level, compared to 3.4 times for gasoline, 4.7 for distillate and 4.4 for the total product barrel. Jersey still has the largest share of the European market, over 13%.[72]

During the postwar period in the United States Jersey filled in the gaps in its coverage. A 1959 reorganization (R59b) merged Humble into Jersey and Carter and other U.S. affiliates into Humble in preparation for extension of marketing to all 50 states.[73] The next year it entered five states to bring the total number in which it sold products to 40. When Socal bought Standard Oil of Kentucky, long a marketer for Jersey, it immediately entered the five states lost on its own and built 1200 stations in two years. In 1963 it entered the West by purchasing 3900 stations and a California refinery (135 TBD) from Tidewater and 1500 Signal stations in 1967. Like others it overexpanded and in 1977 sold off its stations in four states.[74] Still its share of the retail gasoline market was at least four percentage points less than its 10.8% of the U.S. market for all products, attesting to its strength in wholesale markets (PL8) like home heating oil and industrial fuels.[75] Besides being the largest crude producer in the U.S., it also has the largest natural gas reserves (R58).[76]

The other major postwar change in its markets came in late 1960 when it signed a consent decree with the U.S. Justice Department (EC60) to split up the marketing operations of Stanvac between the two partners. It settled some longstanding legal action by antitrust authorities seeking to apply American standards to the foreign activities of the Majors that went back to the Farben investigations and the "As Is" agreements of the 1930's. Since its marketing subsidiaries were in the 1960's able to stand alone it agreed to the decree to accomplish ends it welcomed anyway. Mobil opposed the split-up and never signed the decree.[77]

4.6 Changes in the Postwar Period

There were two administrative changes and three new entries that were of strategic importance in recent years: the appointment of outside directors, changing its name to Exxon, development of the chemicals business, development of alternative energy sources, and entry into venture capital.

The appointment of the first two outside directors in 1966 (R66, PL17) was part of a reorganization begun in 1965 that created regional organizations for Latin America, Africa, and the Far East (PL24,R65).[78] The intent was to decentralize authority. This was not only a matter of policy (PL11) but also of pressure. As an organization grows rapidly at its extremities the number of decisions required multiplies. If they continue to be made at headquarters the pressure at the top of a pyramidal organization, in this case Jersey's directors and committees, becomes unbearable. Relaying local input for each decision becomes impossible. Thus the consequence of growth is the continual need to push decision making outward to the local level through decentralization (PL11). Adding outside directors brought in outside perspectives, relieved some of the burden on directors, and changed the function of the board to less specialized pursuits. The Executive Committee continued to meet from 11 a.m. to 1 p.m. each day, as the board always had, while the full board now met once a week. Directors were expected to meet with the Executive Committee when they were in town.[79]

The name change to Exxon in November 1972 was not accompanied by any reorganization; it was a change of name only. It meant giving up the Standard Oil name. Rights to the Standard Oil name in different geographic markets by other former Standard affiliates had posed a barrier to geographic expansion for Jersey ever since 1911, especially in the retail gasoline market because it made building a national brand name and advertising nationally impossible (PL2, 8). Outside the U.S. Exxon continued to use the Esso brand name. Jersey and, before 1911, Standard Oil had always sold products under many brand names because they were valuable parts of acquired businesses in the 1880's when Standard was formed and because of the policy of secrecy (PL13) which hid the company's actions from competitors and the public. It was also easier for subsidiaries to act independently under a separate brand name (PL11). A single name gave greater unity of identity to the organization. Exxon was the result of an extensive search for a name beginning with "E" that had no meaning in any language and that could, therefore, be protected as a trademark.[80] The "Esso" brand name had been adopted by many affiliates after World War II in a move toward uniformity in brand names.[81] The change to Exxon extended this to corporate names as well.

The period of rapid expansion of Jersey's chemical business (R63) did not begin until after 1958. Petrochemicals had played their first important role during World War II but slipped into obscurity after it.[82] Jersey and Shell had emerged as the largest producers but the market for most petrochemicals was still small compared to that for oil. The policy question for Jersey after the war was not whether to expand capacity but whether to integrate forward to produce end products. It decided not to do so at the time. It remained a producer of basic and intermediate chemicals sold in large volume (PL1,8) at the refinery gate to chemical companies. Chemicals until the 1960's were viewed by oil executives not as a business but as an outlet for crude, as the 1958 Annual Report noted: "The company's chemical business now provides a market outlet equivalent to 40,000 barrels of crude oil a day."[83] Because of their small crude requirements they did not compare favorably with oil when measured by the yardstick of dollars of investment per barrel of crude moved then in use.[84]

Then demand began to grow rapidly. Petrochemical plants were added to Jersey refineries all over, sometimes as bait for selling crude oil. In 1958 Jersey added polypropylene and in 1960, ammonia to its seed chemical businesses in alcohol, additives, and synthetic rubber. In 1962 chemicals were split out of the oil businesses abroad and Esso Chemical Company was formed the following April, marking the recognition of chemicals as a separate business (PL28).[85] Jersey tried a number of joint ventures to integrate downstream in chemicals to acquire technology it did not have but with little success. Its chemical business today, second largest after Shell as a producer of basic and intermediate products in large volume, remains close to its strengths in large-scale refining and wholesale marketing (PL1,7,8).

Alternative energy sources and venture capital, although not yet financially significant, are interesting strategic changes because they are Jersey's response to the appeal of diversification of the late 1960's and 1970's. They have in common a policy of grass roots diversification (PL29)—of building new positions from scratch. This approach was partly forced upon it by the inability to acquire anything large enough to make a noticeable contribution to one of the world's two largest companies both for antitrust reasons and for lack of attractive opportunities.

Jersey has positions in most of the alternative energy source industries. It first ventured into the area by acquiring Colorado oil shale lands in the early

1920's after the post-World War I (PL21) shortage scare. In fact its original interest in I.G. Farben concerned patents for turning coal into liquid fuel, a process it hoped to apply to shale.[86] In 1959 Imperial bought a 30% interest (R59c) in the Athabasca tar sands whose Syncrude Canada project may soon begin producing 125 TBD.[87] It began hunting for uranium in 1967 (R67) and opened a mine in Highland, Wyoming in 1970. It became active in coal in 1970 and operates three mines in Wyoming and Illinois. Exxon's coal production is only perhaps 37th largest in the industry, but it has (PL21) amassed the fifth largest reserve position. It has large reserves of heavy oil (10 degrees API) at Cold Lake, Alberta. It has also invested in the processing technology for these assets (PL7) as in its Syncrude Canada, nuclear fuel fabrication and reprocessing, coal gasification, and uranium enrichment projects. One of its venture capital units makes solar cells and panels for home-heating systems. As in oil this long-term, asset-minded company is stockpiling assets (PL21) and developing technology (PL27) for whatever the future may hold.

Its venture capital activities (R64) began in the 1960's with an industrial gas process, European hotels, and several petrochemical spinoff applications like bricks made from oil and dirt, a petroleum soil mulch, and synthetic protein, not to mention forward integration into fertilizers and fibers.[88] In 1977 its ventures were in three areas, solar energy, new materials, and information processing, and were doing well by venture capital standards—i.e., generating revenues but not yet profits.

But in another sense Exxon found that it could not diversify; it is simply too big to hide. These grass roots efforts absorb only a few hundred million dollars each (venture capital $100 million, Syncrude $150 million so far, coal conversion $200 million, for example). Its real diversification from oil has been into cash and other liquid assets as it maintains its financial strength until more billion dollar opportunities come along.

Exhibit EX-1
Exxon Policies

PL 1	Large scale in every activity: (a) S; (b) 1870-present; (c) PL4.
PL 2	Participants in every geographical market worldwide: (a) G; (b) 1880's-present; (c) PL3.
PL 3	Hold as large a share of the market as possible: (a) H; (b) 1870's (refining), 1880 (pipelines, transportation), 1880's (marketing)-present; (c) PL31.
PL 4	Little new field exploration: (a) V; (b) 1880's-late 1950's; (c) PL11.
PL 5	Supply by purchase of crude, crude reserves, block leases when available: (a) V; (b) 1885-present; (c) PL27.
PL 6	Strength in/control of low cost transportation to offset weakness in crude production: (a) V; (b) 1870's-present; (c) PL5, 26, 28.
PL 7	Refining as the central part of the business: (a) V; (b) 1865-present; (c) PL1.
PL 8	Specialization in wholesale rather than retail marketing: (a) V; (b) 1865-present; (c) PL28.
PL 9	Export supply of foreign markets from large-scale domestic refineries: (a) G,V; (b) 1865-World War II; (c) PL3, 28.
PL 10	Kerosene as principal product: (a) P; (b) 1865-World War I; (c) PL2.
PL 11	Quality and service rather than price competition overall: (a) A; (b) 1865-present; (c) PL8.
PL 12	High prices in some markets to subsidize price wars to gain or hold market share elsewhere: (a) G,P,A; (b) 1865-World War I.
PL 13	Decentralize authority and allow autonomy in subsidiaries: (a) A; (b) 1880's-present; (c) PL6, 23-24.
PL 14	Management through the committee structure devised by Rockefeller to resolve problems: (a) A; (b) 1882-present; (c) PL25.
PL 15	Secrecy and strictly legalistic public responses: (a) A; (b) 1880's-1942 (greater openness after then); (c) PL22.
PL 16	Slowness in innovation and competitive response but capable of massive reaction: (a) S,A; (b) 1880-present except innovation (World War I); (c) PL18-20.
PL 17	Maintain financial strength (little or no long-term debt, large liquid balances) and the willingness to use it to make large strategic investments: (a) A,S; (b) 1870's-present; (c) PL13.
PL 18	Cooperative management by persuasion and consensus: (a) A; (b) 1880's-present; (c) PL18.
PL 19	Active, inside board of directors as senior management: (a) A; (b) 1880's-1966; (c) PL23.
PL 20	Build and maintain excellence in its system of management and executive development: (a) A; (b) 1880's-present; (c) PL32.
PL 21	Take a long-term view of the industry; build for the future; stock-pile resources: (a) A; (b) 1870's-present; (c) PL7, 15.
PL 22	Form and maintain relationships with former affiliates and competitors; U.S. marketing—expand volume within existing territories (1912-27): (a) A,G; (b) 1870's-1950's; (c) PL9.

PL 23	Shift in product mix from kerosene to fuel oil and gasoline but no shift in price or marketing policies: (a) P; (b) World War I-present; (c) none.
PL 24	Expansion of foreign activity including foreign production and foreign export refineries: (a) G,V; (b) 1911- present; (c) PL34.
PL 25	Expansion of controlled production by purchase: (a) V; (b) post-World War I-1954 (last opportunity); (c) none.
PL 26	Regional organization of affiliates: (a) A,G; (b) 1927-present; (c) none.
PL 27	Develop competitive strength in the application of scientific research to the petroleum ndustry: (a) A; (b) 1920-present; (c) none.
PL 28	Recognition of petrochemicals as a separate business: (a) P,A; (b) 1962-present; (c) none.
PL 29	Grass roots diversification—from start-ups rather than acquisitions: (a) P,A; (b) 1960's-present; (c) none.

Notes:

(a) Type of strategic choice (see Ch. 1, Fig.2)

	S—Scale	PL1,4,15
	G—Geography	PL2,10,11,20,22,24
	H—Horizontal concentration	PL3,20
	V—Vertical integration	PL4-9,20,22,23
	P—Product diversification	PL10,21,26,27
	A—Administration	PL10-20,24-27

(b) dates policy was in effect

(c) Cross reference to Standard Oil policies in Exhibit 3-5

Exhibit EX-2
Exxon Resources

R11	Inheritance from Standard Oil.
R11a	Size—$285.4 million in net assets, largest oil company in world.
R11b	Financial strength—no long-term debt, huge assets.
R11c	Production—11 TBD in Appalachia, Louisiana, Romania.
R11d	Refineries—101 TBD, almost all in U.S.
R11e	Markets—43 TBD in eight states in U.S., Canada, the Caribbean, and continental Europe, natural gas, and the wholesale trade.
R11f	Tankers—31 retained after divestiture.
R11g	Excellent management.
R12	NKPM formed through Dutch affiliate to explore and produce in the Dutch East Indies.
R13	Three British producing companies in Peru purchased, including refinery at Talara.
R14	Refinery completed at Tampico, Mexico.
R15	Carter Oil Company subsidiary leases land in Oklahoma.
R18	Transcontinental Oil Company in Mexico purchased.
R19	Humble Oil Company purchased: production in Texas and an able producing organization.
R20a	Purchase of DeMares concession in Colombia.
R20b	Venezuelan concessions acquired.

R 20c	Isopropyl alcohol petrochemical process developed.
R 21	Ellis tube and tank continuous cracking process patented.
R 27	Paraflow lubricant additive developed.
R 28a	Interest in 23.75% of Turkish Petroleum in Iraq acquired through 25% interest in Near East Development Corporation.
R 28b	Purchase of the Creole Syndicate.
R 28c	Purchase of Trinidad Oil Fields Operating Company.
R 30	Acquisition of Anglo-American in England.
R 32	Purchase of foreign properties of Pan-American Petroleum and Transport Company from Standard Oil of Indiana.
R 33a	Purchase of proved reserves and block leases by Humble in Texas, Louisiana, and New Mexico.
R 33b	Standard-Vacuum Oil Company (Stanac) formed.
R 39	Huge U.S. reserves from Humble properties.
R 42	Development of butyl rubber.
R 47a	30% interest in Arabian American Oil Company purchased (Saudi Arabia).
R 47b	Discovery at Le Duc, Alberta, Canada.
R 50	Large additions to refinery capacity after World War II.
R 54	7% interest acquired in Iranian consortium.
R 58	Largest U.S. gas reserves.
R 59a	Discovery of Zelten field in Libya.
R 59b	Reorganization of 1959 preparing Humble to market nationwide.
R 59c	30% interest in Athabasca tar sands.
R 60	Discovery in Colombia.
R 62	Discovery of Groningen gas field in Netherlands with Shell.
R 63	Esso Chemical Company formed—chemicals split out from oil as separate business.
R 64	Early venture capital efforts.
R 65	Reorganization creating regional organizations abroad.
R 66	Appointment of first outside directors.
R 67	Entry into uranium.
R 70	Entry into coal.

Exhibit EX-3
Exxon Handicaps

H 1	Lack of controlled production: 1912-1950.
H 2	Lack of controlled tanker tonnage: 1912-1917.
H 3	Weakness in retail marketing (vertical and geographical): 1912-present.
H 4	Lack of success in exploration: 1912-present.
H 5	Diseconomies of scale—difficulties of co-ordinating a large complex, and decentralized organization.
H 6	Red Line agreement: 1928-1948.

Exhibit EX-4
Exxon Environmental Conditions

EC 1	Divestiture—1911.
EC 2	Competition from Shell.
EC 3	Hostility toward Standard.
EC 4	Commercial production in Mexico—1912.
EC 5	World War I.
EC 6	Decline of kerosene in U.S.; rise of automobile and electricity.
EC 7	Feared crude shortage—post World War I.
EC 8	Upswing of government control worldwide—post World War I.
EC 9	Red Line agreement—1928.
EC 10	Tariff barriers on refined products in Europe and South America—1920's.
EC 11	Shift of U.S. production from Pennsylvania and Ohio to mid-continent and Gulf coast—1920's.
EC 13	Burton-Humphreys thermal cracking process.
EC 17	Jones Act restricts U.S. coastal trade to U.S. tankers—1917.
EC 29	Crude glut—1930's.
EC 40	World War II.
EC 42	I.G. Farben affair.
EC 48	Europe rebuilds—Marshall plan.
EC 50	Heavier share of fuel oil demand in total consumption in Europe.
EC 51	DeMares concession in Colombia not renewed.
EC 57	Crude glut—1957-69.
EC 60	Consent decree splitting up Stanac's marketing operations.

Exhibit EX-5
Exxon Personal Values

PV 1	Walter C. Teagle—interest in international business.
PV 2	Walter C. Teagle—interest in all vertical segments-general management.
PV 3	Everit J. Sadler—interest in production.
PV 4	Everit J. Sadler—interest in international business.
PV 5	Heinrich Riedeman—interest in European business, lifelong desire to crush Shell.
PV 6	Heinrich Riedeman—interest in marketing.
PV 7	Heinrich Riedeman—interest in tankers.

Exhibit EX-6
Exxon Social Responsibilities

SR 1	Medical Department established to improve sanitation and health of employees, especially in foreign producing subsidiaries.
SR 2	Careful and considerate management of personnel.

5

The Strategy of Gulf Oil Corporation

5.1 Overview

Four keys to understanding the strategy of Gulf since and even before its first incorporation in 1907 are presented with their origins as an overview of this chapter.*

The first is that it identifies with upstream in preference to downstream activities. Gulf considers itself an exploring and producing organization (PL 1).[1] The Mellons' interest, even before Gulf's origins at Spindletop in 1901, has always been in finding and producing crude oil. Gulf's principal founder, William Larimer Mellon, first invested less than $2.5 million of the family fortune in Pennsylvania producing properties in 1889 at the age of 21 (PV1).[2] Transportation, refining, and marketing came later. Gulf, Socal, and BP have been the most successful organizations in the world at the most distinctive task of the industry, exploring for new sources of crude oil.

There is a paradox here, however. Gulf is an extraordinarily successful international producer, but it has been crude short in the U.S. almost continuously since 1905, only four years after its origin.[3] The desire to build up its domestic production has had a great influence on its strategy and capital expenditures.

The second key, vertical integration (PL 2), also came from the Mellon family. They believed that the best way to run a business was to develop it from end to end, from raw material production to distribution, a philosophy they followed in financing Alcoa as well (PV2). This also produced a paradox. While Gulf spent vast amounts of money to develop downstream operations in the U.S. to be vertically integrated at home, it was until the mid-1960's almost completely nonintegrated abroad, except for Canada (PL3, PV3). In two key decisions in Gulf's history, its executives entered long-term sales contracts to dispose of crude they could not market. In 1937 they sold half the present and future output of their Venezuelan subsidiary, the Mene Grande Oil Company,

*Policies (PL3), Environmental Conditions (EC#), Resources (R#), Handicaps (H#), and Personal Values (PV#) are keyed to exhibits at the end of the chapter.

to the Canadian subsidiary of Jersey Standard for $100 million, and in 1947 they sold half of their Kuwait production (up to 750 TBD) to Shell under a long-term contract (PL8,PV3).[4]

These sales and the lack of vertical integration abroad demonstrate a third facet of Gulf's strategy, a preoccupation with the U.S. market (PL4,PV3). Gulf was and is the least aggressively international of the seven Majors. BP and Shell were international from their inception due to the colonial aspirations of their governments and the geographical distance between their producing properties in the Middle East and Far East and their markets in Europe. Exxon is the principal successor to Standard Oil, which tried to establish a world oil monopoly in the 1890's. Texaco and the two companies that became Mobil built international marketing organizations well before they were significant producers abroad. Socal joined Texaco in Caltex to market its crude. It did not sell it as Gulf did under long-term contract. While Gulf was not completely without small distributing and marketing affiliates abroad, little was done to develop them until a catch-up push in the mid-1960's. Gulf built its first foreign refineries in Venezuela under protest in 1947-1950 and in Kuwait in 1950. Aside from participations in refineries in France and Iran and its 1956 acquisition of vertically integrated BAOC in Canada, Gulf had no other refineries abroad until the 1960's when a whole string of refineries, petrochemical facilities, and acquired marketing organizations were put together in Europe and Asia. Gulf viewed itself as a U.S. company with foreign production until the 1960's, a policy reflecting the patriotism of the Mellons as well as an unusual aversion to political risk (PV5, PL5).

The fourth key to Gulf's strategy is the continuing role of the Mellon family in Gulf's affairs (PL6, PV4). Gulf is the only Major and one of the few large U.S. corporations where, until it merged with Socal in 1984, a substantial if not controlling interest, 20% in 1976,[5] was still held by the founder's family or foundations. More important than stock ownership, however, has been the Mellon family's continuing influences on policy. Though no family member has been an officer of the company since W.L. Mellon retired in 1948, control or at least influence has continued through the board of directors and through private discussions between Mellon representatives and the chief executive officers whom the family was influential in selecting and retaining (PV4, PL6). In 1976 two of the 13 directors represented Mellon interests.

This overview describes two geographic policies (preoccupation with the U.S. market, aversion to political risk) (PL4,5), three vertical integration policies (to find and produce crude oil as the principal business, vertical integration in the U.S., little or no downstream activity abroad) (PL1-3), and one administrative policy (ownership and control of strategy by the Mellon family) (PL6), which were characteristic of Gulf for many years. The narrative to follow will trace the origin and development of these and other policies which make up the strategy of Gulf.

5.2 1901-1911: Establishment of a Strong Producing Department

It may properly be said that the Texas oil industry began with the discovery at Spindletop on January 10, 1901 (EC3) by the company that became Gulf. The $300,000 loaned by the Mellon brothers (R1b), Andrew W.[6] and Richard B., to James M. Guffey and John H. Galey (R1c), the best oil prospecting partnership of the day (PV6), was almost if not all the capital required to finance the drilling of Lucas No. 1 with the new rotary drilling rig which local residents called a "spindletop." The discovery on January 10, 1901 was the most sensational gusher ever in the United States and established Texas as a major new oil source. The idea of drilling an oil well on "Round Mound" south of Beaumont belonged to Anthony F. Lucas (PV9), a former Austrian Navy officer, who had found traces of oil in salt wells he had drilled along the Texas and Louisiana coasts. Lacking funds, he turned for help to the Pennsylvania operators, Guffey and Galey, who came to the Mellons for the money (R1b,c).[7]

After Spindletop gushed out at the rate of 40 TBD[8] (R1a) the J.M. Guffey Company found itself short of funds, rather than rich. The Mellons' loan was only a fraction of what was needed to finance construction of refining facilities, pipelines, barges, tugs, tankers, and a sales organization since there was no oil industry in Texas (EC2).

Guffey intended to build a large-scale (PL7, PV7), vertically integrated (PL2, PV8) oil company overnight. By May 1901 he had accumulated (R1d-h) one million acres of oil leases in Texas and Louisiana, four producing wells at the Spindletop field, four more which were being drilled, one million barrels of tankage of which 65,000 were at Spindletop and the rest at Port Arthur, 375 acres of land at Port Arthur, 100 rail tank cars of 160 barrels each, and a 16-mile, 6-inch pipeline from Spindletop to Port Arthur.[9] Guffey's strategy was to produce (PL1, PV6), transport, and refine (PL2, PV8) whatever could be refined from Texas crude. Since there were no large markets for oil in Texas (EC2) vertical integration was one way to move the oil to markets elsewhere as refined product. But the scale (PV7) of entry Guffey chose conveyed a clear signal to Standard that a new competitor was on the scene.

Guffey must have congratulated the Mellons on their excellent investment while he outlined his requirements for funds to develop Texas oil. Indeed, Guffey, who was basically a persuasive promoter rather than an oil man, was so busy borrowing money in Pittsburgh that he set foot in Texas only once, about May 16, 1901 when the J.M. Guffey Petroleum Company was granted a Texas charter and established its principal office at Beaumont. This new company was the successor to the J.M. Guffey Company which in reality had been (R1b,c) a partnership among Guffey, Galey, and Lucas financed by the Mellons' $300,000. The Mellons contributed $600,000 for 13.3%, six other Pittsburgh investors invested $900,000 for 20%, and a few weeks later some of the Carnegie instant millionaires put up $2 million for 20%. Guffey owned the rest. His first move as the president of the new company was to buy out Lucas (for $400,000 plus 2/3 of 1% the shares) and Galey (for $366,000 plus a hatful of mining stocks Guffey had in his safe).[10] His second move, on June 28, 1901, was to sign a contract with the Shell Trading and Transport Company (which merged in 1907 with Royal Dutch Petroleum to form Royal Dutch/Shell) to supply 4,500,000 barrels of oil during 20 years at 25 cents a barrel, the company's first huge long-term crude sales contract (R1i, PL8).[11]

The Pittsburgh investors had agreed that a separate company (PL2,PV2,8) should be formed to refine and sell the oil produced by the J.M. Guffey Petroleum Company. Sometime near June 27, 1901, Guffey wrote to his general manager in Beaumont, affirming this plan and saying, "The executive committee propose the name 'Texas Oil Refining Company.'" He added, "If this name has been chosen by others, you will have to change it." The name had been preempted by a group, headed by former Texas governor James S. Hogg, J.W. Swayne, and J.S. Cullinan, which, with the backing of John W. ("Bet-a-Million") Gates, became The Texas Company (Texaco). Desiring to identify the company with Texas oil, company officials borrowed the word "Gulf" from the nearby Gulf of Mexico (R1j). On November 10, 1901 the Gulf Refining Company of Texas was organized (R1k) with $750,000 in capital, 150,000 shares at $5 par each. Guffey held 41,673; the Mellons, 32,966; and others, 24,500. Guffey was president of both companies.[12]

A picture of the original Port Arthur refinery (R2a) would appear today to be more suited to the production of moonshine whiskey than petroleum products. The initial purpose of the Port Arthur refinery was to reduce the sulfur content of the asphaltic Texas crude to meet the requirements of the British Navy, an important customer under the Shell contract (R1i).[13] That the new refining company had been unable to make a quality kerosene from Texas crude and violent price fluctuations from 10¢ to 50¢ per barrel and back in a few weeks concerned the Mellons, but not so much as the fact that Spindletop stopped gushing in August 1902 (EC-5).[14]

They had recently floated a $5 million bond issue for the Guffey company. They had bought $2.5 million for the Mellon bank and had persuaded the Old Colony trust of Boston to take $1.5 million. The rest was unsold. They and their friends had supplied most of the funds for Guffey's two companies. Not only several million dollars (R1b) of their money but their business reputations were on the line.

They turned to their nephew, W.L. Mellon (R2b), the oil man in the family, to see what could be done. He went to Texas in the hot August of 1902 to examine the properties. He found that the wells would resume producing if put on pumps. He also found that Guffey had botched the operation. He had not gotten all of the Spindletop field in spite of his tremendous advantage over competitors. Properties which should have been leased were missed and thousands of acres that were taken should not have been. Money had been spent lavishly while necessities were ignored. The company had not been aggressive in production. The refinery problem was acute. A sales organization was desperately needed. He reported to his uncles that it would take a lot of hard work and $12 to $15 million to make the Guffey-Gulf companies pay out.[15]

Faced with this prospect their first thought was that they might be able to unload the problem on Standard as they had their Pennsylvania properties a decade before (related below). A.W. and W.L. Mellon traveled to New York to meet with Standard's Henry H. Rodgers and John D. Archbold. But Standard had encountered difficulties in the state of Texas both with the Waters-Pierce refining and marketing subsidiary it could not control and with the application of the state antitrust laws to the practices of Waters-Pierce, which resulted in a Texas court decision banishing the unit from Texas in 1900 (EC1).[16] Rodger's response was unequivocal, "After the way Mr. Rockefeller has been treated by the state of Texas, he'll never put another dime in Texas."[17] Standard was out and the Mellons in the fall of 1902 were back in the oil business.

At the next meeting of directors, with Guffey present, W.L. Mellon (R2b) was made executive vice president with complete administrative (PL6) authority. Mellon hired three able lieutenants.[18] George H. Taber (R2d), manager of Standard's Point Breeze, Pennsylvania refinery, was hired away to conquer the problems of refining Texas crude. Gale R. Nutty (R2c) was to set up a sales organization. Frank A. Leovy (R4a), hired in April 1904, was a railroad man who first kept track of the company's tank cars but soon found a new career in leading the search for oil.

In the summer of 1903 Andrew W. Mellon was successful in renegotiating the long-term contract with Shell (R1i) which became disastrous when Spindletop declined and crude prices rose above the 25¢/bbl in the contract.[19]

By early 1903 Nutty had established sales offices (R3b-d) in New Orleans (R6a), New York, and Philadelphia. In 1907 an office was opened in Atlanta to head a southeastern division in the Carolinas, Georgia, and Florida (R7a).[20] By this time the first element in Gulf's marketing strategy, marketing in the South, Southeast, and East coast, was fixed (PL9). These are still the areas in which Gulf is competitively strongest today.

The marketing organization sold what Taber's refiners were able to produce from the Texas crude (PL10). In a memorable exchange of letters in August 1903 Nutty wrote to M.H. Warren, superintendent of the Port Arthur refinery, complaining of the poor quality of Gulf's lubricants. In a response typical of refiners Warren suggested that the products might sell better if a five dollar bill were nailed to the head of each barrel.[21]

Also in 1903 Gulf adopted the trademark, the orange disc (R3a), which is its visible identity throughout the world. On a visit to New Orleans W.L. Mellon and Nutty found that their customers identified suppliers by the color of their tankwagon, not by brand. They adopted a single color on a company-wide basis: orange.[22] It was the second important step, after the selection of the name Gulf, in binding the Guffey-Gulf companies together everywhere under a single identity as an integrated company (PL11).

Following Spindletop came discoveries at Sour Lake, Saratoga, Batson, and Humble in east Texas and Jennings and Welsh in Louisiana (EC6). Mellon responded with a Standard Oil strategy by quickly laying pipelines to the new fields (R4b,c). Two barges and three tankers were ordered. The refinery was expanded to 12 TBD by 1906 (PL2,7). The problem from 1902 to 1905 was how to dispose of the crude.[23]

By 1906 the problem had changed for two reasons. The first was that for two years after 1904 no new fields were found (EC7) and Texas production fell from 77 TBD in 1905 to 34 TBD in 1906. The second was that although Taber and his associates had not solved the problems of refining all Texas crudes, they found they could get better products by using only the better crudes, less than half of those available—about 12 TBD in 1906 (PL12). The Guffey-Gulf companies were able to obtain only about half of that. With refining capacity of 12 TBD and an expensive pipeline and marine transportation system they suddenly did not have enough crude (H2).[24]

This stimulated the search for new sources of refinable crude. The most promising prospects were in Oklahoma where sweet paraffinic crude had been discovered at Glenn Pool on November 22, 1905 (EC8).[25] But it would require a 413-mile, eight-inch pipeline to connect it to their system. The pipeline would cost $1.2 million, which would have to come from additional outside financing, not to mention the cost of leases (eventually $3 million). The problem was that with Guffey still president and largest stockholder the Mellons were not willing (PL6) to invest additional money and Guffey would not agree to any reorganization which would further diminish his power. The Gulf Pipe Line Company of Texas (capital $3.5 million) was organized separately in 1906 (R6a) as the Gulf Refining Company had been.[26]

The Mellons also found that they had the votes to reorganize the companies without Guffey's approval and they did (PL6). On January 30, 1907 the Gulf Oil Corporation, with $15 million capital and $5 million bond authority, was organized in New Jersey for the purpose of acquiring the majority of stock in the three Guffey-Gulf companies. A.W. Mellon became president until 1909 when he was succeeded by W.L. Mellon (PL6). Guffey sold out. In August 1907 the pipeline was finished (R7b) and Gulf was ready to do business in Oklahoma.[27] The 1906 decision to enter Oklahoma (PL13) shows Gulf's strategy of seeking production (PL1), vertical integration (PL2), and sources of refinable crude (PL12) on a large scale (PL7).

Mellon sent Frank Leovy to Tulsa to form a company to buy leases for Gulf. The Gypsy Oil Company (R7c), named after Leovy's favorite sailboat, spent $625,00 for one and $459,299 for another lease at Glenn Pool, enormous sums for unproven leases even though they eventually produced more than 5-

1/2 million barrels. Buying in after the initial strike eventually cost Gypsy more than $3 million at Glenn Pool. The high cost of buying leases struck Leovy as poor business and he resolved that Gulf would become a discoverer rather than a purchaser of leases (PL14). [28]

To this end in 1911 he hired M.J. Munn (R11a), a geologist who had developed a technique of three dimensional mapping of subsurface contours of coal lands while with the U.S. Geological Survey, to establish a geological department at Gulf. In so doing Gulf was the first of the Majors to recognize the importance of geology in exploration. In March 1912 a strike was made at Cushing, Okla. by another company. Using Munn's maps Leovy quickly acquired comparatively cheaply the areas which looked most productive. They ultimately produced 11 million barrels of oil and established the importance of geology at Gulf (PL15). [29]

As Oklahoma production grew after 1907 refining and marketing fell behind. In 1908 the Port Arthur refinery was expanded. In 1911 Gulf built its second refinery, a 5 TBD plant at Ft. Worth (R11b), Texas to divert Oklahoma production from Port Arthur so that it could handle crude from the new Caddo, Louisiana fields. [30] The Ft. Worth location illustrates the local or regional nature of the company at this time. It was content to produce crude oil and expand in its own backyard where Standard was not present.

Gulf expanded its production in Louisiana (1906) and east Texas (1910) and conducted the first drilling operation over water in the U.S. at Ferry Lake during 1909-1911, increasing its exploring expertise (R6b,10,PL15). [31]

By the end of its first decade in 1911 it was apparent that W.L. Mellon had turned the Guffey venture into a solid company. It earned $2.5 million, had assets of almost $28 million, and capital expenditures of $4.4 million. [32] It was a large-scale enterprise (PL7). Its production of 21 TBD was 10% of Texas, Louisiana, and Oklahoma production, 3.4% of U.S. production, and was exceeded only by Standard's 90 TBD among the Majors. [33] Geographically, it produced in those three states, refined in Texas, and marketed in the South, Southeast, and East coast states, but not abroad (PL 3,4,7,9,13) except for crude sales (PL 8). Horizontal concentration apparently played little or no part in its policies. Its market shares were significant but not dominant. Gulf was vertically integrated with exploration and production as its principal activities (PL1,2,14). Its refinery runs (27 TBD) exceeded its net production because it used its pipelines to purchase royalty crude at the wellhead as well as transport its own. In 1911 its capital expenditures were allocated 54% to exploration and production, 18% to refining, 22% to pipelines, 1% to marine, and 2% to marketing. [34] It would discover rather than purchase leases and had established a geological department to use scientific advances to gain a lead over competitors in finding oil (PL14,15). [35] Refining and marketing volumes would be basically set by the level of gross production to dispose of Gulf's production without having to deal with Standard (PL1,2). Refiners would refine only the better grades of crude and marketers would sell the products refiners could produce (PL10,12). The company would be known everywhere by the name Gulf and its symbol, the orange disc (PL11). The Mellon family would exercise ownership and control of strategy (PL16).

The shaping of Gulf's strategy in response to Standard Oil (EC-10) is a factor that must not be underestimated. The entire strategy was carefully shaped to avoid Standard wherever possible. Gulf located in and did not stray far from an area Standard would not enter until 1919. It specialized in the vertical segment of the business, production, where Standard was least active, and the subsegment, exploration, where Standard was absent.

Two of Gulf's most important policies, pursuit of production and vertical integration (PL1,2), were partly a response to the discovery of oil in Texas (EC3), and the necessity of vertically integrating to bring it to market since there had been no oil industry (EC2) there before, and partly due to the beliefs and experiences (PV1-5) of the Mellons in devising a strategy to compete with Standard's strategy of building pipelines to every new field and becoming practically the sole purchaser of crude. W.L. Mellon described it: "I concluded that the way to compete was to develop an integrated business which would first of all produce oil. Production, I saw, had to be the foundation of such a business."[36] A.W. Mellon held a strong belief in the merits of vertical integration, advising his nephew that "the only real way to make a business out of petroleum was to develop it from end to end."[37]

These beliefs came in part from an experience W.L. Mellon had had with the Pennsylvania Railroad in 1892 when he tried to dispose of his crude to a customer other than Standard Oil. He had begun acquiring producing properties in western Pennsylvania in 1889 at the age of 21, financed by his uncles. For a year and a half he sold to Standard, but then began supplying some of his crude to a French firm, taking the business away from Standard. This export crude was initially shipped in tank cars to New York over the Pennsylvania Railroad, whose largest customer was Standard. In 1892 the railroad raised shipping rates just enough to make the export business of Mellon and other independent producers profitless. Mellon fought back by forming the Crescent Pipeline Company and built a line from Pittsburgh to Marcus Hook on the coast 271 miles away to supply the export trade. Nearly every crossing of the railroad's right-of-way required a battle. Lines laid during the day would be torn up at night. Railroad employees would be arrested and a court order would be obtained to allow the work to proceed. There would be no further trouble until the next crossing. In 1893 Standard made a generous offer for the properties and the Mellons sold out.[38]

5.3 Geographic Expansion: 1911-1930

Between 1911 and 1930, the remainder of W.L. Mellon's administration, Gulf began producing abroad, refining on the East coast and in the Midwest, and marketing in the Midwest and to a minor extent in Europe. These policy changes were all geographical; other policies changed little.

Mexico Gulf Oil Company, Gulf's first foreign investment, was formed in 1912 to explore and produce in Mexico (PL1,14,EC11).[39] Gulf sent its crews down the Gulf coast to Tampico in 1913 and made its first strike a year later (R13).[40] Tampico was actually closer to Port Arthur than some of Gulf's U.S. fields. The intention was to import the crude to Port Arthur for refining (PL16). Gulf and Shell were the first Majors to explore in Mexico, both about the same time. Mexico Gulf became one of the largest producers in Mexico and was the only large oil company which was not nationalized in 1938.[41] Political troubles prevented further investment but Gulf continued to produce there until 1951.

Its second foreign investment was in Venezuela in 1924. It was not the first Major into Venezuela: Shell had been there since 1912 (EC12) and Jersey (unsuccessfully) since 1920.[42] Its exploration was successful here, too, obtaining production in 1925 (PL1,14,16,R25a). By 1929 it was the second largest producer after Shell.[43] In 1925 in a related venture it bought a controlling interest in the Barco concession (R25b) on the Venezuelan-Colombian border (PL1,14), a third foreign investment.[44]

In its fifth foreign investment (the fourth, European lube oil marketing, is discussed below), it began looking for oil in the Middle East. On November 30, 1927 Gulf purchased through Major Frank Holmes a concession he had acquired on Bahrain Island and in a second transaction his options in Kuwait, the Neutral Zone, and al-Hasa (the eastern province of Saudi Arabia) (R27a-d).[45] Gulf sent out a geologist, Ralph O. Rhoades, later chairman of Gulf in 1959 and one of its greatest explorers, to map the area. From the air above Bahrain he spotted a textbook dome. He did not know if it contained oil but he knew where to drill before he even set foot on the ground.[46]

Gulf never did get to drill that particular dome. Other events (H3) intervened. On July 31, 1928 the Near East Development Corporation (R28a) (owned 1/6 by Gulf and its sixth foreign investment) obtained a 23.75% interest in the Turkish Petroleum Company (renamed Iraq Petroleum Company in 1929) which had struck oil in northern Iraq on October 15, 1927 (EC17). On the same day they signed the "Red Line" agreement, agreeing that all stockholders in Turkish Petroleum would operate solely through it (H3) in an area delineated by a red line, which included Gulf's options (R27a-d) in Bahrain, al-Hasa, and the Neutral Zone, but not Kuwait.[47] On December 21, 1928 Gulf reluctantly sold the Bahrain option to Socal, a non-participant in the Red Line agreement, for $50,000. It offered the Neutral Zone and al-Hasa options as well, but Socal, fearing they were invalid from delinquency on payment of protection fees, declined and Gulf formally let them lapse on April 1, 1932.[48]

Thus Gulf was the first American Major to obtain a concession in the Middle East and the first Major to invest in the Arabian side of the Gulf. The amount of oil found on Holmes' concessions is staggering. Had Gulf retained and developed them, it would have nearly locked up the Middle East and could well have been the largest Major today. As it was, with its sixth foreign investment it got a share of production in Iraq. Five of its first six foreign investments were to explore for oil (PL1,3,14). All found oil. By 1930 three were successful, one was not, and the fifth was Kuwait. It was an extraordinary record of successful exploration.

In the U.S. Gulf was similarly successful as it expanded geographically in the mid-continent with strikes in Kansas (1916), Kentucky (1918), Arkansas (1921, EC15, the Smackover field), and New Mexico (1929) (R16-29). Oklahoma, Texas, and Louisiana provided the bulk of its U.S. production. When oil was discovered in west Texas and the Panhandle in November 1920 (EC14) Gulf's geologists moved quickly to the counties southwest and northeast of the strike to blanket promising areas, to which other prospectors had not been attracted, with leases. In 1926 the first wells came in, including shares in three giant fields, the Yates, McElroy, and Hendrick. Gulf's aggressive moves nearly sewed up west Texas (R26a). It was still the largest producer there by mid-century (PL1,14).[49]

Both the capacities and geographic locations of Gulf's expansion of downstream facilities reflected its strategy, which was to dispose of its discoveries of crude (PL1,2,17). As Gulf's Mexican production boomed from 1920 to 1922, matching its U.S. production of 66 TBD in 1922, all Gulf's refinery expansion took place at Port Arthur. By 1923 it was the world's largest refinery with a capacity of 125 TBD.[50] The intrusion of salt water caused a sharp decline (EC16) in Mexican production not entirely offset for Gulf until 1927 when west Texas and Venezuela boomed. More capacity was needed, but geographically it made little sense to ship crude west from Venezuela to Port Arthur for refining and then ship the products back east to the Atlantic coast (EC18) when Venezuela was almost directly south of the U.S.

Gulf built new East coast refineries in Philadelphia and Bayonne, New Jersey (R26b,c) (moved to New York after a fire) between 1925 and 1927 at a cost of $20 million, adding 60 TBD to its refinery capacity. It built a new tanker fleet to bring crude from Venezuela (PL16). It also (R29a) built a 5 TBD refinery at Sweetwater, Texas to handle some of its west Texas production.[51] These refineries were built to dispose of Gulf's new production (PL1,2,17).

The locations on the East coast also reflected aversion to political risk (PL5,PV5) and the economics of refineries near markets which had been made attractive by the rise of the automobile and the demand for gasoline (EC13). Not only was the gasoline market profitable, but it also provided a marketable product for the whole barrel, and it offered an opportunity to invade Standard's territory in a product (gasoline) and a vertical segment (retail marketing) where it was weak. Gulf's product mix changed from 4% gasoline in 1907 to 18% in 1912 to 43% in 1924 as it became a strong U.S. gasoline marketer (PL18,R29c). By 1926 it was the fourth largest U.S. gasoline marketer and second largest among the Majors (next to Socony) with 7% of the market.[52]

With the discoveries in Venezuela and west Texas and peacetime prosperity Gulf's net income rocketed from $9 million in 1921 to $44 million in 1929, a level it would not reach again for 16 years.[53] With money to spare it made new investments. In its fourth foreign investment it spent $5 million in 1926 to acquire 25% of the Nobel-Andre-Good group, marketers of lubricating oils in Western Europe (R26d). In 1928 for another $5 million it increased its ownership to 75% (100% in 1934) at the time it began marketing its first premium quality motor oil, Gulfpride, one of the first oils to be sold in refinery-sealed, non-returnable containers (R29d,PL18).[54] In 1930 it began building new headquarters in the center of Pittsburgh opposite the Mellon bank, its central depository and ally (PL6). It was completed in 1932 at a cost of $5 million (R32a).[55]

Gulf Production Company (Gulf's Texas producing subsidiary) entered a 50-50 joint venture with Texas Gulf Sulphur (R27e) Company in 1927 to commercially extract sulphur from its oil properties at Boling Dome, Texas. This venture was an application of its exploration expertise (PL14,15). In 1952 Thompson notes that with minor exceptions all significant sulphur deposits in the United States were on lands originally leased by Gulf and assigned by it to the sulphur companies.[56]

But Gulf's major new investment was an expansion of its downstream facilities in 1929 into Ohio, Michigan, Indiana, and Illinois (R31a-e), which already had 1/6 of the U.S. population. It built refineries in Cincinnati (12 TBD) and Pittsburgh (6 TBD). It bought the Paragon Refining Company with 344 bulk plants and retail stations and an 8 TBD refinery in Toledo for $22 million. It built a pipeline from Tulsa, Oklahoma to Lima, Ohio to bring crude from its new fields in west Texas and elsewhere. It completed the Midwestern expansion at a total cost of $88 million.[57] This and the Nobel-Andre-Good companies' European lube oil marketers (R26d) were its first substantial geographic expansions downstream by acquisition (PL19).

This investment was an overreaction to short-term pressures, demonstrating two strategic handicaps, a tendency to overreact to pressure (H4) and a tendency to take a short-run view of the business (H5), policies which have been characteristic of the company's strategic responses more than once. Due to the rule of capture (EC19) Gulf was unable to control the surge of production after 1927; its refining and marketing organization could not dispose of it (PL17); and it expanded downstream unwillingly to capture its benefits.

The expense of the investment so strained Gulf's finances (H6) that a separate subsidiary, the Union Gulf Corporation, was formed to raise $60 million through the issue of 5% gold bonds. A.W. and R.B. Mellon had to personally provide collateral as security.[58] This shows a third recurring strategic weakness in the adminstrative area (H7), vulnerability in financial management, obliquely confirmed by Drake's reforms below. This reinforces the image of Gulf's strategy of greater strength as operators in finding and producing oil than in refining, marketing, and making a profit from it; that is, of being better oil men than businessmen. It is an ironic weakness in an organization whose founders were investment bankers.

In the geographic expansion from 1911 to 1930 Gulf began exploring and producing abroad, becoming the largest U.S. producer (137 TBD—5% U.S. total) and third largest (214 TBD—5.25% world total) worldwide. Refining expanded geographically to the East coast and Midwest, but not abroad, making Gulf the second largest U.S. refiner (3rd worldwide)[59] (PL2,3,4,5,7, 16,17). It became a strong gasoline marketer (PL18) and made a premium lubricant, responding to market demand, making policies 10 and 12 less important. Downstream activities were strengthened but the Midwest expansion was to impose great financial strain on Gulf.

5.4 Retrenchment in the 1930's

Due to a crude glut and price fall, the imposition of prorationing and gasoline taxes, the Depression (EC20-23), and financial overextension, the roof fell in at Gulf during the 1930's, more so than at any other Major. Its strategic response was the sale of interests in three foreign producing properties (PL4,5,R25a,b, 28a).

It was a case of Murphy's law. With the flood of east Texas crude (EC20) in 1930 the wellhead price dropped from $1.25 to $.60/bbl. With most of its production in Texas and Oklahoma, the first states to institute prorationing (EC21), Gulf was hit hardest of all the Majors. Having overextended itself in the Midwest, the production cutbacks left it with excess capacity, high fixed charges, and no money to drill wells to offset the decline. The refinery price of gasoline fell by 2/3 and the retail price fell to 12¢ a gallon. Gasoline taxes were imposed to raise money for unemployment relief. Racketeers bootlegged gasoline like alcohol, undercutting prices. In 1931 Gulf lost money on gasoline on record volume. Consumption fell 25% from 1929 to 1932 (21% for Gulf). Production dropped 20% in the U.S. but 36% for Gulf. It sustained the only two losses in its history in 1931 and 1933 and lost $30 million from 1931 to 1934.[60]

W.L. Mellon, now in his early 60's, and Frank Leovy (R2b4a) found this to be one too many battles to fight. They stepped aside to become chairman and vice-chairman of the board in April 1931. An outsider, James Frank Drake, one of Andrew Mellon's political lieutenants at the U.S. Treasury, who had just turned around Standard Steel Car, a Mellon company, and sold it, was brought in (R31f), demonstrating another administrative weakness, a lack of management development (H8), as well as continuing control by the Mellons (PL6). Drake's contribution was the "inauguration of decentralization," a reorganization of the company into producing, refining, marketing, and transportation departments whose profitability could be judged individually.[61]

Gulf was in difficulty during the 1930's but not in danger. The greatest cost was opportunity lost. Operating losses and the trust indentures of the Union Gulf bonds limited funds for capital expenditures to increase U.S. production by drilling or purchase of the many excellent producing properties available at bargain prices due to low crude prices and the lengthened payout from prorationing. On November 27, 1934 Gulf retired these bonds and extinguished Union Gulf. Whether Gulf, the great explorer (PL14), would have purchased these proven properties is debatable, but the loss of the opportunities was a matter of keen regret to Gulf executives for many years thereafter.[62] They were correct in assessing it, at least in hindsight, as a once-in-a-lifetime opportunity denied to them by the Union Gulf overextension (H6).

After the bonds were retired, capital expenditures for production and marketing increased but no proven properties of consequence were purchased. There was even a debate as U.S. production declined as to whether profit could be made on purchased crude. For a time Gulf cut back its refining and marketing operations, following its crude disposal policy (PL17), but in July 1935 it decided to run the refineries and pipelines as close to capacity as marketing would permit, becoming a net purchaser of U.S. crude as needed (PL20), ending the disposal policy (PL17) in the U.S.[63]

Gulf's other strategic response to financial stringency in the 1930's was to sell some foreign producing properties. It sold its (R28a) interest in Iraq to Jersey and Socony-Vacuum in 1934. In 1936 it sold the disappointing Colombian Barco concession (R25b) to a Texaco Socony-Vacuum joint venture. Then on December 15, 1937, having a large increase in Venezuelan production for which it had no markets, it sold half the Mene Grande Oil Company, its Venezuelan subsidiary (R25a), to International Petroleum Company, a Jersey subsidiary, for $100 million (R37), $25 million in cash and the rest over a 15 year contract. It made a profit of 20-30¢ per barrel on the Venezuela oil under the contract.[64] This was its second large crude sale contract (PL8) and showed its preoccupation with the U.S. market, foreign production for import, and lack of interest in refining and marketing abroad (PL4,3,16). What it could not import it sold.

5.5 Kuwait

Through the difficulties of the 1930's few Gulf executives must have suspected that Gulf's greatest discovery (PL14), Kuwait (R38), lay just around the corner. After difficult negotiations with the British (EC24), Gulf obtained a concession with BP as its equal partner in 1934 (R34a) and discovered oil in 1938. To dispose of its Kuwait crude Gulf signed the largest crude sale contract in the history of the industry with Shell in 1947 after the war (PL3,5,8).

The intervention of the British government (EC24) greatly altered the outcome of events in Kuwait. Kuwait was the only concession of the four (R27a-d) Gulf had purchased from Frank Holmes in 1927 that lay outside the Red Line so that agreement was not a factor in these events. But Kuwait was a British Protectorate and the British Political Agent made it clear to Holmes, representing Gulf, in 1928 that no non-British company could get a valid concession in Kuwait. In March 1929 Gulf complained to the U.S. State Department and protests were made but to no avail.[65] But in March 1932 the new ambassador to Britain was none other than Andrew W. Mellon, illustrious former Secretary of the Treasury. On April 9, 1932 the British agreed to drop the nationality clause from any oil concession. On May 26, 1932 Gulf submitted a draft concession. BP,[66] whose geologists did not believe there was oil on the Arabian side of the Gulf, was not interested in Kuwait but it was not interested in Gulf having it either. Prodded by the British government it submitted a rival concession bid. In June 1932 Socal struck oil in Bahrain, proving its existence on the Arabian side of the Gulf and BP became more interested.[67] For more than a year the negotiations were a standoff, Gulf claiming it offered better terms while BP insisted on preferential treatment. Archibald Chisolm, BP's negotiator, had British government support, but Holmes, acting for Gulf, was close to the Sheikh, who wanted American participation. On December 14, 1933 they agreed to cooperate and formed the Kuwait Oil Company (KOC) (R33), owned equally by the two parent companies, to resolve the impasse. Following some devious negotiations and a third bid by a British group, KOC secured a concession on December 23, 1934.[68]

Without the intervention of the British government, Gulf would probably have secured sole possession of the Kuwait concession (R34) in 1927. Also whether due to the joint venture preventing the price from being bid up or to the pessimism of BP's geologists and British influence on the Sheikh, royalty payments were 13¢/bbl. compared to roughly 22¢/bbl. for concessions in Saudi Arabia, Iraq, and Persia negotiated at similar times. Much of the difference is due to the acceptance by the Sheikh of Indian Gulf rupees rather than gold in payment and the effect of subsequent currency devaluations.[69]

Drilling began 18 months later on the coast but produced a dry hole.[70] Later in 1936 Gulf sent out a team to do geophysical research, the first such party equipped with magnetometer, gravimeter, and seismograph (PL15), according to Thompson.[71] The drillers were moved 28 miles south to the Burgan district where there were traces of oil. Under the direction of Ralph O. Rhoades they brought in the first well of the fabulous Burgan field in April 1938 (R38).[72] The wells were capped during the war and commercial production did not begin until mid-1946 (R46).[73]

The size of Gulf's reserves greatly altered its geographical and vertical balance. Its share of Kuwait reserves (R38) was estimated at four billion barrels in 1946 and 5138 million a year later, compared to 1100 million in the U.S. (R3,14,17,23,25) and 575 million in Venezuela (R25a) including the 50% committed to the long-term contract (R37).[74] Suddenly Gulf had foreign reserves five times as large as its U.S. base and was literally more an international producer with U.S. downstream activities than a vertically integrated U.S. company with foreign production (PL4). Even in 1950 its foreign product sales amounted to only 26 TBD, 5.5% of its total and 0.6% of non-U.S. consumption.[75] A near purchase of Petrofina in Belgium was its only attempt to expand downstream after the war (PL1,3,4).[76]

Even though its assets had changed greatly with Kuwait, its policies did not. Only PL16, developing foreign production for import, required modification. Rather than forming a Caltex or Stanvac crude for markets type of joint venture, in 1947 Gulf entered a long-term contract to sell half of its Kuwait production (R47), up to 750 TBD, to the Royal Dutch/Shell group.[77] Profits from the crude were to be split equally, giving Gulf participation in Shell's downstream operations without investment, reaffirming PL1,3,4,5,8,17 of its strategy and indirectly the Mellons' values (PV1,3,5). All it wanted were outlets for its crude.

Neither did its weaknesses change. The Shell contract was an important strategic move that relieved Gulf of the necessity of refining and marketing abroad, either alone or by joint venture, for 15 years, after which Gulf was at a disadvantage to earlier entrants. It was a move that met its short-term needs (H5) at the expense of perpetuating its weakness downstream (H1).

5.6 After the War: 1947-1960

Once its direction had been reaffirmed by the Shell contract it was business as usual at Gulf until the 1960's. Gulf clung to its strategy of first finding and producing crude oil (PL1) and second integrating downstream only in the U.S., not abroad (PL2-5,8), while the world changed.

The changes in Gulf's world, which reflected those in the industry, were (1) its U.S. production could not keep up with the postwar surge in U.S. consumption, falling from 60% to 45% of its downstream requirements in the five years after the war, while (2) foreign production soared far beyond its downstream capabilities causing (3) a crude oil glut (EC31) and price cuts which jeopardized its control of foreign sources (EC34) as well as their profitability, and (4) its frustration at being unable to match its glut in crude supply abroad with its deficit at home due to import controls (EC30).

Chief executives came and went while the board dominated by Mellon supporters, led from 1945 to 1968 by Richard K. Mellon, retained control of strategy (PL6). J.F. Drake became chairman in 1947 and retired in 1952. Again Gulf went outside its own executive ranks to select Sidney A. Swensrud for its top position (H7). His tenure proved briefer than he no doubt had anticipated. In 1953 William K. Whiteford, another outsider (H8), was brought in. By all accounts he was an autocratic executive (PV10) who ruled rather than managed the company until 1964, first as president and later as chairman.[78] His legacy was the swamp of political troubles over illegal and unethical payments which would disgrace his two successors after the 1974 Watergate scandal. Their guilt is unclear but their dismissal came from the concealment of these activities from the Mellons (PL6).[79]

In late 1950 Gulf established a crude oil purchasing department (PL20) for domestic operations (R50a).[80] But there is no doubt that Gulf hoped to import its cheap Middle Eastern crude. Its policy (PL16) of seeking oil abroad for import as a guide to foreign investment was a factor in the decision to invest in Kuwait. But it was continually frustrated by import controls (EC30). Whiteford's program to increase Gulf's influence in Washington through individual payments was a response to this frustration in part.[81] In 1958 its New York refinery, which had been equipped to process heavy Venezuelan crude, was shut down and dismantled due to import controls.[82] Gulf financed the construction of two refineries (R60b,66a) in Puerto Rico and Hercules, California by Edwin A. Singer, then an independent and in 1976 a member of Gulf's board, to take advantage of the more favorable treatment granted to small refineries under import restrictions.[83] A refinery site in Charleston, South Carolina purchased in 1956 was never used.[84]

Net income rose fivefold in the decade after the war, but the U.S. portion fell from more than 3/4 to 1/4 of the total at the most extreme points in 1948 and 1958.[85] Gulf took two actions to counter and exploit the trend. It tried to boost its U.S. business (PL4) and it continued to explore abroad (PL1).

To boost its North American earnings base (PL22) Gulf took several steps. In 1954 it began exploring offshore in the Gulf of Mexico.[86] Since it had been so successful in finding domes in the Louisiana Delta it is curious that Gulf did not take to deep water for a full decade after Magnolia (part of Mobil) and Socal. But it has been (R54a) successful there. Late in 1955 it acquired the Tulsa based Warren Petroleum Corporation, a natural gasoline and gas liquids company (R55a) from the founder, William K. Warren, for 1.5 million shares (5%) of Gulf common worth $86 million.[87] In 1956 it acquired 58% (later increased to the legal maximum 68%) of the British American Oil Company, Limited (R56a), Canada's second largest oil company, in exchange for Gulf's production (24 TBD) and pipeline assets in Alberta.[88] BAOC's four refineries with a capacity of 100 TBD total gave it an outlet for its Canadian crude and BAOC had some U.S. production as well.[89] It became Gulf Oil Canada, Limited. Also in 1956 Gulf bought $120 million of 3-1/4% convertible debentures of the Union Oil Company which it sold in 1961, passing up that possible acquisition (R56b).[90]

Between 1947 and 1960 Gulf explored in 21 countries in Europe (4), Africa (7), and Latin America (10). It had been exploring in Canada since 1943, struck oil in 1948, and began production in 1950 (R50b).[91] In late 1954 it was admitted to the Iranian consortium and obtained (R54b) another significant source of production. Even so it was becoming steadily more dependent on Kuwait. After the Suez Crisis in 1956 it began to acknowledge a policy which had been guiding its wider search for oil in the postwar decade.[92] The goal of its foreign exploration was not only to find oil but also to diversify its sources of production (PL21,26) to reduce its dependence on Kuwait and any one foreign investment in general (PL5).

Diversifying production geographically while avoiding political risk (PL21,5) meant exploring in the least politically risky places, rather than where it might find the most oil. Before 1956 these were (1) the small countries of Central and South America, which were in the U.S. sphere of influence if anything went wrong and where any surplus of crude would be closer to home, (2) Europe, which was reasonably hospitable, and (3) Canada. After the closing of the Suez Canal in 1956 demonstrated the vulnerability of the Middle East to the interruption of marine transport, Gulf like others turned its attention west of Suez to north and west Africa. Aside from Iran, it never sought additional production in the Middle East as the independents were then doing and its efforts in the Far East have been minimal.[93]

5.7 The Glut: 1960-1973

After the Suez Crisis (EC26) in 1956 threatened the world with shortage, its resolution in the form of larger tankers (EC35) and rapidly rising Middle East production ended the ten year postwar shortage. A glut appeared and with it falling prices (EC34,31). As Gulf's Annual Report described it, "Oil operations in 1958 were conducted in a setting of rising costs, declining prices, and relentless rivalry for markets."[94] Both U.S. and worldwide markets were affected by this rivalry. In the U.S. gasoline price wars broke out. Abroad Exxon cut the posted price in Libya in 1959 and OPEC was formed as a result (EC31,32).

Gulf responded in the U.S. by starting (along with Sun Oil Company) the bloodiest of the price wars and by expanding its marketing network to 48 states by 1966 (PL29,30). In the Eastern Hemisphere Gulf found markets for its crude by integrating downstream (PL23). In a crash program it built or acquired a network of refining, marketing, and transportation facilities from Europe to Asia (PL24,25,27). It also continued to explore abroad to reduce dependence on Kuwait (PL14,16,21).

Though Gulf did not produce all the crude its refineries processed or its marketers sold, it faced a surplus of crude supplies in the U.S. beginning in 1957. As noted in its 1957 Annual Report, "Shortly after the middle of the year our Company found that the total allowable production of all the wells connected to its pipelines was a greater quantity than, looking to the future, it could use, sell or store."[95] This was partly due to its recent success in west Texas, Oklahoma, and the Gulf of Mexico (U.S. drilling activity hit an all-time high [EC27] in 1957) and partly due to lower consumption than expected from the impending recession (1958-1959). In response to the surplus Gulf reduced its liftings and sold its Oklahoma gathering system, whose volume was only about 1/4 from Gulf at the time.[96] In spite of the surplus in 1959 its U.S. net production of 333 TBD was only 74% of its 450 TBD refinery runs and 57% of its 580 TBD product sales.[97] Purchases, imports, and refining by others for Gulf made up the differences. Product sales volume in the U.S. declined 4% in 1960, triggering a response.[98]

It was in November 1961 that Gulf began marketing Gulftane (R61), a new economy gasoline priced below its regular grade (PL29). The price differential was 2¢/gallon, the same margin by which independent marketers sold their private label products below regular in most markets.[99] The move was aimed at boosting product volume (PL28) and countering the inroads of the private brand marketers who bought their supplies from the surpluses of large refiners. Large refiners had begun the practice of dumping excess products at low prices to maximize their own volume and profits from the crude produced and refined. The private brand industry grew up from these surpluses. Gulftane was introduced in San Antonio where these so-called independents held 50% of the market in the summer. By the end of the year Gulftane was sold throughout Gulf's market area from the Midwest to the Atlantic coast. The Sun Oil Company introduced a similar grade, Sunoco 190, priced 1¢ below its regular at approximately the same time.[100]

Wherever Gulftane was introduced it set off a cutthroat price war as Gulf tried to increase its market share while the independents cut their prices further to maintain their price advantage. They also had a quality advantage since their products, though private label, were 94 octane (the regular grade) while Gulftane was only 91. Prices frequently dropped 4-6¢/gallon. The other Majors with Shell in the lead followed suit by cutting their price structure, though none introduced a subregular gasoline of an octane similar to Gulftane. After about 18 months a new equilibrium was established with Gulftane selling 1¢ below regular and the independents 2¢ below. It is the general consensus that everyone lost in this most famous of U.S. gasoline price wars.[101]

Gulf's other important U.S. move during the 1960's was the push to market nationwide (PL30). This geographic expansion began in 1960 with the acquisition of the Wilshire Oil Company, an independent marketer (PL19) in the Los Angeles area with 700 stations and a 32-TBD refinery at Santa Fe Springs (R60).[102] Gulf previously supplied its approximately 35,000 stations from its four refineries in Port Arthur, Philadelphia, Toledo, and Cincinnati. In 1963 Gulf formed a partnership (R63) with Holiday Inns to combine lodging and service station facilities and use Gulf's credit card for both. It also bought a chain of stations in Arizona and some sites near San Francisco.[103] It continued to operate the Wilshire and Blakely (Arizona) chains as independent price discount stations until 1965 when they were converted to the Gulf name and orange disc trademark (PL11).[104] In 1966 Gulf purchased 2300 stations (R66b) in ten Rocky Mountain states, achieving its goal of 48-state marketing.[105]

In 1958 Gulf's production in Kuwait was 714 TBD, 57% of its net production worldwide. Profits in the Eastern Hemisphere were 71¢ per barrel, an all-time high for the Eastern Hemisphere. They totaled 61% ($202 million) of Gulf's total net income.[106] Kuwait had made Gulf an international oil company whether or not it wanted to be.

Declining prices due to the crude glut (EC31,34) caused profits in the Eastern Hemisphere to fall to 36¢/bbl in 1959 and a low of 16¢ in 1964 while U.S. profits rose to $1.08/bbl in 1964 and a high of $1.50 in 1968 for the decade as Gulf's marketing program succeeded.[107] Gulf refined and marketed only about 125 TBD in the Eastern Hemisphere and perhaps 350 TBD through (R47) the Shell contract. Including its other Eastern Hemisphere production (R54b,55b,56c) in Iran (36 TBD), Sicily (20 TBD) and France (4 TBD), this left perhaps 300 TBD of Eastern Hemisphere crude for which it had no secure markets and on which the price was falling.[108] Particularly after Suez demonstrated the vulnerability of the Middle East, a downstream organization abroad equal to the task of handling the flow of crude from Kuwait became a necessity (PL23). This meant refineries, tankers, and marketing outlets and less money for exploration.

The factors which propelled this decision were political as well as economic. (1) Consent decrees with the FTC from antitrust action broke up Stanvac and eliminated the provision in Gulf's 1947 contract with Shell which forbade it from entering Shell's territories.[109] (2) The influence of the British and French in the Middle East had suffered a mortal blow from Suez (EC26). Mossadegh in Iran was a recent memory of the threat of nationalization. Supplies in the Middle East looked less secure after Suez. (3) Crude prices turned down (EC31) in 1957 and did not turn up again until 1969.[110] The disadvantage of having only half the profits from downstream activities under the Shell contract widened as the crude price fell. (4) After the cut in the posted price the formation of OPEC was a defiant gesture of the bitterness the Arabs felt over the price cut and the arrogance the companies demonstrated by lack of consultation (EC31,32).[111] (5) The Russians had made new discoveries and began to export oil to Western Europe for the first time since the 1920's and at cut-rate prices (EC33). Averse as ever to political risk (PL5), Gulf executives feared a day, lurking over the horizon, when Kuwait crude would no longer be theirs to sell, and downstream facilities would be required to capture its benefits.

Gulf responded in characteristic fashion. It embarked on a massive charge downstream reminiscent of its Midwestern move in the late 1920's. Like that move, it ended in disaster. It was an overreaction (H4) to political risks (PL5) and the recent (H5) margins between U.S. and Eastern Hemisphere profitability that was based on the assumption that downstream profits would be present when Gulf arrived.

Between 1962 and 1972 it built 12 refineries in the Eastern Hemisphere with 669 TBD of capacity.[112] This was 60% of total capacity added in the decade. From a low of $28 million in 1960, Gulf's total refinery capital expenditures soared to $341 million at the 1970 peak. The refineries were located at Stigsnaes, Denmark (30 TBD, 1962, Scandinavian market), Europoort, Netherlands (1962, 30 TBD, near Rotterdam), in 1962 the Philippines (10 TBD, 44% Gulf with Philippine nationals), Ulsan, Korea (1963, 35 TBD, 25% Gulf with Korean government), Taiwan (lube oils, 75% Gulf, 10 TBD, 1963), Neuchatel, Switzerland (1967, 25% Gulf, 60 TBD), Milford Haven, Wales (78 TBD, 1968), Huelva, Spain (40 TBD, 1968, 40% Gulf with Rio Tinto mines), Emden, Germany (1970, 50 TBD, acquired with the Frisia Refining, Marketing, and Transportation Company), Okinawa (1971, 92 TBD, 45% Gulf), Bilbao, Spain (112 TBD, 40% Gulf), and Milan, Italy (80 TBD), both in 1972 (R62-72).

In the Western Hemisphere it added 249 TBD capacity in the U.S. between 1962 and 1972, 155 TBD in Canada, and 49 TBD in Latin America, a total of 453 TBD capacity. New refineries were built or bought in Black Creek, Mississippi (1963, 19 TBD, purchased), Venice, Louisiana (1967, 20 TBD), Hercules, California (1966, 26 TBD), Alliance, Louisiana (1971, 155 TBD), Edmonton, Alberta (1971, 80 TBD), Pt. Tupper, Nova Scotia (1971, 80 TBD), and Ecuador (1968, 5 TBD, 50% Gulf with Texaco, purchased) (R63-71).[113]

Before this push Gulf had foreign refineries only in Kuwait (1949, 50% with BP), Venezuela (67%, 1950), Iran (consortium 7% share), and an 18% (33 TBD) interest in a French refinery from a crude supply contract with Exxon, and six Canadian refineries (R49-56).[114] This excludes the offshore refinery in Puerto Rico built in response to import quotas.

To connect this network with supplies of crude in Kuwait, since it had neither tankage nor marine terminals abroad, it built three large marine transshipment facilities, unique in the industry, from which crude could be transferred from the new 300,000 dwt (deadweight ton) supertankers to 100,000 dwt coasters which could enter existing harbors (PL27).[115] The first was begun at Bantry Bay, Ireland, in 1966 and began operation in 1967 (R67a). Two more followed in 1970 at Pt. Tupper, Nova Scotia and on Okinawa (R70a,b). As an example of the grand design, since Gulf was unable to import crude into the U.S. it built the Nova Scotia terminal and refinery as an offshore facility which could use the 300,000 dwt tankers to run heavy Kuwait crude into North America. Its principal product, low sulfur fuel oil for Canada's maritime provinces, could also be sent to the U.S. Gulf was the first company to order the 300,000 dwt ships. These radically different ideas attracted a lot of industry attention.[116] The purpose was to overcome the cost disadvantages of 5-10¢/bbl. built into its downstream network by the small-scale, geographically dispersed refineries (PL24).[117] For a time it overcame the transportation losses it once had had to absorb.[118]

The objective of all this downstream investment was a network of geographically diversified assured outlets to dispose of its crude (PL24) and convert it into dollars which could be sent home to the Mellons. Kuwait, a British protectorate, was in the sterling area, an advantage for crude sales since purchasers could pay in plentiful pounds rather than scarce dollars. But at some point Gulf had to convert the crude or products or currencies into hard dollars. It was not always an easy task. Sterling was blocked (i.e., nonconvertible) after Suez, one of the original stimuli for the downstream expansion. The partial ownerships in many of these refineries were often less important to Gulf than the crude supply contracts for the entire refinery which came with them (PL25). To "upgrade" a barrel of heavy Kuwait or Iranian crude into fuel oil as Gulf did could be done with relatively simple and inexpensive refining units. It took partners to minimize (PL25) its investment, but not with other Majors, all of whom were better positioned downstream. As in the U.S. the goal was to move the maximum volume of crude (PL28) and it cut prices to do so.[119] It did not enter any giant new crude sale contracts like the 1947 Shell contract (R47). Gulf chose to build small-scale refineries (30-60 TBD when others were building 100 TBD) and scatter them over a wide area (PL24) for diversification and to minimize political risk (PL5,21). Having no refineries anywhere in Europe and Asia, it entered everywhere within a decade. Or almost everywhere. Close study shows that it built refineries only in areas of low political risk (Canada, Western Europe) or where there was a large U.S. military presence (Western Europe, S. Korea, Okinawa rather than Japan, the Philippines) or commitment (Taiwan) (PL5). Of the refinery additions, none were in Africa and only one local market refinery was in Latin America. The marketing network received even less attention. The main product was fuel oil which did not require extensive marketing facilities. But Gulf also built up at the 1970 peak 13,000 retail stations, half in Canada and half in Europe. For a time marketing kept pace, but in the 1960's it fell far behind refining, as refining fell behind production, under the relentless push of crude.[120]

The crude came from old sources and new: Venezuela, Iran, Bolivia, Colombia, Ecuador, Nigeria, Angola, and above all Kuwait.[121] Some new sources were small but those in Nigeria, Angola, and Ecuador were substantial. Gulf continued to have success in finding oil (PL1). It struck oil on Christmas Eve, 1963, in Nigeria (R63c).[122] In Cabinda, an enclave in the Portuguese colony of Angola, where it had been searching onshore since 1957, it discovered oil offshore in 1966 (R66b). Legend has it that the Cabinda concession was obtained after Gulf executives were summoned to Portugal by the dictator Salazar and Gulf was the only company asked to bid on the territory, so highly regarded were its oil finding abilities by that government. Small strikes in southern Colombia led Gulf, in a joint venture with Texaco, to look over the border into Ecuador, another area where no one else had been looking (R68c). Gulf missed out on Prudhoe Bay or rather it obtained only two small positions when management overruled its geologists who urged the project.[123] It has participated in four North Sea fields: Thistle, Statfjord, Dunlin, and Murchison.[124]

With a surplus of crude it is surprising that Gulf continued to invest in exploration for higher cost sources (PL26). The reason of course is that Gulf feared the loss of Kuwait and it was first of all an exploring company (PL1,21). Men like K.C. Heald, Ralph O. Rhoades, P.H. Bohart, and Hollis Hedberg kept the emphasis on exploration and had, as one of Gulf's explorers put it, "a deep appreciation of the desirability and necessity of a worldwide exploration program."[125] This had become the policy by the mid-1960's. Hollis Hedberg led Gulf into west Africa in Cabinda and Nigeria in 1961.[126] A converted fishing trawler, the Gulfrex, was commissioned in 1967 to conduct marine geophysical surveys worldwide. It was joined in 1974 by a second geophysical research ship, the *Hollis Hedberg*, designed for arctic waters.[127]

The entire logistical system Gulf built from 1962 to 1972 for disposing of crude (PL17) was a thing of beauty if not a marvel of efficiency. It shuffled huge volumes of crude through the system, fulfilling Gulf's policy of maximizing profits on crude (PL28), a policy quite in contrast to that of maximizing profits on the total barrel or marketing products as, for example, Texaco did.

What it did not shuffle through with equal efficiency were profits. Eastern Hemisphere profits remained at about 20¢/bbl. through 1972, never regaining the 60¢/bbl pre-1959 levels much less the $1.00-plus U.S. level.[128]

The problem was that while Gulf was finally arriving downstream, the industry was going elsewhere. Refinery scale was increasing rapidly from an efficient scale of 100 TBD to giant facilities three to six times that size. Harbors were deepened and channels widened to accommodate the huge new tankers. The multiple stop 100,000 dwt tankers and 100 TBD refineries gave way to huge refineries served by huge tankers at a cost saving of 35-40¢/bbl. over the smaller facilities like Gulf's.[129] What Gulf had done with its small-scale refineries and expensive transshipment facilities was to build itself permanently into a high cost position in the Eastern Hemisphere. It had entered as a marginal competitor while consumption was growing rapidly. As the industry overbuilt and costs declined due to scale economies in refining and marine transportation, it remained a marginal competitor with high costs, extremely sensitive to any decline in demand or price. History obliged shortly after in 1973. In 1975 it lost $67 million on downstream operations in Europe.[130]

5.8 Diversification and Retrenchment

Gulf's nonpetroleum operations have always been viewed as byproducts of its oil business, just as its downstream activities have been byproducts of its upstream business (PL1,17), except for the years 1972-1976.[131] Twice in recent years Gulf has tried to solve long-standing coordination problems through reorganization without notable success. These efforts at diversification and reorganization have been part of a program of retrenchment occurring since 1972 to adjust to the end of the era of cheap Middle Eastern crude.

Gulf's early diversifications were simply offshoots of its oil business like the Texas Gulf Sulfur joint venture in 1927 (R27e). Gulf's first experience with petrochemicals came in 1942 with its participation in the Neches Butane Products Company (see Chapter 8) (R42) which made synthetic rubber and for which Gulf supplied butylene from its Port Arthur refinery.[132] A Gulf-Goodrich joint venture (R54c) bought the Port Neches plant in 1954 and another in 1955.[133] Gulf began to manufacture ethylene at Port Arthur in 1950 (R50d) and was the leading producer by 1959.[134] In the 1950's it built up a large business in bulk petrochemicals like ethylene, benzene, oxo-alcohols, and the like (PL31), sticking to the large volume building block petrochemicals (Figure 4) as it stuck to the upstream large scale crude oil production segment of the petroleum business (PL1,7). Acquisitions (R55a,R63f) of Warren Petroleum (1955) (LPG and gas liquids) and especially Spencer Chemical (1963), a former ethylene customer, boosted its product line and pushed it downstream into plastics, fertilizer, explosives, and olefins for detergents and fibers. The Spencer acquisition was Gulf's first major move downstream in petrochemicals and brought marketing and financial expertise as well as products.[135]

In the 1960's it added petrochemical facilities to its foreign refineries: fertilizer plants in Korea (1964) and Kuwait (1966, 20% Gulf), benzene-cyclohexane in Wales and Spain (1966), plastics fabricating plants in Ecuador (1967), Singapore (1970), Taiwan and Indonesia (1971), and a bulk petrochemical complex in Rotterdam (1970) (R64-71).[136] Host governments lobbied for the facilities. In the Eastern Hemisphere, where lighter products were produced at low marginal cost as byproducts to fuel oil, petrochemicals offered a high value outlet. Until 1970 the chemical businesses were managed as part of the oil businesses in each geographical region and were not thought of as diversification.[137] Gulf was then caught by the price declines (EC41) of basic chemicals due to excess industry capacity, especially in fertilizer, similar to crude oil.[138]

With Spencer Chemical Company in 1963 came a subsidiary (R63f,g), the Pittsburg and Midway Coal Mining Company, headquartered in Kansas City like Spencer. It was then the tenth (now seventeenth) largest coal producer and one of the three most profitable.[139] Thus Gulf was unintentionally the first Major to enter coal production.

W.K. Whiteford retired on December 1, 1965 and was succeeded by the first insider ever. E.D. Brockett, a production engineer, was chairman until the end of 1971 and was succeeded by another insider, Bob R. Dorsey, a refiner, until January 1976. Brockett and Dorsey began to make investments outside the petroleum industry for the first time (PL32,34,PV11,12).

For the intentional diversification (PL32) one must credit at least two sets of factors: political and economic uncertainties in the Middle East and Libya (EC39-45) and the conglomerate diversification mania (EC40) of the late 1960's. The crude glut, the fact that it was not getting rich from its downstream charge in the Eastern Hemisphere, the Six Day War in 1967, nationalizations in Bolivia (1969) and Libya (1970), and the Teheran agreements beginning nationalization in 1971 (EC34-45) made Gulf with its aversion to foreign political risk (PL5) less than confident about its future in the oil industry. Diversification offered the possibilities of increasing its North American earnings base (PL22) while entering promising new businesses (PL32).

Gulf responded to diversification in its usual wholehearted fashion, overreacting with enthusiasm to immediate prospects (H4,5) and underinvesting in financial prudence (H7). In 1967 it began drilling for uranium deposits and the following year created (R67d) Gulf Mineral Resources Company "to direct a worldwide search for uranium, coal, oil shale, sulfur, potash, phosphate, and other minerals related to the company's business."[140] The relation it had in mind must have been the application of its expertise in exploration (PL1,15) rather than the prospects of their development. Also in 1967 it acquired (R67e) from General Dynamics its General Atomic subsidiary, "which is internationally recognized as a pioneer in the development of high temperature gas-cooled reactors (known as the HTGR)."[141] In 1974 a half interest was sold to Shell. Sales of the nuclear reactors were good: 35 utilities had signed up at the end of 1968.[142] Gulf General Atomic also began selling in 1968 a reverse osmosis water purifying system and nuclear fuel rods for light water reactors in 1971 (R68a,71d).[143] In 1970, in a joint venture with Allied Chemical, Gulf began constructing a light water reactor fuel reprocessing facility near Barnwell, South Carolina (R70d)in the district of Mendel Rivers, a prominent U.S. Congressman.[144] Gulf made large uranium discoveries at Rabbit Lake, Saskatchewan and two locations in New Mexico in 1968-1969 (R68e,f). It added to them in 1971 (R71b). It would become "a major producer of uranium for the world's power markets," and participate in "almost all sectors of the rapidly growing nuclear energy business that are open to private enterprise" (PL33).[145] Its nuclear strategy to develop a vertically integrated enterprise in the U.S. and Canada built upon exploration and production was identical to its petroleum strategy (PL1,15).

It approached alternative fuel sources in the same fashion, concentrating first on the raw materials. In addition to new coal (R71c) reserves in Montana and New Mexico Gulf took a 10% interest in 1971 in Syncrude Canada Limited (R71e), which would produce crude from the Athabasca Tar Sands in Alberta, and oil shale leases in Utah (R71d).[146] In 1972 it greatly expanded its tar sand reserves to 340,000 acres (R72a). In January 1974, with Amoco, Gulf bid the highest price offered for any oil shale land, $210 million, and won the richest oil shale tract in Colorado offered by the U.S. Department of the Interior (R74a). In 1974 Gulf began exploring for geothermal steam, zinc, and copper.[147]

Gulf's diversification efforts were not limited to energy and minerals. In 1964 it took an interest in a real estate venture to build a new city from scratch in Reston, Virginia, near Washington, D.C. (R64). Successful by 1968, it expanded in 1971 to add a new headquarters for the U.S. Geological Survey. A second residential-industrial-commercial complex near Disney World in Orlando, Florida was begun in 1971 (R71e).[148] A new subsidiary, Gulf Oil Real Estate Development Company, was created to seek more creative solutions for surplus Gulf properties, many of them former service station locations (R71f).[149] In 1969 it took an interest in "Venture Out in America," a firm formed to develop and operate high quality camp grounds for recreational vehicles (R69).[150]

Dorsey began to correct the mistakes of the 1962-1972 period. In 1972 Gulf took a $250 million write-off on losing properties (PL30,R66a, c,70c,62d).[151] It divested all its 3500 marketing outlets in 20 upper midwestern and northwestern states and a similar number elsewhere in the U.S. The Hercules, California refinery was scheduled to be sold but was retained until February 1976 when it was sold to Coastal States Gas. Gulf exited completely from Germany, selling its refinery and all marketing facilities. It also disposed of its interest in the Philippine refinery and production in Colombia. By 1976 Gulf had closed 40% (2500) of its stations in Europe, 1/3 (2000) in Canada, and almost 40% (11,000) of its U.S. stations. Long-term debt was reduced from $2.1 billion in 1971 to $1.1 billion in 1976. Since the oil business did not look attractive short term (H5) in the 1970's Gulf retrenched downstream and abroad. Downstream, at least, more retrenching remained to be done.

Dorsey's solutions (PV12,13) were reorganization and conglomerate diversification. In 1973 he brought in Jurgen Ladendorf, a consultant and Harvard Business School professor, and an elite staff of outsiders to institute from the top down the long-term planning Gulf had always lacked. These non-oil men were also supposed to push Gulf into (PL34) diversification.[152] In 1974

Gulf's 16 regional operating companies (R68c), which had been reorganized geographically in 1968, were shuffled (R74b) into seven "strategy centers" based on functional activities and product lines. Gulf, like Sun and Mobil, dropped the geographic organization. Coordination among functional units continued to be a problem, however, and may remain until the problem of centralization of decision making and strategy in Pittsburgh on Gulf's board (PL6) is resolved. The legions may not cooperate if all roads lead to Rome.

The seeds of diversification bore fruit in the 1970's.[153] Gulf established a strong position in coal and uranium. Their profitibility is not publicly known. From 1970 to 1976 its mineral activities lost $85 million, though some were going through start-up periods. Oil shale (R74a) turned sour. Gulf wrote off $310 million from 1970 to 1976 on the HTGR (R67c), the Barnwell facility, and nuclear fuels and reserved another $250 million for contingencies. The HTGR has never gotten through the pilot stage. With Shell's share the HTGR ranks with the half dozen largest corporate losses of all time. General Atomic cancelled its last contract in October 1975. The conglomerate diversification ideas were never consummated. They included the acquisition of CNA Financial, a large Chicago insurance and real estate company, in 1973; the Ringling Bros.—Barnum and Bailey Circus—and Pittsburgh-based Rockwell International Corporation, an aerospace conglomerate, in 1975.

Murphy's law has prevailed for much of the last decade at Gulf. By 1976 most of its foreign production had been nationalized (EC48). Its Eastern Hemisphere refining and marketing network was deep in the red. Chemical prices had taken a beating until 1973 (EC41) and Gulf lost money on them from 1970 to 1972.[154] Fifty-state marketing was unprofitable. Gulf lost through a 1973 Supreme Court decision some of its valuable west Texas leases (EC46).[155] After Watergate in 1974 came charges and evidence of illegal contributions to political figures and bribery in many places. No sooner had it cleaned house thoroughly in January 1976, dismissing Brockett, Dorsey, and others, than two employees were charged with "providing illegal gratuities" to an Internal Revenue Service agent. This was followed in 1977 by a swamp of lawsuits alleging that Gulf participated in a Canadian uranium cartel.[156] In addition to the public censure for its own shortcomings it made a perfect villain for politicians to chastise for the energy crisis (EC47), high petroleum prices and profits, and low corporate ethics—all the excesses of big business to which Americans objected in the 1970's. It was a case of mismanagement (H5,7) added to misfortune.

Diversification was deemphasized when Dorsey departed (PL32,34, PV12). His successor, Jerry McAfee, returned Gulf to its traditional (PV14) strengths (PL1-4,14,15,17): "What I hope to be remembered for, what I *will* be remembered for, is for strengthening our basic businesses—oil and gas, and related ventures in chemicals, coal, and uranium."[157] This means further retrenchment downstream and renewed emphasis on finding and developing oil and other resources. His $10 billion five-year capital spending program called for $1.2 billion each year for oil and gas exploration, mostly in the U.S., $200 million/year for coal and uranium and a similar amount for chemicals (PL1,4,5,31). Never a company to do anything in a small way (PL7,H3), Gulf's level of capital spending has been greatly boosted to $3 billion in 1977, up from $1.75 billion in 1976, and is second only to Exxon's $4 billion, though Exxon is three times as large.[158] *Business Week* reported that Nathan W. Pearson, a Gulf director and financial adviser to the Mellons, "says that he is pleased with the direction Gulf is taking under McAfee" (PL16).[159] After the attempts in the 1960's and 1970's to change its basic strategy failed, Gulf returned to what it does best.

It appears to have paid off.[160] The decline in Gulf's U.S. production was arrested. The 12 leases in the Gulf of Mexico for which it paid top dollar in June 1977 appear promising. It committed very little to offshore California and Alaska and the Baltimore canyon, areas that so far have produced little for other companies. With crude oil prices controlled in the U.S. Gulf is exploring for gas. There is today, three-quarters of a century later, no better description of Gulf's strategy now and over the years than the one W.L. Mellon gave in 1901, "I concluded that the way to compete was to develop an integrated business which would first of all produce oil."[161]

5.9 Summary and Analysis

Gulf, whose strategy was formed in response to the opportunities presented by the absence of Standard Oil in Texas (EC1) and the rise of the U.S. gasoline market (EC13), took a different path than the other Major, Texaco, which had been formed in similar circumstances. Texaco was a balanced organization with low cost production that sold products. Gulf is an integrated business that first of all found, produced, and often sold crude oil. Texaco's and Gulf's vertical and product strategies developed differently in similar environments. This was not entirely due to Gulf's early commitment (PL1,14) to find and produce crude oil, for Texaco had a similar commitment until 1913. It was partly Texaco's early commitment to sell products, not crude, and to the tug of war between Texas and New York directors over upstream and downstream policy from 1913 to 1933. At Gulf the role (PL6) played by the Mellons—the founder, the uncles, the heirs, and now the Mellon directors—has been crucial to the direction and execution of strategy. Their personal values have been and are particularly important to Gulf: the emphasis on crude, vertical integration, and family ownership and control (PV1-5). Their position is perhaps unique among large U.S. companies. Once both owners and managers, not unlike the Ford family at Ford, their closest counterpart, they own and control but do not manage. An interesting question is whether, in addition to the documented effects on strategy, their role may have contributed to Gulf's organization problems (H1,4,5,8) of falling behind (neglect), followed by overreaction, the inability to decentralize, and lack of management development. If so, as a model for greater involvement and responsibility by boards of directors of large corporations, they raise questions about the effect of active directors on management's ability to manage.

With regard to the models of the firm it appears that Gulf's successful profit maximizing and oligopolistic responses took the form of selecting geographical and vertical areas of the industry where it could outperform competitors. In support of the business policy model the effects on strategy of its resources, handicaps, environment, and personal values are clearly important. The event of the discovery in Kuwait, whose production took off and simply ran away with the American company, forcing it to be an international Major, was one with which Gulf never appeared completely comfortable. Of special note to the difficulty of changing corporate strategy is Gulf's experience of spending literally billions of dollars fruitlessly since 1960 in attempts to adapt its strategy to the new environment by integrating downstream in the Eastern Hemisphere and by diversifying, only to return to doing what it does best. It has proved at least as difficult and costly for Gulf to become a worldwide refiner and marketer as for Exxon to become an explorer.

Exhibit G-1
Gulf Policies

PL 1	Principal business is upstream (exploring, producing) rather than downstream (refining, marketing, transporting activities: (a) V; (b) 1901-present.
PL 2	Desire to be vertically integrated in the U.S.: (a) G,V; (b) 1901-present.
PL 3	Little or no downstream activities abroad: (a) G,V; (b) 1901-62; 1972-present.
PL 4	Preoccupation in downstream activities with the U.S. market— identity as a U.S. company with foreign production: (a) G,V; (b) 1901-62; 1972-present.
PL 5	Aversion to political risk: (a) G,A; (b) 1901-present.
PL 6	Ownership and control of strategy by the Mellon family: (a) A; (b) 1902-present.
PL 7	Large scale entry: (a) S; (b) 1901-present.
PL 8	Disposal of crude by long-term, large volume contract: (a) V,P; (b) June 28, 1901-1962.
PL 9	U.S. marketing in the South (1903), Southeast (1907), and East coast (1903): (a) G,V; (b) 1903, 1907-present.
PL 10	Marketers sold the products refiners could produce—i.e., marketing controlled by refining to dispose of its output: (a) V,P; (b) 1902, 1907-present.
PL 11	Adoption of the name Gulf (1902,1907) and the orange disc trademark (1903) to give all parts of the company one identity: (a) A; (b) 1902, 1903, 1907-present.
PL 12	Refine only the better grades of Texas crude: (a) V,P; (b) 1903-1906(?).
PL 13	Entry into Oklahoma (1906) to produce and transport refinable crude on a large scale: (a) G,V,S; (b) 1906-present.
PL 14	Discover rather than purchase leases: (a) V, A; (b) 1911-present.
PL 15	Use scientific advance to gain a technological lead over competitors in finding oil: (a) P,A; (b) 1911-present.
PL 16	Develop foreign crude production for import, not for sale abroad: (a) G,V,P; (b) 1912-present.
PL 17	Purpose of downstream facilities is to dispose of crude: (a) V,P: (b) 1902-July 1935; 1939-present.
PL 18	Become a strong gasoline marketer in the U.S. (1926 or before) and abroad (Canada and Europe) (1956), 1962-present: (a) V,P; (b) 1926 or earlier, 1956, 1962-present.
PL 19	Geographical expansion downstream by acquisition: (a) V,A; (b) 1929-1966.
PL 20	Purchase crude in U.S.: (a) V; (b) July 1935-present.
PL 21	Diversify production geographically to reduce dependence on Kuwait and spread political risk: (a) G,H,A; (b) 1950's-present.
PL 22	Boost North American earnings base: (a) G; (b) 1954-present.
PL 23	Expand downstream in the Eastern Hemisphere: (a) V,G: (b) 1960-72.

PL 24	Build small-scale, simple refineries in many locations to move the maximum volume of crude at minimum investment expense: (a) S,H,G,P; (b) 1962-1972.
PL 25	Enter joint ventures downstream and supply crude for them: (a) V; (b) 1962-present.
PL 26	Worldwide exploration for higher cost crude sources: (a) G,H; (b) 1960-present.
PL 27	Marine transshipment facilities like Bantry Bay built; (a) S,V,G,H; (b) 1966-72.
PL 28	Maximize profit on crude and crude volume: (a) S,P; (b) 1901-present.
PL 29	Sell Gulftane, a subregular grade gasoline, at cut-rate prices: (a) P; (b) 1961-73.
PL 30	Market in 48 states: (a) G; (b) 1961-72.
PL 31	Enter bulk petrochemicals: (a) P,V; (b) 1950-present.
PL 32	Intentional diversification outside petroleum: (a) P; (b) 1967-present.
PL 33	Vertical integration in nuclear industry: (a) V,P; (b) 1968-74.
PL 34	Conglomerate diversification: (a) P; (b) 1972-6.

Notes:
(a) Type of strategic choice

S—scale		PL 7,13,24,27,28
G—geography		PL 2-5,9,13,16,21-24,26,27,30
H—horizontal concentration		PL 21,24,26,27
V—vertical integration		PL 1-4,8-10,12-14,16-20,23,25,27,31,33
P—product diversification		PL 8,10,12,15-18,24,28-29,31-34
A—administration		PL 5,6,11,14,15,19,21

(b) duration

Exhibit G-2
Gulf Environmental Conditions

EC 1	Standard banished from Texas: 1900.
EC 2	No oil industry in Texas before 1895.
EC 3	Discovery at Spindletop: January 10, 1901.
EC 4	Asphaltic character of Texas crude.
EC 5	Spindletop stops gushing: August 1902.
EC 6	Additional discoveries in Texas and Louisiana: 1903-4.
EC 7	No new fields discovered: 1905-6.
EC 8	Discovery at Glenn Pool, Okla. 1905.
EC 9	High cost of leases in 1910.
EC 10	Standard Oil before 1911.
EC 11	Discovery in Mexico 1912.
EC 12	Discovery in Venezuela 1916.
EC 13	Rise of the automobile and the demand for gasoline, 1908-24 but especially after World War I.
EC 14	Discoveries in west Texas 1920.
EC 15	Discoveries in Arkansas (Smackover), 1921.
EC 16	Decline in Mexican production 1921.

EC 17	Discovery in Iraq, October 15, 1927.
EC 18	Location of Venezuela relative to U.S. East coast and Texas.
EC 19	Rule of capture.
EC 20	Discovery of east Texas field 1929.
EC 21	Coming of prorationing 1930's.
EC 22	Great Depression 1930's.
EC 23	Imposition of gasoline taxes.
EC 24	British resistance to American participation in the Middle Eastern oil, World War I—1956 (Suez).
EC 25	Huge reserves and remoteness of Kuwait.
EC 26	Suez Crisis, 1956.
EC 27	All-time peak in U.S. drilling.
EC 28	Surplus in U.S. production.
EC 29	U.S. price structure breaks.
EC 30	Mandatory import quotas.
EC 31	Cuts in posted prices.
EC 32	Formation of OPEC.
EC 33	Consent decree with FTC changing Shell contract (see text).
EC 34	Worldwide crude glut.
EC 35	Scale changes in refineries and tankers in 1960's.
EC 36	Rapid growth of Eastern Hemisphere demand in 1960's.
EC 37	Lack of supertanker ports in Europe.
EC 38	Trend toward 50-state marketing in Europe.
EC 39	Six Day War in Israel.
EC 40	Conglomerate diversification mania.
EC 41	Chemical prices decline.
EC 42	Nationalization in Bolivia.
EC 43	Nationalization in Libya.
EC 44	End of crude glut.
EC 45	Teheran agreements.
EC 46	Supreme Court decision terminating some of Gulf's west Texas leases.
EC 47	OPEC embargo and nationalization.
EC 48	Foreign production nationalized.

Exhibit G-3
Gulf Resources

R 1a	Production at Spindletop on the Gulf coast, January 10, 1901.
R 1b	Management of J.M. Guffey. 1901-2.
R 1c	Mellon's money. 1901-34.
R 1d	Guffey's 1 million acres of leases in Texas and Louisiana. 1901-7.
R 1e	Tankage at Spindletop and Port Arthur. 1901-present.
R 1f	Pipeline from Spindletop to Port Arthur. 1901.
R 1g	Refinery site at Port Arthur. 1901-present.
R 1h	100 rail tank cars. 1901.
R 1i	Shell contract for 4.5 million barrels of crude at 25¢/bbl. June 28, 1901-3.

R 1j	Name—Gulf. November 10, 1901.
R 1k	Gulf Refining Company of Texas. 1901.
R 2a	Refinery at Port Arthur. 1902.
R 2b	Management skills in oil business of W.L. Mellon. August 1902-April 1931.
R 2c	Managerial skills of Gale R. Nutty. 1902.
R 2d	Skills of George H. Taber in refining. 1902.
R 3a	Orange disc trademark. 1903-present.
R 3b-d	Sales offices in New York, Philadelphia, New Orleans (all 1903).
R 4a	Skills in production of Frank H. Leovy. April 1904-April 1931.
R 4b	New production in Texas and Louisiana from discoveries at Sour Lake, Saratoga, Batson, Humble, and Jennings. 1901-4.
R 4c	Pipelines to new fields (R4b), 2 barges, 3 tankers, and refinery expansion to 12 TBD. 1904.
R 6a	Gulf Pipe Line Company of Texas. 1906.
R 6b	Discoveries at Caddo, La. 1906.
R 7a	Sales office in Atlanta to head Southeastern sales division covering the Carolinas, Georgia, and Florida. 1907-present.
R 7b	Pipeline from Tulsa to Port Arthur. 1907.
R 7c	Gypsy Oil Company—production in Oklahoma (Indian territories). 1921.
R 10	Discovery in east Texas at Ferry Lake (first to drill over water). 1910.
R 11a	Geological skills of M.J. Munn. 1911.
R 11b	Ft. Worth refinery. 1911-54.
R 13	Production in Mexico. 1913-51.
R 16	Production in Kansas. 1916.
R 18	Production in Kentucky. 1918.
R 21a	New production in Oklahoma (Indian territories). 1921.
R 21b	Production in Arkansas (Smackover). 1921.
R 25a	Production in Venezuela. 1924-76.
R 25b	Barco Concession in Colombia. 1925-36.
R 26a	Production in west Texas. 1926-present.
R 26b,c	Refineries at Philadelphia (30 TBD, 1927) and Bayonne/New York (30 D, 1926).
R26d	Purchase of interest in Nobel-Andre-Good, lube oil marketing company in Europe. 1926.
R 27a,d	Concessions on Bahrain (11/30/27-12/21/28), Kuwait (11/3/27-1976), Neutral Zone, and al-Hasa (both 11/30/27-4/1/32).
R 27e	Joint venture with Texas Gulf Sulfur to extract sulfur from oil properties.
R 28	1/6 interest in Near East Development Corporation, production in Iraq. 1920, 1927-1934.
R 29a	West Texas refinery at Sweetwater (5 D). 1929-54.
R 29b	Production in New Mexico. 1929.
R 29c	Strong U.S. gasoline marketing. 1920's-present.
R 29d	Gulfpride, premium motor oil. 1929-present.
R 31a-c	Refineries in Cincinnati (12 TBD, 1931-present), Pittsburgh (6 TBD, 1931-54), Toledo (12 TBD, 1931-present).
R 31e	Marketing in Ohio, Michigan, Indiana, and Illinois.
R 31e	Pipeline from Tulsa, Okla. to Lima, Ohio. 1929-31.

R 31f	J.F. Drake.
R 32	New headquarters building in Pittsburgh. 1932.
R 33	Kuwait Oil Company formed as joint venture with BP.
R 34	Concession for KOC in Kuwait. 1934-76.
R 37	Contract to sell half the present and future production of the Mene Grande Oil Company, Gulf's Venezuelan producing subsidiary, to International Petroleum Corporation, a Jersey subsidiary, for $100 million. 12/19/37-1950's.
R 38	Discovery in Kuwait. 1938.
R 46	Production in Kuwait (50% Gulf). 1949-76.
R 47	Long-term contract to sell half its Kuwait production, up to 750 TBD to Shell
R 49	Refinery in Kuwait (50% Gulf). 1949-76.
R 50a	Establishment of a crude oil purchasing department for domestic operations. 1950.
R 50b	Production in Canada. 1950.
R 50c	Refinery in Venezuela (67% Gulf).
R 50d	Manufacture of ethylene at Port Arthur. 1950.
R 52	Interest in Neches Butane Products Company 1942.
R 54a	Production offshore in Gulf of Mexico. 1954-present.
R 54b	7% share in Iranian consortium production and refining. 1954-71.
R 54c	Goodrich-Gulf joint venture to manufacture butadiene. 1954-70.
R 55a	Acquisition of Warren Petroleum Company. 1955-present.
R 55b	Production in Sicily. 1955.
R 56a	Acquisition of 58% of BAOC, Canada's second largest marketer, refiner, producer.
R 56c	Production in France; refinery (18% interest) in France.
R 60a	Caribbean Refining Company (Puerto Rican refinery); 100% in 1962. 1960-present.
R 60b	Acquisition of Wilshire Oil Company (700 stations + 32-TBD refinery near Los Angeles). 1960-present.
R 61	Introduction of Gulftane. San Antonio in July; nationwide, November—December 1961-73.
R 62a	30 TBD refinery at Stigsnaes, Denmark. 1962-present.
R 62b	30 TBD refinery at Europoort, Netherlands. 1962-present.
R 62c	10 TBD lube oil refinery in Taiwan (70% Gulf). 1963-present.
R 63a	Gulf-Holiday Inn partnership. 1963.
R 63b	35 TBD refinery in Korea (25% Gulf). 1963-present.
R 63c	10 TBD lube oil refinery in Taiwan (70% Gulf). 1963-present.
R 63d	Purchased 19 TBD refinery in Black Creek, Miss. 1963-71.
R 63e	Discovery in Nigeria. 12/24/63.
R 63f,g	Acquisition of Spencer Chemical and a subsidiary, Pittsburg and Midway Coal Mining Company.
R 64	Reston, Va. real estate venture.
R 66a	Refinery at Hercules, Cal. 1966-76.
R 66b	Discovery in Cabinda.
R 66c	Purchase of 2300 stations in 10 Rocky Mountain states from Cities Service 1966-72.
R 66d	48-state marketing 1966-72.
R 67a	Bantry Bay marine transshipment facility.
R 67b	Refinery at Neuchatel, Switzerland (60 TBD, 25% Gulf).

R 67c	Refinery at Venice, La. (10 TBD).
R 67d	Gulfrex geophysical research ship commissioned.
R 67e	Gulf Mineral Resources Company organized.
R 67f	General Atomic acquired.
R 68a	Refinery at Milford Haven, Wales. (78 TBD).
R 68b	Refinery at Huelva, Spain (40 TBD, 40% Gulf).
R 68c	Discovery in Ecuador (50% with Texaco).
R 68d	Reverse osmosis water purifying system.
R 68e	Uranium discovery at Rabbit Lake, Saskatchewan.
R 68f	Two uranium discoveries in New Mexico.
R 68g	Reorganization into 16 geographical operating companies.
R 69	Interest in Venture Out in America.
R 70a	Pt. Tupper, Nova Scotia marine transshipment facilities.
R 70b	Okinawa, marine transshipment facility
R 70c	Acquisition of Frisia Refining, Marketing, and Transportation Co., including refinery (50 TBD) at Emden, Germany) 1970-72.
R 70d	Nuclear fuel reprocessing facility built at Barnwell, S.C.
R 70e	Gulf's first North Sea discovery.
R 71a	Gulf produces nuclear fuel rods for light water reactors.
R 71b	Uranium discoveries.
R 71c	Syncrude in Canada (10% Gulf).
R 71d	Shale leases in Utah.
R 71e	Refinery on Okinawa (92 D, 45% Gulf).
R 71f	Refinery at Alliance, La. (155 TBD).
R 71g	Refinery at Pt. Tupper, Nova Scotia (80 D).
R 71h	Major Center near Disney World (real estate development).
R 71i	Gulf Oil Real Estate Development Company formed.
R 72a	Additions to tar sand reserves in Canada.
R 72b	Refinery at Bilbao, Spain (112 TBD, 40% Gulf).
R 72c	Refinery at Milan, Italy (80 TBD).
R 74a	Colorado oil shale tract "A" (50% with Amoco).
R 74b	Reorganization into seven strategy centers.
R 74c	Hollis Hedberg geophysical research ship commissioned.

Note: R## is generally the year 19## when Gulf acquired this resource or it became of strategic importance.

Exhibit G-4
Gulf Handicaps

H 1	Lack of downstream organization—weakness downstream. 1901-present.
H 2	Crude short in the U.S. 1906-present.
H 3	Red Line agreement, July 31, 1928.
H 4	Tendency to overreact to pressure 1929-present.
H 5	Tendency to take a short run view of the business.
H 6	Financial stringency—Union Gulf bond indentures, 1929-late 1930's.
H 7	Weakness in financial management.
H 8	Lack of management development.

Exhibit G-5
Gulf Personal Values

PV 1	W.L. Mellon: interest in production.
PV 2	W.L. and Andrew Mellon: belief in vertical integration.
PV 3	Mellons: interest in U.S. market; lack of interest in foreign investment.
PV 4	Mellon family: control of strategy as well as ownership of Gulf.
PV 5	Mellon family: aversion to political risk.
PV 6	J.M. Guffey and J.H. Galey: interest in discovery of crude.
PV 7	J.M. Guffey and J.H. Galey: belief in large scale.
PV 8	J.M. Guffey and J.H. Galey: belief in vertical integration.
PV 9	Anthony F. Lucas: belief that there was oil along the Gulf coast.
PV 10	Wm. K. Whiteford: arrogant autocrat.
PV 11	E.D. Brockett: commitment to diversification.
PV 12	Bob R. Dorsey: commitment to conglomerate diversification.
PV 13	Bob R. Dorsey: commitment to planning and reorganization.
PV 14	Jerry McAfee: return to traditional strengths upstream.

6

The Strategy of Mobil Corporation

6.1 "Long on Brains; Short on Crude"

The strategy of Mobil Corporation can be traced to origins in the roles of its components within Standard Oil before 1911, demonstrating the greater importance of early circumstances, resources and values over those more recent in the formation of strategy.* Only two changes of more recent origin have significantly affected Mobil's strategy to date: a 1959 reorganization cut costs, retired the hidebound, installed long-term planning, and brought bright, new people with aggressive attitudes to top management(PL21,22); and in 1968 the board made a commitment to conglomerate diversification resulting in the 1974 acquisition of a retailer and a paper company and the near acquisition in 1976 of a real estate firm (PL25). Thus Mobil has the distinction of being one of the few Majors to make important changes in its basic strategy successfully in recent years.

The two former Standard Oil units which united in 1932 to form Socony-Vacuum, renamed Mobil Oil in 1955 and Mobil Corp. in 1976, were both primarily marketers (PL1). Socony, the former Standard Oil Company (New York), had been the principal export unit and headquarters (PL4,5) for Standard Oil until an 1893 reorganization shifted on paper the control of many assets to the Standard Oil Company (New Jersey), now Exxon. Since Standard's headquarters remained at its 26 Broadway offices in New York even after the divestiture it still thought of itself as the headquarters for many years (PL5), though its assets had in fact been reduced to refining and marketing in New York and the New England states and the export of products, principally to the Far East (PL2,4). Vacuum Oil had been the premium lubricants specialist (PL15) within Standard and exported to a worldwide network of sales offices (PL19) by the 1890's. Its name was the most valuable product brand name (R31c, PL18) in the world for many years. Neither company had any crude (H1).

*Policies (PL#), Resources (R#), etc. are keyed to exhibits at end of text.

As the strategic comparisons in Table 1-2 (Chapter 1) show, Mobil is still primarily a marketer, still crude short, still worldwide and larger in the Eastern Hemisphere than in the Western Hemisphere, still stronger in lighter and specialty products, and is the most diversified Major (PL1,3,4,11,19,23,25,H1). Not shown is that in petrochemicals Mobil has followed its marketing instincts and specialized downstream in more consumer products like Hefty trash bags than in bulk chemicals like Exxon, Shell and Gulf. It is also still a lubricants specialist, now selling its new synthetic Mobil "1" oil at $3.95/quart (4-5 times the price of most competing brands) (PL15).

By maintaining its strategy it has emphasized its strategic differences from other Majors in being for many years shortest on crude and strongest in marketing (PL1,3,H1). Between 1911 and 1976 it became a strong nationwide U.S. gasoline marketer and acquired some production, though never enough (PL9-11). In the 1970's it turned its scrappy new (PL22) attitudes to public relations to emphasize these differences from other Majors (and to affirm its identify as Mobil rather than the headquarters of Standard, its self-image for so many years) by advertising its corporate opinions, doing battle with its critics in print, and taking the highest public image profile of the Majors (PL5,22). The phrase "long on brains; short on crude" is its own description of its new identity.

6.2 SOCONY: Inheritance 1911

One can describe Mobil's strategy from 1911 when it was separated from Standard (EC11) until 1959 when it reorganized and began to achieve its own identity (PL22) by saying that its strategic response (PL8) to the divestiture (EC11) was to try to put Humpty-Dumpty back together again. Mobil inherited from Standard Oil both its principal parts and its policies. Both parts came from Standard Oil: Standard Oil Company (New York), known by its cable address SOCONY, and the Vacuum Oil Company.[1]

Socony had been one of the principal refiners and marketers (PL1,2,14) in the Standard combination with five refineries in New York City and one in Buffalo with a total capacity of 6 TBD. In 1911 it was principally a marketing company (PL1). It had been Standard's marketing arm in New York and New England, handling some 85% of the petroleum products in the area, and in the Far East (PL2,4,R11). With $90 million in assets it was a large company, the second largest piece after Jersey Standard split off the Standard Oil empire in 1911, and the third largest of the Majors just as Mobil is today.[2]

It owed much of its success to shrewd retail marketing (PL1). In 1893 it sent salesmen and agents to China, India, and Japan to market (PL4) Standard's kerosene in the Orient. To promote the new product it designed a small, efficient lamp with a tin bowl and a glass chimney that became known as the Mei Foo ("Admirable Company") lamp. Millions were sold or given away with the first case of kerosene. As a marketing program and a social contribution, it was a stroke of genius. Both the lamps and Standard's blue tin kerosene cases became necessities. With kerosene they were among the first industrial products to reach large numbers of Asian consumers. When Shell first tried to sell kerosene in bulk by itself it found no demand because customers wanted the cases as much as the kerosene and would not buy the product alone. The company became known as Mei Foo, the name was registered as a trademark, and marketing "oil for the lamps of China" became part of Socony's identity (PL1,R11d).[3]

But more important to its strategy than the obvious assets and markets it inherited in 1911 was the position it had formerly occupied within Standard Oil. In 1865 William Rockefeller, brother of John D., went to New York to set up the export trade (PL4). In 1882 Standard's headquarters were moved from Cleveland to New York. From 1882 to 1899 New York Standard was the principal unit of Standard oil (R11e, PL5). It was the export center, the financial center, and the administrative center for the entire combination.[4] It was the banker which lent funds to subsidiaries, invested their short-term funds in New York, and settled foreign accounts (PL6). It allocated capital expenditures of the group. The recalcitrance of the state of New York and the willingness of the state of New Jersey to pass a law permitting Jersey Standard to become a holding company in 1899 and take over many Standard assets was merely a legal maneuver, a charade. Did not the managements manage from the headquarters at 26 Broadway in New York for years after the divestiture much as they had before?

In the view of its executives, and the public, New York Standard was not a subsidiary of Standard Oil. It was Standard Oil. And its policies and strategy (PL5,6) reflected this view for many years thereafter. In 1911 Socony began life on its own with small refineries in New York, a strong marketing organization in the Northeast, and the romantic business in the Orient as its assets (R11). It was a large company which thought itself larger still. But it had not one drop of crude (PL3,H1).

6.3 Response: 1911-1931

During its first years of independence Socony was at least as preoccupied with other changes in its environment as with divestiture: the rise of the automobile market and gasoline demand (EC12) (Tables M-1,2 below); resulting changes in refinery technology and product mix (EC13) (Table M-3 below), and invasion of its market by competitors (EC14) (Table M-4 below).

The decade between 1912 and 1922 was a period of great change in marketing. World War I demonstrated the superiority of motor vehicles over animal power in transportation. Armies went into the war on horses and mules and came back in tanks and trucks. The same revolution (EC12) affected civilian transportation after the war (Table M-1,2). Henry Ford introduced the Model T in 1908. The first motor tank truck was built in 1911.[5] Because its delivery range was so much greater than the horse-drawn tank wagon, fewer and larger bulk delivery and storage plants were needed. Many marketing properties became obsolete. By 1915 the gasoline filling station had begun to replace the garage.[6] Until about 1921 they were largely owned and operated by independent dealers, as garages had been. But intense competition among the large companies forced marketers to seek greater control over retail outlets through ownership or lease. Large new investments in marketing facilities were required. Socony borrowed $50 million in 1921, its first long-term debt and first notable departure from Standard Oil strategy (PL7). Socal borrowed $25 million. Gulf also made its first major borrowing that year and Texaco doubled its long-term debt. Socony's gasoline sales exceeded its kerosene sales in 1912 for the first time.[7] It continued to market Jersey's products. Wholesale and retail distribution channels and product mix were all profoundly and permanently changed by the rise of the automobile (EC12-16). At the start of this decade Socony decided to become a leading gasoline marketer (PL11).

Gasoline demand and technical change, like the Burton-Humphreys thermal cracking batch process developed at Indiana Standard in 1913 (EC13), improved efficiency and shifted product mix in refineries (Table M-3). Fear of U.S. crude shortage after World War I (EC18) prompted Socony to build a refinery at East Providence, Rhode Island in 1919 (R19) to take advantage of South American supplies.[8]

Socony also had to contend with invasions of this territory (EC16). In 1911 Texaco, Gulf, Cities Service, and Pure Oil Company, all independents, had footholds in Soconyland.[9] By 1916 other former Standard affiliates had begun to enter its market.[10] By 1922 Tidewater, Sinclair, Sun, Atlantic Refining, and Continental had extended their businesses into New York and New England.[11] As they grew Socony's volume grew but its market share declined (Table M-4).

After World War I Socony began to respond to the increased competition with two new policies: (1) it sought to obtain crude production to reduce its total dependence on purchased crude (PL9) (vertical integration); and (2) it began to expand geographically to counter the expansion into its territory (PL10). In the U.S. it acquired three vertically integrated oil companies before 1931 and, after its merger with Vacuum in 1931 became the leading U.S. gasoline marketer in the 1930's (Table M-5) (PL1,11).

In its first independent years Socony had continued to purchase (PL3) crude and products through the former Standard Oil affiliate network. After World War I crude prices rose sharply from $2 in 1918 to $3 per barrel in 1920[12] as consumption expanded and fear of a worldwide (EC18) shortage spread. The source of the fear was a narrow escape from scarcity during 1917-1918 in the midst of the war. Had the war been prolonged an actual scarcity might have developed.[13] High crude prices and unsure supplies placed Socony at a disadvantage compared to integrated competitors just as it was beginning to make heavy investments in its own retail marketing network. The risk forced it to integrate (PL8) into production. In 1918 Socony made its first acquisition, a controlling interest in Magnolia Petroleum (R18).[14]

Magnolia, named after the flowering trees in Texas in 1911, was the first refining company in Texas, having begun operations in 1898 as the J.S. Cullinan Company at Corsicana where the first oil in Texas had been discovered in 1894 by men drilling for water.[15] It was founded by J.S. Cullinan, who founded Texaco in 1902, and financed by two Standard executives, Calvin N. Payne and Henry C. Folger, who was president of Socony from 1909 to 1922. Standard's secret interest through its executives had been divested in 1909 when the U.S. Supreme Court upheld a state antitrust case in Texas banning Standard from the state.[16] In 1911 Magnolia consisted mainly of refineries (PL14) at Corsicana and Beaumont, but it soon began to acquire extensive (PL9) proven producing properties in Texas, Oklahoma, Arkansas, and Louisiana. Three more local market (PL14) refineries were added in 1914, 1921, and 1923. Socony acquired the remainder of its stock in 1925. By 1926 it was the leading gasoline marketer (18% share) in both Texas and Oklahoma.[17] Magnolia gave Socony crude production, vertical integration, and expanded geographical coverage (PL9,10).

In 1926 Socony acquired for stock worth approximately $110 million 100% of the General Petroleum Corporation of California (R26). GP, as it was called, was the successor to a number of companies organized by Captain John Barneson, a retired sea captain who had been among the first to experiment with oil as a marine fuel in 1898. As late as 1912 it was solely a producing company which had been supplying 2 to 3 TBD of heavy crude to Standard Oil of California. In 1913 it built a refinery on the outskirts of Los Angeles at Vernon and later a larger one at Torrance (PL14). By 1916 it had 235 producing wells, pipelines, tankers, and sold its products from Alaska to Chile and from the Rockies to Hawaii. It was one of the eight largest oil companies in California but was primarily a fuel oil producer, refiner, and marketer. In 1923 it began gasoline marketing and had 1500 outlets in 400 towns by 1925. In 1925 GP produced 27 TBD (= thousand barrels per day), purchased 41 TBD more, and refined 49 TBD. Perhaps more important to Socony than the crude and geographic expansion, GP provided a source of supply on the Pacific coast for Socony's marketing business in the Far East (R11d) (PL4,9,10). When Socony acquired it, Captain Barneson's son, Lionel, was president of the company. Following acquisition, GP like Magnolia continued to operate independently under the same management (PL12).[18] Its relationship with these new subsidiaries, engaged in production, an operation it did not understand, was perhaps less owner and manager than the banker it had been for Standard Oil (PL6).

Socony now had operations in three widely separated areas: the Northeast, Southwest, and West coast. In 1930 Socony expanded into the Midwest by acquiring the White Eagle Oil and Refining Company (PL9,10). It had a small amount of mid-continent production, refineries in Augusta, Kansas (1916) and Casper, Wyoming (1923) and most importantly was the second largest marketer in 11 states from the Mississippi River to Montana and from Arkansas north. Its book value was $14.5 million (R30).[19]

Socony followed the same strategy abroad as at home (PL9,10,R13-26). It began marketing in Bulgaria in 1913 and Greece in 1915. It formed the Colombia Petroleum Company (Colpet) in 1917, a joint venture with Texaco, to market in South America. It expanded in the Far East. It re-entered Australia in 1922 and South Africa in 1926. Its most important expansion abroad from 1911 to 1931 was the 1928 acquisition of a 25% interest (50% by 1934) in the Near East Development Corp. (R28), which held 23.75% of Turkish Petroleum Company, renamed Iraq Petroleum Company Limited in 1929 (Table M-6 below).[20] It was strategically important as Socony's first foreign crude source but even more important as its entree into the Middle East. Without it Socony would never have had the opportunities to join Basrah or Mosul Petroleum (other Iraq, Abu Dhabi, and Qatar properties), Aramco, or the Iranian consortium. Without these it would not be a Major today, as are none of the other companies which did not obtain production in the Middle East.

In 1931 Socony marketed in 29 states through 37,077 outlets (PL1,11) (23,521 in Soconyland). By itself it was probably the second largest gasoline marketer after Indiana Standard. Its business was 81% U.S. (PL2,4) and 19% foreign.[21] It had grown rapidly in its first two independent decades, but neither as rapidly or as profitably as the industry.[22] Its market share in New York and New England shrank from 85% to 34% and in the U.S. from 19% to 7% (Table M-4). In size Gulf, Texaco, and Socal had nearly caught up to it. It had expanded geographically (PL10) and integrated vertically (PL9) by acquisition following the Standard pattern (PL8). It maintained its financial strength (PL7), large liquid assets (PL7), contractual relationships with former Standard affiliates (PL5,8), an active inside board of directors (PL13), headquarters at 26 Broadway (PL5,6,8), and subsidiaries even more autonomous than under Standard Oil (PL12).

6.4 The Vacuum Oil Company

A major opportunity appeared in January 1930 when the board of directors of the Vacuum Oil Company approved the terms of a friendly merger with Socony. Vacuum was a more famous marketer than even Socony (PL1), having been the premium quality lubricants specialist (PL15) in the Standard Oil organization. It marketed worldwide (PL19) and its trademarks were the most valuable in the industry. On July 30, 1931 the two companies formed the Socony-Vacuum Corporation. Vacuum (R31) contributed products and policies which are characteristic of Mobil today (PL15-19) and reinforced the central concern of both companies with marketing (PL1).

The Vacuum Oil Company was founded in 1866 by Hiram Bond Everest, a grocer, and Matthew Ewing, a carpenter, to exploit a method of distilling crude under a vacuum to produce kerosene. Kerosene production was unsuccessful and Ewing sold out. But Everest found a use for the oily residue produced by the vacuum process in the manufacture of leather. The vacuum process and each improvement were patented and Everest began peddling Vacuum Harness Oil in used oyster cans from a pushcart (PV1) to farmers and businessmen in the Rochester, New York area. He looked for new applications (PV2) and in 1869 patented the famous Gargoyle 600-W Steam Cylinder Oil, one of the most famous products ever in the industry. He improved his process and developed a line of specialty products: hoof dressing, coach oil, axle grease, and tanner's products. Since his products replaced animal and vegetable lubricants like tallow, operators of machinery had to be convinced by tests in actual use that a mineral oil would be a better lubricant. Practical tests (PV3, PL17) for each product became a policy at Vacuum. Everest advertised his products (PV4) from the beginning often in ads extolling Vacuum Oils while commemorating the performance of famous racehorses of the time, probably precursors of the Flying Red Horse trademark.[23]

These early practices became fixed policies in Vacuum's strategy: (1) reliance on purchased crude (raw material)—no interest in production (PL3); (2) manufacture and marketing of high quality specialty lubricants—kerosene, gasoline, naphtha, and other fractions—regarded simply as byproducts (PL15); (3) patent positions based on research (PL16); (4) practical testing in actual use of very product (PL17); (5) direct retail marketing (from the pushcart) to persuade customers of the superiority of its products (PL1); and (6) consumer advertising of its products (PL18). Today, a century later, Mobil is regarded as one of the premier makers of lubricants in the world with its new motor oil "Mobil 1" made from synthetic crude being the first of its kind.

The business grew rapidly and Vacuum acquired a reputation for quality and know-how (PL15,17) in the field of lubrication. In 1879, after refusing several offers, Hiram Everest sold 75% of Vacuum to Standard for $200,000, retired and installed his son, Charles, who had the confidence of Standard, as chief executive. He retained a 25% interest and the title of president until 1900 when he sold the rest to Standard. Affiliation with Standard brought Vacuum a dependable supply of crude at favorable prices, financial backing, and Standard's widespread warehousing facilities for its distributing network.[24]

Charles Everest continued to operate the business as a lubricating oil specialist relatively independently within Standard. He established a technical department and a fully equipped laboratory to study every new form of machinery and devise the right lubricant for each (PL16). Much of Vacuum's success was due to this policy of first determining a customer's particular requirements and then manufacturing a product specifically for those needs, rather than the common practice of trying to find customers for the mix of products the crudes and refineries produced (PL1,16,17).[25]

Rather than expand by integrating vertically through production or horizontally by adding products like kerosene Vacuum chose to expand geographically to market its products worldwide beginning in 1886 (PL19). This commitment to international sales by 1911 had made Vacuum the world's leading marketer of lubricants and the most multinational business within Standard and, with the possible exception of Singer Sewing Machine, anywhere else in its day or since. Over 80% of its business was done abroad.[26]

The advertising of products continued to be an important policy (PL18) at Vacuum. Since the word "Vacuum" could not be registered in all countries, the name "Gargoyle" (a mythical creature symbolizing the victory of good over evil) became the official trademark in 1906. The name "Mobiloil" was coined about 1899 and appeared in the first advertisement for motor oils anywhere in the Saturday Evening Post on September 15, 1906. The two were used together until the 1930's when "Mobil" began to appear by itself. The "Flying Red Horse" symbol first appeared in a South African advertisement about 1908 describing Vacuum's gasoline as "Pegasus Motor Spirit."[27] Many of Standard's subsidiaries were acquired for their brand names and their continued use demonstrates not only their value but also Standard's policies of decentralization and secrecy about holdings in other companies. Perhaps because of its emphasis on advertising (PL18), most of the corporate trademarks used by Mobil came from the Vacuum side of the family tree. Trademarks identified not only products; they were and are the most important symbols of corporate identity. In 1911 the divestiture left Vacuum with assets of $26.5 million, the best trademarks in the industry, refineries at Rochester and Olean, N.Y., but no crude supply, ships, or distribution system in the U.S. It retained its own marketing properties and received Standard's assets in Portugal, Egypt, South Africa, west Africa, and Australia.

From 1911 to 1931 Vacuum had to cope not only with the divestiture (EC11) but also with the rise of the automobile (EC12). Its business grew rapidly before World War I. Like Jersey it acquired a fleet (R15) of seven tankers in 1915 and three more in 1919. To meet export demand it erected a new refinery on the coast at Paulsboro, New Jersey in 1918 (R18b, PL14) and built a modern research laboratory (still one of Mobil's main laboratories) nearby (R18c). Completing the recognition of the international (PL19) character of its business it moved its headquarters from Rochester to 61

Broadway in New York City in 1918. As at its other refineries the Paulsboro facility was designed to produce lubricating oil, with the gasoline and kerosene byproducts sold in bulk in the U.S. or exported to Portugal, Africa, or Australia where it marketed them through the Standard subsidiaries acquired in the divestiture (PL115).[28]

After the war Vacuum expanded its foreign marketing network (PL1,19). Between 1926 and 1929 its investments in foreign companies tripled from $21 to $64 million. In 1929 its products were sold in every civilized country except Russia where its properties had been confiscated in 1917. In 1929 Vacuum's lubricating products were on sale at more than 230,000 places throughout the world, including 100,000 in Europe and 90,000 in the United States.[29]

Its efforts at vertical integration were more modest. In response to the fears of crude shortage after World War I (EC18), it acquired some leases on producing property in Texas and Louisiana in 1920 (R20) and achieved production in 1924 (PL9). But in 1931 Vacuum's crude production was only 5% of its product requirements. It refined less than half of the products it marketed, although it did refine nearly all its lubricating products (PL15).[30]

Prior to 1920 competition in Vacuum's field of specialty lubricants was not especially strong, coming mainly from the animal and vegetable oils being replaced by petroleum. After 1920 companies began vigorously building their own retail distribution networks to push their products. In this new specialized channel of distribution, the filling station, a full line of products of one company was sold. The filling station, the first architectural monument to the automobile age, changed the landscape of America nearly as much as did the automobile itself. Without its own U.S. distribution facilities, the company-controlled filling station presented a strategic problem for Vacuum. Many companies continued to sell Vacuum products as a premium supplement to their own lubricants, but competition was increasing.[31] As their quality improved Vacuum would slowly be squeezed out of the automotive market, now becoming the largest market for its products. Strategically, it was a tail looking for a dog. It had to either change its strategy and become primarily a gasoline marketer in the U.S. when it had limited outlets, refineries, and crude or it had to find them through merger.

In the late 1920's Vacuum acquired five refining-marketing companies with refineries near Detroit, Chicago, and St. Louis and 2800 outlets in New York, Wisconsin, Michigan, Illinois, Indiana, Missouri, Ohio, and Ontario (Canada). But these were only a drop in the bucket compared to its lubricating outlets.[32] Still, by 1931 only 2/3 of its business was abroad.[33] Socony, the second largest gasoline marketer, had national distribution and 37,000 outlets. The two companies were complementary geographically (Socony's large U.S. business; Vacuum's large international interests) and strategically (both primarily high quality marketers). Socony, with strong marketing organizations everywhere but Canada, Latin America, and Western Europe, was closer to becoming Standard Oil again and Vacuum had found its new dog.

6.5 Crude Acquisitions

If Socony and Vacuum had similar strengths in marketing, they also had similar weaknesses (H1,2) in refining and most of all, lack of crude. That their major move in the 1930's was a marketing merger demonstrates the strategic preoccupation of both companies with the marketing function (PL1). In the 1930's with a crude oil glut and marketing outlets at a premium, Socony-Vacuum's preoccupation with strengthening its marketing network was wholly reasonable, logical, and rational, but wrong, especially when compared with the success of Jersey's policy of buying proven crude reserves at depressed prices. Neither of these two first rate marketing organizations wanted to change long-standing policies favoring purchased (PL3) rather than owned crude and lack of emphasis on vertical integration. They wanted to buy crude as cheaply as possible and balance purchases (PL3) among alternative sources. With high value added in refining and marketing the cost of crude seemed even less important when it was cheap. Vacuum in particular, with its relatively small crude requirements, low volume, and premium quality and price simply did not see itself as being in the fuels business, like it or not (PL19). It had, after all, merged to avoid that problem. The unwillingness to invest heavily in crude properties and to make vertical balance an important element of strategy was the consequence of the policies (PL3,19) both companies had followed since their founding and of the attitude (PV5) of financial conservatism toward the riskier end of the business practiced by the chief executives which these strategies pushed to the top of the organization. Thus Socony-Vacuum missed the first of only two major opportunities (it missed the other one also—offshore leases in the Gulf of Mexico described below) to improve its vertical balance in the United States. Abroad, these attitudes set the background for the decision in 1947 to take only 10% of Aramco (R47), the largest strategic error in its history. In retrospect, and especially since the Vacuum merger had added to rather than decreased cash, the company should have viewed the merger as freeing it to pursue the improvement of vertical balance through the acquisition of crude reserves cheaply available in the 1930's.

Socony-Vacuum did not entirely ignore the issue of crude ownership or vertical balance but, compared to the opportunities available and to its financial resources, it did not pursue them vigorously. Its policy was to purchase supplies from the most advantageous sources (PL3), a policy which worked well in times of glut but poorly in times of shortage. The risks of crude ownership it wanted to avoid were spending money on exploration without finding oil, taxation or nationalization of resources, wells going dry, and capital tied up in high cost crude sources. By purchasing, for example, in the

Far East it had its choice of supplies from Romania, Russia, Persia, Burma, the Dutch East Indies, and California crude dumped on the market by Socal or Union Oil, often at low prices.[34]

Socony did seek crude briefly in China in 1914-1915. Talks to found a joint Socony-Chinese government company foundered over political difficulties, principally Chinese indignation because Socony had not raised a loan for China in New York.[35] On July 31, 1928 Socony had purchased 1/4 of the 23.75% of Turkish Petroleum held by the Near East Development Corporation (R28,PL9), thereby becoming party to (PL8) the Red Line agreement and in 1932 to the "Heads of Agreement for Distribution" arrangement for sharing foreign markets in the depression.[36] By 1934 when oil began flowing through the pipeline from Kirkuk, Iraq to the Mediterranean it owned 50% of Near East as Atlantic Refining, Gulf, and Pan American Petroleum and Transport sold out in the depression (EC30) (Tables M-6 A and B).[37] Because of the depression and the world oil glut the Iraq Petroleum partners had been in no great hurry to develop Iraqi production, especially Anglo-Persian, the operator, since the pipeline would give Iraq oil an advantage over Iranian crude where Anglo-Persian had 100% ownership instead of 25% (Table M-6). Socony's 11.875% amounted to 7.4 TBD.[38]

In June 1933 Socony-Vacuum joined Jersey in a joint venture (R33), the Standard-Vacuum Oil Company (Stanvac), through which Jersey's producing properties in Indonesia would be merged with Socony-Vacuum's marketing facilities throughout the Far East (PL9). Since 1927 all of Jersey's products from Indonesian crude had been sold to Socony under contract, not at the customary 5% commission, but at the opportunity cost of California supplies less a transport adjustment—a very low price in accord with Socony's purchase policy (PL3). The Dutch government, which under Jersey's concession was entitled to 20% of net profits from its oil, felt cheated and demanded more revenue. Jersey was in a tight spot. It might lose the chance to obtain new concessions or even its present concession in Shell's home territory and the power to underprice Shell there. Socony was not interested in revising a good deal. So Jersey bought a refining and marketing company in Australia and New Zealand, forcing Socony-Vacuum to see the advantages of the merger, compared to the alternative of competing with Jersey in the Far East, and Socony-Vacuum somewhat reluctantly agreed. On December 14, 1933 it received half ownership in Stanvac and $13 million in bonds for its Far East marketing assets with book value of $76 million.[39]

In the U.S. Magnolia and General Petroleum did somewhat accelerate their leasing programs in 1933 and in 1934 Socony-Vacuum made its only substantial acquisition of U.S. crude reserves during the 1930's (PL9).[40]

In May 1934 Socony signed options for concessions in Colombia (R34a) and in December 1934 a Venezuelan subsidiary (R34b) was formed to acquire concessions there. These were the first exploration subsidiaries Socony-

Vacuum had formed anywhere in 20 years since the China venture. In April 1936, in a joint venture with Texaco, Socony-Vacuum purchased 80% of the Barco concession (R36), a partially proven tract in Colombia, from Gulf for $12.5 million and the other 20% for $2 million. Gulf's judgment in selling the field eventually proved correct since it never produced the exportable surplus sought.[41]

Socony-Vacuum also sought to bolster its refining position abroad (PL14). Due to tariffs on products similar to those which led to Vacuum's purchase of the French refining and marketing company, Compagnie Industrielle des Petroles, in the late 1920's Socony-Vacuum purchased a refinery at Brasov, Romania and two Italian companies, "SIPOM" and "BENIT" with one refinery and plans to open another in Naples in 1937 (R29). In 1933 Socony-Vacuum joined Texaco in the Ultramar Petroleum Company which opened a refinery in Buenos Aires in 1934[42] (R34c). All these efforts in the acquisition of production and refining taken together did not significantly change Socony-Vacuum's vertical imbalance and did not amount to a major effort to do so. It continued to rely on purchases and, in the U.S., its position as the second most active Major (after Exxon) in pipelines (R38, PL20).[43]

In the 1940's two opportunities arose which offered the possibility of substantially increasing crude production. The first was the development of offshore drilling techniques about 1946. Magnolia and Socal were the first companies to develop them. Magnolia acquired its first offshore leases in Louisiana in 1946 (R46).[44] A year later it drilled a well 28 miles off the coast.[45] But it was not as successful as Socal in finding oil and by the mid-1950's the effort declined. It was not revived until almost 1970 when Mobil spent more than anyone on offshore Gulf leases from 1970 to 1975.[46] Socony-Vacuum's second opportunity to boost its crude production in the 1940's (PL9), the purchase of 10% of Aramco in 1947, was by far its most successful and its most reluctant move (R47).

The opportunity did not result from its own initiative. Jersey had led the push all through the 1920's to open the British and Dutch colonial hegemonies to American companies, resulting in participation in Iraq Petroleum in 1928.[47] Jersey discussed a merger before and more intensely after Socal obtained its concessions in Bahrain and Saudi Arabia.[48] It was Jersey who had been approached in 1946 by Socal and Texaco.[49] The partnership between Jersey and Socony-Vacuum began with Near East Development in 1928 and was further cemented by the creation of Stanvac in 1934 (PL8). They were partners in many of the fruitless discussions of the 1930's about how to get out of the Red Line agreement.[50] But Jersey was the leader.[51] Socony-Vacuum sneaked, or rather was dragged, into Aramco by Jersey.

The Flying Red Horse[52] was led to water. But it was most reluctant to drink. There was intense debate within Socony-Vacuum's board about every aspect of the Aramco deal.[53] There were doubts about the size of the reserves.

The price was too high. There would be antitrust problems. How could they get out of the Red Line agreement? Factions arose on the board and there were bitter arguments. Harold F. Sheets, the chairman, and Laurence B. Levi, a vice-president, were in favor, thinking it a once-in-a-lifetime opportunity. Brewster Jennings, the president, was opposed, feeling the price was too high (10% for $75 million)[54] and not willing to incur debt in the uncertain postwar period (PV5). George V. Holton, a vice-president, worried about the antitrust aspects.[55]

There was a question of how much oil Socony-Vacuum could handle. Its prewar foreign sales were only about 120 TBD and it produced nearly 40 TBD.[56] The prewar crude glut was expected to continue. It was negotiating favorable cost plus a shilling per barrel contracts with Anglo-Persian in Iran and Kuwait.[57] At the opportunity cost of 1 shilling (24¢) per barrel, $75 million would buy over 300 million barrels, which was 80 TBD for 10 years, and no cash had to be laid out now. Reserves in Iran and Iraq were estimated to be 5-6 billion barrels each and Kuwait somewhat near five billion.[58] Each of these areas could produce 500 TBD for 30 years, a total about equal to Eastern Hemisphere consumption at the time. In the depression consumption had dropped more than 20%. The larger the Saudi Arabian reserves, the greater the glut and the cheaper long-term purchases would be. With the great political and economic uncertainty after the war (Russians trying to Balkanize the Middle East and postwar depression in Europe until the Marshall Plan revived it and stopped the Russians in Greece and Turkey) and no end in sight to the glut, why lay out so much cash for such a prospect? Jenning's view was not without merit (PV5,PL3).

Socony-Vacuum decided to take half of Jersey's share if one-third of the Aramco concession were offered but less if one-half were offered. In the end they took 10% while Jersey picked up 30%. Sheets and Levi retired. Holton became chairman and Jennings remained chief executive until 1954. The victors reigned; the losers retired. But what the decision amounted to was the inability of Socony-Vacuum's management to make a substantial strategic commitment of funds for any purpose (PV5) and, in particular, to change their crude position.

The consequence of these decisions on crude acquisition through the 1930's and 1940's was that the second half of the beginning phrase (H1), "long on brains; short on crude," was assured for at least the three decades from 1947 to the present. It would be a decade and a palace revolution later before the way was cleared for executives who possessed the former characteristic.

6.6 Strains after the War: 1946-1957

The Aramco blunder was soon apparent. The predicted glut did not materialize. In the U.S. conversion to a peacetime economy was made quickly.

Europe began to recover with the Marshall Plan in 1948. By 1951 Eastern Hemisphere consumption surpassed 2.5 million barrels per day, about double prewar levels.[59] Saudi Arabia was producing 500 TBD by 1950.[60] When Iran (660 TBD) was shut down in 1951, Iraq, Kuwait, and Saudi Arabia picked up the slack. Mobil was making profits of 60¢/bbl, 40-50¢ of which were from crude.[61]

The Aramco decision was a turning point for Socony-Vacuum. It led to bitter recriminations on the board. Jersey's objective in including Socony-Vacuum was less than totally successful since its unequal status was a constant point of friction. More than anything the Aramco decision made undeniable the need for changes in leadership and strategy at Socony-Vacuum. Albert L. Nickerson became president in 1955. Brewster Jennings became chairman until early 1958 when he retired. Then changes began to take place.

The need for change was more apparent internally than externally. The company was not unprosperous. They had lucked into Aramco because they were partners with Jersey in Iraq and Stanvac. They were included in the Iranian consortium in 1954. They had done some of the right things: merger with Vacuum, acquisition of Magnolia and General Petroleum, and the U.S. marketing network. But ever since World War I, Mobil, the Major with the least crude, had ranked consistently last among the American Majors in net income, return on assets and equity, and growth of assets and equity.[62] But it was still clearly a major international oil company with perhaps the widest marketing network of any Major.

There were heavy internal strains (H3) from 1946 on within the organization. Profit and loss was calculated theoretically by function based on indexes from Platt's (Oilgram/Press Service) tankcar prices. Separate functions like refining and marketing bickered over pricing, and as less product moved by tankcar and the series became less accurate, they bickered more. There was some truth to the criticism by refiners that as much profit as possible should be priced at the refining level because the marketers would just give it away. Indeed, there were occasions when there was little difference between retail prices and the Gulf Coast tankcar plus freight price.[63]

Various functions had little in common. Each function under the system was an independent business and its head was a director (PL13). The parent was a holding company. The cost accounting system of each function was similarly designed to run at arm's length so that it should make no difference whether refining bought its crude from Magnolia's production or from outside. One executive related an incident when the Vice President of Manufacturing wanted to know whether to run Colombian crude produced by Mobil in the Paulsboro (N.J.) refinery or buy more crude in Texas. There was no consideration by the refiners of the profit the producing function would earn if Mobil ran its own crude compared to buying outside.[64] They were concerned only with the cheapest source of supply to their function. With such cost systems strategic decisions such as the size, siting, and equipping of a new refinery were impossible to analyze. Besides, refiners were unwilling to give cost data to other functions or to their superiors for fear it would adversely affect their transfer pricing negotiations.

Not only were the functions independent, there were geographical separations as well. Magnolia on the Gulf Coast and General Petroleum in California had only dotted line relationships to the functional vice-presidents in New York, and most of the time the relationships were more space than dots. Magnolia was a bastion of independence (PL12) and General was even worse. At one point the president of G.P. came to New York to meet with the Mobil board of directors and took the position that he would not tell them anything about the business. They could tell him how much they wanted him to make or replace him if they didn't like the results and that was the limit of their authority.[65] Functional groups had similar problems with their subsidiaries. The Beaumont refinery was just as hard for Dallas as for New York to handle (PL12).[66]

The same lack of unity permeated Mobil's international business (PL12,H3). The Socony and Vacuum businesses had never been effectively merged. The European organization was a Vacuum organization and still thought of itself as such. Stanvac was handled out of New York. In addition to the Far East, east Africa came under Stanvac. North and south Africa were Vacuum and west Africa was Socony.[67] The geographical separations extended to the country level as well. The marketing affiliates in each country were all different, depending on the personalities of the general managers.[68]

The Socony-Vacuum organization was as hyphenated as its name (PL12). It was a company of pieces, decentralized into geographical and functional fiefdoms, some of which were strongly managed and some of which were not. Nor was it an accident. It was intentionally so, and had been ever since the separation from Standard Oil. Brewster Jennings' philosophy of organization (PV6,PL12) was, "Pick a good man and tell him what you want."[69] What good men wanted, first of all, was to run their own show. The decentralization of authority and the emphasis on management through people rather than legal and organizational structures like a chain of command was a policy inherited from Standard Oil. In the nineteenth century when communications were poor and the complex legal structures of trust, holding companies, hidden companies made clear lines of authority impossible, the logical alternative was to manage through people and functions. In addition, marketing in particular had always been a localized business, just as producing, refining, and transportation were centralized. A premium was paid for local market expertise. New Yorkers were thought, possibly with some justification, to be as ignorant of the local conditions in the California market as they were of darkest Africa. The company was managed as a bank manages its portfolio (PL6).

Socony-Vacuum in the 1950's was a nineteenth-century organization. Like the shortage of crude oil, Socony-Vacuum's situation was the result of its strategy and heritage. The decentralized philosophy was not the Standard Oil policy run amok so much as it was Standard's policies gone to seed. The coordinating committees used by Standard to overcome the disadvantages of decentralization were absent at Socony-Vacuum. The only place the functions met was the board and it was not equipped to handle such problems. Little thought had ever been given to the strengths and weaknesses of different forms of organization. As Albert Nickerson compared it, "Jersey tried harder to coordinate—to extract the strengths out of its problems. Mobil just lived with them."[70]

It lived with them as long as it could. Some changes began to appear. It named its first outside director in 1950 and more in 1959 (PL13).[71] Albert L. Nickerson became president in 1955.[72] The name of the company was changed to Socony-Mobil, the company's main marketing trademark, and Vacuum was dropped.[73] Vacuum distillation no longer had the sophisticated technological ring it had in 1866 and the company was no longer primarily a lubricants specialist. With its face lifted to erase the wrinkles of time the company moved from the legendary 26 Broadway address, the Standard headquarters (PL5), to a new skyscraper at 150 E. 42nd Street.[74] But by 1956 the postwar shortage was over. Markets were glutted and prices began to slide. Having less crude and consequently thinner margins than others, Socony-Mobil could not afford to run downstream losses to make profits by moving crude. In two years profits went from $270 million to $157 million.[75] It was the final impetus needed to force restructuring the organization.

6.7 Reorganization

The process of reorganization began in March 1958, when Jennings retired and Nickerson took over.[76] That the collapse of profits was the sign of a major problem was obvious, but the solution to it was not. The disturbance of the Suez Crisis (EC56) served to obscure deeper problems. Jennings, with disinterest in marketing and organization, had not understood. Neither had the financial and banking community, which continued to view Socony-Mobil as a fine, progressive company.[77] Fine, yes. But progressive? Perhaps only compared to bankers.

In fact the old guard at Socony-Mobil was blue-blood conservative and socially elite, a "good, gray Republican company" as Rawleigh Warner put it.[78] Whether blue or gray they had long practiced a policy of hiring and promoting their own kind, a policy which filled the ranks with men whose conservatism substituted for ability. There was a distinct snobbery within the company (H4) which was to play a role in shaping the policies to come. Whatever else they were, progressive they were not. The departure of 12,000 of their ranks to join the unemployed, as a consequence of the reorganization, occasioned a glut in the antique markets of John D. Rockefeller roll-top desks and cuspidors which were removed upon their departure (H4).

As at Shell at the same time and for many of the same reasons, Socony-Mobil brought in the same outside management consultants, the McKinsey company, for a dispassionate view. McKinsey worked closely with the company through the mid-1960's.

The first step in the reorganization was conceptual (PL21). It was the realization that all the functions—production, refining, marketing, transportation, and the lot—were parts of *one* business.[79] They were not separate profit center entities—some Socony, some Vacuum, some Magnolia, some G.P., some production, some manufacturing, some marketing—in a holding company as had been the case since 1911. Managers throughout the company had to recognize that actions taken in their subsidiary affected the organization as a whole and such effects had to be carefully considered. They were all part of one entity—a worldwide, integrated oil company—whose identity had yet to be defined.

To recognize this unity required the most difficult step in the reorganization—the elimination of the domestic subsidiaries, Magnolia and General Petroleum, in 1959.[80] Two divisions of Socony-Mobil were created: the Mobil Oil Company would handle all operations in North America (U.S. and Canada); and the Mobil International Oil Company would conduct operations abroad, except in the Far East.[81] Mobil Petroleum was created in 1960 to manage Stanvac assets from the 1962 dissolution (EC62).[82]

Another major step of the reorganization was a design to glue the functional and geographical pieces of the company into a cohesive whole. In Standard Oil the committee system had performed this function of coordinating functional differences that naturally arose from decentralization. At Socony-Mobil the coordinating organization (R59a) was to be called Supply and Distribution. It was to be responsible for the logistics and cost effectiveness of operations from producing/purchasing through crude transportation, refining location, product transportation, and distribution (R59a).

It was to be the heart of the new operating organizations, controlling how products flowed through the organization—which sources of supply were used and where they went. This included the functions of the old Traffic (transportation) Department and those of Purchasing but with an important difference. It was accountable for the profitability of these decisions for the organization as a whole. Its goal was to use joint ventures, purchases, swaps, transportation, and capacities at a variety of locations to attain lower delivered product costs than competitors like Texaco, Gulf, Exxon, and Shell.[83] Supply and Distribution was to be the connecting organization from E&P (exploration and production) at one end of the organization to Marketing, which was still composed of geographical entities, at the other. It was to use Socony Mobil's strengths in marketing and pipelines (U.S.) to overcome its lack of crude (PL1,20,H2, R11,30,38).

The new Supply and Distribution organization addressed the operating problems of coordination, making it run more efficiently with existing assets, but another structure was required to address the strategic problems of capital expenditure for new assets, facilities type, scale, and location, balance of capital expenditure among functions, and long run strategy for the company.

This was the Planning Department (R59b). It would consider the long range issues of where the company would be in the coming years. It became the intellectual center of the organization—the brain—as S&D was the heart.

Planning became the most powerful function in the company, usurping the role once held by marketing. It has been responsible for much of the conscious strategic change at Mobil, including diversification. In conjunction and often in combination with the Middle Eastern area, Planning became the route to the top at Mobil, a required stop for aspiring executives. With the exception of Herbert Willets, who was already near the top at the time of reorganization, all of Mobil's chief executives since Albert L. Nickerson have come through the Planning and Middle Eastern departments. Rawleigh Warner, Jr., the diplomatic current chairman whose father had once been president of the Pure Oil Company in Chicago, was a vice-president for international affairs in the Middle East at the time of reorganization.[84] Later he served in Planning and Economics before becoming president in 1964.[85] The vice-chairman, Herman J. Schmidt, a tax lawyer and accountant with experience in Mobil's Middle Eastern affairs, was the first president of the international division, Mobil International Oil Company.[86] William P. Tavoulareas, now president, was a Middle East tax accountant who became the first Director of Planning.[87] The top planning job rotated at one or two year intervals for several years until it was occupied by the current holder, Lawrence M. Woods, ten years ago,[88] making him effectively the number four man at Mobil, after Warner, Schmidt, and Tavoulareas.

To entrust a group of executives with planning the future course of the company required a personnel policy which promoted the most able regardless of social background rather than the comfortable, prestigious elite.[89] In 1956 during the Suez Crisis (EC56) a young tax accountant of Greek extraction, Brooklyn born and educated, was laboring obscurely in the Middle Eastern Department. The Suez Crisis and the Arab-Israeli war had closed the canal, leaving the industry short of tanker capacity and burdened by minimum lifting commitments for oil they could not move. Through the maze of overlift/underlift pricing and profit sharing agreements, he discovered that Jersey's actions had the effect of picking its partner Mobil's pocket of $5 million/year. He got nowhere in his department with his discovery. The hidebound could not believe that someone who lacked an Ivy League education could make a discovery they had not. And how could one branch of Standard cheat another (PL8)?

He put his idea in the company suggestion box, every company's paper clip saving device, which at Mobil promised 20% of first year savings to the author of suggestions adopted. At the other end of the box were the same people, now forced to give his idea a hearing. It aroused considerable consternation among the ranks, since 20% of $5 million was not a run-of-the-mill size suggestion, and became an all or nothing (unemployment) gamble for William P. Tavoulareas. Finally he demanded that his idea be taken to someone who was smart enough to determine whether he was right or wrong. The matter went to Herman J. Schmidt, accountant, tax lawyer with expertise in the international area, and General Counsel at the time. He won his point and the recognition of top mangement while demonstrating the lack of acumen of several layers of the old guard, who then tried to block his $1 million reward. He compromised, demanding personal apologies from the men involved. They were told by top managment to apologize or be fired, and did. The incident revealed several shortcomings in company thinking: (1) ability was not being rewarded (H4), often on the basis of prejudice; (2) little attention was being paid to foreign (H5) especially Eastern Hemisphere operations; and (3) the interests of Socony-Mobil and Standard Oil of New Jersey, the real heir of Standard Oil, were not identical (PL8). Management decided the company would become "long on brains" even if it remained short on crude. The realization that they were not Standard Oil, that the time of gentlemanly competition was over, plus the vigor of new men and ideas, launched the company on a quest for a new identity: to be scrappy, aggressive, and intelligent (PL22). It would get the best of competitors, not just live with them.

6.8 Foreign Operations: 1946-Present

Two other necessary shifts in strategy (PL23,24) were recognized concurrently with reorganization: (1) the need to expand foreign, especially Eastern Hemisphere, operations (PL23); and (29) the need to develop a fuels business abroad (PL24). These shifts had been occurring gradually since World War II, but it was with the reorganization that they were recognized, endorsed, and pursued vigorously. These were the first shifts in strategy in response to postwar changes. All the reorganization and strategic changes discussed previously were in response to pre-World War II problems, many of which went back to 1911.

After World War II Socony-Vacuum's foreign business was in a shambles. One of the two areas in which it had been strongest, Eastern Europe, lay behind the iron curtain.[90] In the other, civil war raged in China, and it was lost in 1948. Japan had been lost as the war broke out.[91] American companies did not reenter Japan until 1951 when Caltex was the first.[92] These difficulties were imposed on top of the fragmented structure—some Socony, some Vacuum—of the company's subsidiaries. In addition the primary product in many foreign areas was still lubricating oil, not fuel (H6). One of the purposes of the Socony-Vacuum merger was to minimize investment in the fuels (PL19) business by using Socony's U.S. retail distribution network to distribute Vacuum products. At least this was true from Vacuum's point of view, and most of the foreign subsidiaries were Vacuum organizations. Abroad there was no Socony to turn to, so they didn't.

After the war Socony-Vacuum's foreign marketing network consisted of Stanvac's remaining outlets in Southeast Asia, India, east and south Africa, Australia and New Zealand, approximately 30% shares in the markets of Greece, Austria, Turkey, and Egypt, 5-7% shares in England, France, Germany, and Italy, plus small businesses, primarily lubricating oil distributorships (gasoline in Portugal), in the rest of Europe, west Africa, Latin America, and nearly everywhere else.[93]

In two ways many of these foreign marketing businesses were anachronisms. It is not that they were simply geograpically distinct, isolated, and tied to conditions in their host country's economies (PL12). Marketing has always been a local business. The moves toward centralization in the reorganization did not change that. But conditions of trade in the prewar period in many countries when there were no telephones and poor communications were based on local custom, personalities, and gentlemen's agreements to split markets (the "As Is" pact signed at Achnacarry, Scotland in 1938), maintain prices in the depression, and meet at the club to play golf and cards together every Tuesday. A sense of community and civility developed in an isolated outpost. It was more important to get along than to feud over markets and profits. In a way it might be accurately named commerical colonialism because the models for its conduct were the rapidly fading British and European colonial empires which still governed in many of these areas until the 1960's (PL8).

The other anachronism was the structure of the foreign distributorship itself. The integration of Standard Oil forward from bulk sales at the U.S. refinery gate into foreign wholesale marketing through foreign and domestic marketing affiliates had begun in the early 1890's in Europe and the United States. But the distributors in many of these commercial backwaters were still relics of the nineteenth-century export business. They were classic importers (PL4), F.A.S. New York, of Mobil oil products—engine and industrial oils like the famous Gargoyle 600-W Steam Cylinder Oil—packaged in the U.S. and shipped in drums and cans to local agents. These local distributors often handled tractors, pumps, or machinery as their principal business with lubricants as a sideline service.[94]

In the United States the revolution in oil consumption (EC12) had begun after World War I with the rise of the automobile. In Europe and Japan a similar revolution based on the benefits of cheap energy (oil) took place after the war and the postwar recovery and then accelerated from the Suez Crisis in 1956 until the OPEC embargo in 1973 (shown in Tables M-7A&B below). It was fueled by a combination of rebuilding from the war, advancing technology, low wages, and cheap Middle Eastern and Venezuelan oil. Economic growth and growth in fuel consumption was even greater abroad than it was in the United States.[95]

It even affected the developing countries. By 1951 Mexico, Brazil, and Argentina were among the ten largest national consumers of petroleum. As the volume of business in these countries through distributorships expanded, they found it increasingly difficult to finance and turned to their suppliers, the oil companies, for help (PL23).

As foreign sales volumes grew and the demand for loans increased, the logical course of action for the oil companies was to set up their own distributors who would handle only the company's petroleum products rather than to lend money to distributors for whom their products were only a sideline. Because wholesale distributors of one company's petroleum products needed higher volumes over which to spread overhead, it also made sense for them to distribute higher volume fuels like gasoline and fuel oil rather than lubricants alone (PL24).

These changes proceeded slowly at Socony-Vacuum in the decade after World War II. Management at the company was oriented toward the United States market and foreign, to them, meant Venezuela, except for the export of lubricants. This is less myopic than it seems when one considers that as late as 1950 the U.S. consumed 60% of all petroleum products, the Western Hemisphere totaled 70%, and consumption in the United Kingdom, the second largest free world consumer, was only 3.4% of the world total, while Western Europe consumed 11.8%.[96] Stanvac took care of the Far East, whose consumption was about equal to that of the U.K.[97] The remainder of Socony-Vacuum's businesses, aside from lubricant sales through general distributors described above, was small. The marketing of exported lubricants in many of the smaller countries, like Rhodesia, though low volume, had always been quite profitable. But management's vision of operating subsidiaries did not go beyond North America or the Western Hemisphere at most.[98] Furthermore, lubricant salesmen, proud of their Vacuum image, had little interest in selling fuels. Obtaining company commitment to the importance of the possibilities of foreign business and of adding fuels to the product line required a change of strategic perspectives that did not occur until approximately 1956 at Socony-Mobil.

The supply side of the equation was at least as important as the vigorous growth in demand after World War II. In addition to the cheap foreign crude reserves it had acquired by purchase or joint venture in Iraq, Indonesia, Saudi Arabia, and the favorable purchase contracts it had signed with Anglo-Persian for crude from Iraq and Iran after World War II and Kuwait in 1951, Socony-Vacuum geologists had developed production on the Barco concession in Colombia by 1944 in the joint venture with Texaco, and they had made significant discoveries in Venezuela which began production in 1945.[99] Fred W. Bartlett, the company's chief geologist in Venezuela, became chairman from 1957 to 1960.[100] The company also shared 50% of a small discovery in Egypt at Sudr in 1947.[101]

With strong demand after World War II profits on cheap foreign crude averaged as high as 60¢ per barrel.[102] Socony-Vacuum, whose lubricant business had always been a low volume, high margin operation, discovered the advantages of high volume operations designed to move the crude. Since fuels, in particular gasoline and not lubricants, were the way to move crude, the possession of low cost crude and favorable purchase contracts placed enormous pressures on the company to develop its fuels business abroad (PL24).

The integration forward into wholesale distributorships and the entry into the fuels business, the rapid growth rate, and the new blood from the reorganization produced vigorous competition for foreign business and, especially, for experienced foreign distributors. Mobil aggressively sought to lure away distributors from Esso, Texaco, and Shell. They were successful in finding loopholes in Esso and Shell contracts with distributors but had little luck with Texaco. Texaco took a dispute over one Puerto Rican distributor to the U.S. Supreme Court, but lost to Socony-Mobil (PL23).[103]

The push into fuels and foreign business (PL23,24) began in the mid-50's, propelled by Albert L. Nickerson.[104] Latin America was first. Socony-Mobil had been shipping gasoline from its Venezuelan production to west Africa. It discovered that 5.5 million Venezuelans were using as much gasoline as 40 million Italians.[105] It began marketing gasoline in Venezuela, Colombia, and Puerto Rico. The typical pattern was to set up its own lubricating distributors and begin marketing gasoline a year later.[106] A great deal of Mobil's success in breaking into the fuels market can be attributed to the qualities brought out by the reorganization and the new look it turned loose upon the competition (PL22). By being vigorous, aggressive, and quicker to respond, and sometimes smarter, it outmaneuvered its established rivals. Since the end of World War II its foreign product sales have increased from 65 TBD to 1409 TBD, in 1976 over 2000%, while U.S. sales have increased 87%.[107]

The company's personnel policy in foreign marketing subsidiaries soon reflected the change in its image. Gone were the members of the Old Boy expatriate network.[108] With the exception of the top man in the country who was usually an American in his 50's, the new Mobil managers were bright nationals under 30 who stayed with the company a few years and moved on.

Turnover was 30% annually, but the company felt it was better to hire someone bright and lose him than employ dullards who would stay 35 years. In many foreign countries Socony-Mobil recognized that this policy meant they were running training schools for managers.[109] Besides staffing their subsidiaries with the brightest people, they were making important contributions (SR1) of trained managers to the countries. By staffing with nationals earlier, they obtained an edge on competitors. They were able to rebuild more quickly after the war and, more importantly, gain a share of the fuels business before the mid-1960's when actions by consuming country governments and market conditions limited entry.[110] It was for these reasons, curiously, that Mobil never developed a fuels business in Canada.[111]

In 1959 Socony-Mobil began marketing fuel (PL1,23-4) in Germany and the Benelux countries, though more conservatively and less successfully than in Latin America.[112] In Germany it supplied crude to and owned 20% of ERAL, a large marketer whose name stood for full service in Germany. The other owners were chemical companies and other non-oil companies who did not understand gasoline marketing, particularly the trend to self-service stations which were rapidly taking over the market in the 1960's. Without managerial control and bound by rigid rules banning such customary advertising hoopla as signs, posters, and plastic streamers, and by the full service image which prevented the introduction of self-service stations, Socony-Mobil could not make the kind of vigorous campaign that gained it market share elsewhere. In Italy it bought ERAL's stations for their operating permits, which turned out to be worthless. But it did well in France and the U.K., achieving a 7% market share in the U.K. with 3% of the total number of stations.[113]

It is apparent that the strategic change (PL22,24) to enter foreign fuels marketing aggressively was not accompanied by a change of similar magnitude in refining or transportation to achieve scale economies and lower costs. The reorganization did initiate a tanker construction program in 1961,[114] but Mobil was relatively slow (PL20) in moving to larger tankers.[115] Historically it has relied relatively heavily (PL20) on the charter market for transportation.[116] The historical dominance of the marketing (PL1) function shows up clearly in the scale and location of refineries (PL14). Both in the U.S. and abroad Mobil has matched its geographically localized marketing subsidiaries with small-scale refineries located in each geographical market area. Since 1959 it has attempted to move away from this policy and build larger refineries. For years it wanted to expand the Paulsboro, New Jersey refinery to run large volumes of imported crude but was restrained by import quotas. It opened a 175-TBD refinery at Joliet, Illinois (near Chicago) in 1972 but has had problems in obtaining a supply of Canadian crude, even from its own production there. The cost advantage of a 175-TBD refinery over a 30-TBD refinery is about 1¢/gallon or $100 million per year in profit after tax.[117] It has no offshore refineries and only one of its 40 refineries worldwide (Beaumont, Texas) is larger than 200 TBD, as efficient scale moves steadily toward 500-TBD giants. It has used processing agreements and part ownership to gain some of the cost advantages of capacity in larger scale refineries. It boasts more coking capacity than other refiners for upgrading crude into higher value products like lubes and gasoline. But overall it has not under taken a commitment to build the large-scale transportation/refining logistical system to match its marketing program.

6.9 U.S. Marketing, Exploration, Chemicals, and Diversification

Accounting for Mobil's success in international marketing is easier than explaining its lack thereof in other areas. The components of its expertise were (1) knowledge of foreign markets (PL1,19), (2) aggressiveness (PL22), (3) quick response to events (PL22), (4) early entry (PL4,19), (5) concentration on lubes and gasoline (PL11,15), the most profitable products, (6) managerial control over its investments, and (7) early staffing with nationals (SR1). By developing its marketing capabilities it was able to make money by moving crude. To an extent its lack of success in other areas results from practices which were the reverse of these.

In the U.S. the company followed a different marketing strategy than it had abroad. It could not make money by moving crude as it did abroad since (1) it did not produce it (it produced about 40% of its U.S. marketing volume (PL3,H1),[118] (2) it could not import it (due to U.S. import quotas) (EC59), (3) it had no large offshore refineries (PL14) from which to ship in products, and (4) it could not purchase it cheaply since import quotas and output restrictions kept U.S. crude prices artificially high. Its U.S. marketing strategy did not change with the reorganization. Its most prominent policies were (1) no wholesale marketing or private branding (PL1) and (2) full-service stations (PL11).[119] Like other Majors Mobil has continued to lose market share in the U.S. Today it has about 15% in New England and 6.5% nationwide. There has even been trade press speculation that the company might pull out. Overall there have been very few years when domestic marketing was a good business even for a marketer like Mobil.[120]

An increased emphasis on exploration and production was part of the reorganization (PL9). New production men were hired.[121] Capital spending, however, was not noticeably increased until 1968 and more heavily from 1972 to 1974 when the company spent freely at the U.S. offshore lease sales. Mobil has made or participated in significant discoveries in Libya (1961-1962), Nigeria (1964), and the North Sea (1972-1973),[122] but its overall record is one of late entry, lack of aggressiveness, and frustration.[123] Its hard luck in finding oil is as legendary as its marketing expertise. In one confirmed account another company discovered oil on a lease given up by Mobil by drilling a well beneath the campsite of the Mobil exploration team. The biggest addition to Mobil's production is likely to come from a phased increase in ownership from 10% to 15% in Aramco from 1975 to 1979 squeezed out of its other partners after the OPEC Crisis (EC73,PL22).[124]

Mobil was a latecomer to petrochemicals, forming the Mobil Chemical Company in June 1960, as a direct result of the reorganization (R59c).[125] It had not come earlier because of Mobil's weakness in refining (H2), a consequence of the small scale of its market-oriented refineries (PL14), and its lack of a volume-oriented fuels business (H6) to produce the petrochemical raw materials as byproducts. It went through a grabbag of chemical businesses, a number of them joint ventures, which did not pan out before carving out a niche for itself in polyethylene films (Hefty trash bags), polystyrene foam products (egg and meat trays, plastic tableware), phosphate rock mining (third largest in U.S.), and crop chemicals. Mobile got into its two most successful areas, polyethylene film and phosphate mining, relatively early with acquisitions of Kordite Corp. (Hefty trash bags) in 1962 and Virginia Carolina Chemical Corp. (phosphate) in 1963 and was able to survive later shakeouts. It did not go into bulk petrochemicals. Its original Beaumont polyethylene plant was built as much to balance refinery operations as for polyethylene.[126] Except for phosphate, its successes have come at the fabricating/marketing end of the petrochemical industry, a different strategy from other Majors, by applying its marketing skills (PL1) to a new industry, an important precedent for later diversification.

By 1965 all the changes since 1959 were becoming apparent in the form of a new strategy and a new identity. It was still a marketing company but now it marketed fuels as well as lubricants abroad, and it was far more aggressive, intelligent, and responsive (PL1,15,22,24). It was no longer Standard Oil and it changed its name to show it (PL5). Socony, the abbreviation of Standard Oil Company (New York), was dropped. On May 17, 1966, 100 years after its predecessors were founded, it became Mobil Oil Corporation, taking its name, as befits a marketing company, from one of its most valued trademarks. It had taken 55 years, since the divestiture in 1911, to sever its ties to Standard Oil and find its own identity (PL22). Its advertising and high profile public image, begun after Rawleigh Warner became chairman in 1969, which has led it to air its views on television, publish advertisements opposite the editorial page in the *New York Times* and differ publicly with other oil companies, reinforce its new identity (PL22). In effect it is advertising itself (PL1,18).

Perhaps in no way does Mobil differ more from the other Majors than in its most recent strategic move, to diversify outside the oil industry (PL25).[127] It began in 1968 with a policy decision by the board to diversify, spurred by the conglomerate diversification mania (EC68), the Six Day War of 1967 (EC67), and the fear of being too dependent on foreign operations in politically unstable areas (conveniently translated into lower stock prices by the market). It looked for a large, well-managed company that would significantly lift

Mobil's U.S. earnings base.[128] Flush with inventory profits from the 1973 OPEC embargo, in August 1974 at the bottom of the 1973-1974 market crash it tendered for control of Marcor, a combination of Montgomery Ward, a retailer, and Container Corp., a paper company, in a move (PL6) it would not have dared to make (PL7) before the 1959 reorganization. Total cost was about $1.7 billion with 1976 earnings of $144 million, 17% as large as Mobil's 1976 net income. In 1977 it barely lost a bid for the Irvine Company, a West coast land developer with very valuable holdings in the Orange County area near Los Angeles. In 1975, to celebrate its diversification (PL25), it became simply Mobil Corporation, no longer just an oil company.

This concludes the narrative of the development of the strategy of Mobil Corporation from its Standard Oil origins a century ago to the present.

Exhibit M-1
Mobil Policies

PL 1	Retail marketing as central function: (a) V; (b) 1866-present.
PL 2	Refine crude and market products in New York and New England: (a) G,V; (b) 1880's-present.
PL 3	Dependence on purchased crude and products (lack of vertical integration): (a) V; (b) 1866-present.
PL 4	Export refined products to the Far East: (a) G,V; (b) 1890's-World War II (export), present (Far East).
PL 5	To be the administrative headquarters of Standard Oil: (a) A; (b) 1882-present.
PL 6	To be the financial headquarters of Standard Oil: (a) A; (b) 1882-present.
PL 7	Standard Oil financial policy: no long-term debt, large cash reserves, internal finance policy: (a) A; (b) 1866-1921 (first long-term debt in 1959).
PL 8	Put Standard back together again: Attitude of gentlemanly cooperation rather than competition with other companies, especially Jersey Standard: (a) H; (b) 1911-1959.
PL 9	Integrate vertically by acquiring and discovering crude production: (a) V; (b) 1918-present.
PL 10	Expand geographically in the U.S.: (a) G; (b) 1918-present.
PL 11	Become a leading full-service gasoline marketer: (a) H,V,P; (b) 1911-present.
PL 12	Allow subsidiaries extreme autonomy after acquisition: decentralization of management: (a) A; (b) 1911-1959.
PL 13	Active inside board of directors (Standard Oil policy): (a) A; (b) 1880's-1950 (first outside director), 1959 (several).
PL 14	Large U.S. export refineries on tidewater; small local market refineries elsewhere: (a) G,S,V; (b) 1880's-present.
PL 15	Manufacture and market high quality specialty products and lubricants—kerosene, gasoline, naphtha, and other fractions were regarded as byproducts (Vacuuum Oil policy): (a) S,P: (b) 1866-World War II (minimize fuels), present (lubricants).
PL 16	Develop products through research by designing them to customer's requirements and patent them: (a) A; (b) 1866-present (Vacuum Oil policy).
PL 17	Practical testing in actual use of every product (Vacuum Oil policy): (a) A; (b) 1866-present.
PL 18	Consumer advertising of products: (a) A; (b) 1866-present.
PL 19	Expand international sales to market worldwide but not to integrate vertically or become a fuel company: (a) G,V,P; (b) 1886-1959 (fuels), present (worldwide).
PL 20	To be strong in U.S. pipelines, but not in marine transportation: (a) V; (b) 1920's-present.
PL 21	Integrate functional and geographical parts to become a single vertically integrated major oil company through two new functions: Supply and Transportation/Distribution to coordinate operations and Planning to form long range strategy: (a) V,A; (b) 1959-present.

PL 22	To be Mobil: a scrappy, aggressive, response, intelligent competitor: (a) A; (b) 1959-present.
PL 23	Expand foreign, especially Eastern Hemisphere, operations: (a) G; (b) Post World War II-present.
PL 24	Market fuels abroad: (a) G,P; (b) 1900-present.
PL 25	To diversify outside the energy industries: (a) A,P; (b) 1968-present.

Notes:
(a) Type of strategic choice (see Chapter 1, Figure 2)

S—scale		PL6,14,15
G—geography		PL2,4,10,14,19,23,24
H—horizontal concentration		PL8,11
V—vertical integration		PL1-4,9,11,14,19-21
P—product diversification		PL11,15,19,24,25
A—administration		PL5-7,12,13,16,17,18,21,22,25

(b) date of origin and duration

Exhibit M-2
Mobil Environmental Conditions

EC 11	Divestiture from Standard Oil—1911.
EC 12	Rise of the automobile market and gasoline demand—1908-29.
EC 13	Changes in refining technology and product mix to increase gasoline production—1912-14, 1920's on.
EC 16	Invasion of New York and New England by competitors—post World War I.
EC 15	Filling station replaces garage as retail gasoline outlet: i.e., gasoline volume increased to the point of requiring specialized distribution outlets—1915 on.
EC 18	Fear of crude shortage after World War I.
EC 30	Great Depression—1930's.
EC 56	Suez Crisis—1956.
EC 62	Dissolution of Stanvac—1960-62.
EC 67	Six Day War between Arabs and Israel—1967
EC 68	Conglomerate diversification movement—1968-70
EC 73	OPEC Embargo and Energy Crisis—1973.

Exhibit M-3
Mobil Resources

R 11	Standard Oil Company (New York).
R 11a	$90 million in assets.
R 11b	Six refineries in U.S.
R 11c	Markets in New York and New England (Soconyland).
R 11d	Physical headquarters of Standard Oil at 26 Broadway, New York City.
R 11e	Markets in Far East.
R 13	Marketing in Bulgaria.
R 15a	Vacuum acquires a fleet: 7 tankers (1915), 3 more (1919).
R 15b	Marketing in Greece.
R 17	Colpet—joint venture to market with Texaco in South America.

R 18a	Controlling interest in Magnolia Petroleum with refineries at Corsicana and Beaumont, Texas.
R 18b	Paulsboro, N.J. refinery—Vacuum.
R 18c	Paulsboro, N.J. research lab.—Vacuum.
R 19	Refinery at East Providence, Rhode Island begins operation.
R 20	Vacuum acquires leases on producing properties in Texas and Louisiana.
R 22	Marketing in Australia.
R 26a	Acquisition of General Petroleum Corporation of California by Socony.
R 26b	Marketing in south Africa.
R 28	Interest acquired in Near East Development Corp.—Iraq.
R 29a	Vacuum purchases Compagnie Industrielle des Petroles, a French refining and marketing company—late 1920's.
R 29b	Refinery at Brasov, Romania purchased—late 1920's.
R 29c	Two Italian refining and marketing companies, SIPOM and BENIT, purchased.
R 30	Socony acquired White Eagle Oil and Refining Company.
R 31	Vacuum Oil Company.
R 31a	Worldwide retail marketing.
R 31b	Premium quality lubricants and specialty products.
R 31c	The most valuable brand names in the petroleum industry.
R 33	Standard-Vacuum Oil Company formed.
R 34a	Concessions in Colombia—1934.
R 34b	Venezuelan subsidiary formed—1934.
R 34c	Ultramar Petroleum Company—joint venture with Texaco in Argentina—refinery opened 1934.
R 36	Barco concession in Colombia purchased in joint venture with Texaco.
R 46	Louisiana offshore leases acquired—1946.
R 47	Purchase of 10% of Aramco.
R 59a	Reorganization, elimination of domestic subsidiaries, formation of two operating divisions: Mobil Oil company—domestic, and Mobil International Oil Company—international operations.
R 59b	Supply and distribution coordinative departments established.
R 59c	Establishment of the Planning Department.
R 60a	Mobil Chemical Company, a third operating division, formed.
R 60b	Mobil Petroleum Company, a fourth operating division, formed to manage Stanvac assets.
R 61	Discovery in Libya.
R 62	Kordite Corp. acquired (Hefty trash bags).
R 63	Virginia-Carolina Chemical Corp. (phosphate rock mining) acquired.
R 64	Discovery in Nigeria.
R 71	Discovery in North Sea.
R 72	Second discovery in North Sea.
R 74a	Controlling interest in Marcor acquired.
R 75	Interest in Aramco will rise to 15% by 1979.

Exhibit M-4
Mobil Handicaps

H 1	Shortage of crude 1911-present.
H 2	Weakness in refining 1911-present.
H 3	Organizational strain 1946-59.
H 4	Elitism and social prejudice; intelligence not rewarded; managers who were not "long on brains."
H 5	Little attention paid to foreign, especially Eastern Hemisphere operations.
H 6	Lack of foreign fuels business.

Exhibit M-5
Mobil Social Responsibilities

SR 1	Training managers in developing countries.

Exhibit M-6
Mobil Personal Values

PV 1	Hiram Everest—interest in direct retail marketing.
PV 2	Hiram Everest—belief in research to develop new applications and better products.
PV 3	Hiram Everest—practical testing of products in actual use.
PV 4	Hiram Everest—belief in use of advertising.
PV 5	Brewster Jennings and predecessors—financial conservatism.
PV 6	Brewster Jennings—belief in decentralization of authority.

Table M-1
Motor Vehicle Registrations

Year	Cars	Trucks
1908	Model T 1st delivered	
1909	305,950	6,050
1910	458,500	10,000
1911	619,500	20,000
	Both	
1913	1,260,000	
1915	2,450,000	
1917	4,980,000	
1922	12,240,000	
1930	26,500,000	

Source: Mobil II (6,8)

Table M-2
Volume Changes 1912 and 1922 (TBD)

	1912	1922
U.S. Gasoline	68	433
% U.S. Crude	11%	28%
U.S. Kerosene	121	151
% U.S. Crude	20%	10%
U.S. Crude Prod.	611	1527
World Crude Prod.	966	2523

Sources: Mobil II (8); API (1930) p.8

Table M-3
Industry Product Yields 1912 and 1922
From a 42 Gallon Barrel of Crude

Product	1912 gallons	%	1922 gallons	%
Gasoline	6	14	12	29
Kerosene	12	29	4-1/2	11
Gas and Fuel Oil	16	38	21	50
Total Products	34	81	37-1/2	89
Loss	8	19	4-1/2	11

Source: Mobil II (8)

Table M-4
SOCONY's Market Share in
New York and New England

Year	Market Share	Market
1909	92%	gasoline
1911	85.19	all products
1918	60	gasoline
1922	50	all products
1926	46	gasoline
1931	34.7	gasoline

Sources: Mobil I (10), II (8,13,16-17); *Mobil World,* Jul-Aug 1976 p. 8

Table M-5
Shares of the U.S. Gasoline Market

	1926	1933	1934	1935
Standard Oil Company (California)	3.4%	2.7%	2.8%	2.7%
Socony-Vacuum Oil Company	9.6	9.3	8.8	8.7
Standard Oil Company (Indiana)	10.6	8.7	8.7	8.5
The Atlantic Refining Company	4.2	2.7	2.6	2.5
Standard Oil Company (Ohio)	2.4	1.7	1.5	1.5
Standard Oil Company (New Jersey)	5.4	6.1	6.0	6.1
Standard Oil Company (Kentucky)	2.5	1.4	1.4	1.4
Continental Oil Company	1.4	1.8	2.0	1.8
The Texas Company	6.5	7.7	7.8	7.5
Sinclair Oil Corporation	7.8	6.0	6.0	6.0
Shell Oil Company	5.3	6.4	6.4	6.3
Sun Oil Company	(a)	3.2	3.2	3.2
Tide Water Associated Oil Company	3.3	3.5	3.3	3.3
Cities Service Company	(a)	2.5	2.5	2.5
Pure Oil Company	2.3	2.3	2.3	2.5
Phillips Petroleum Company	(a)	1.8	1.9	1.9
Union Oil Company	1.5	1.2	1.3	1.3
Skelly Oil Company	(a)	0.4	0.5	0.5
The Ohio Oil Company	(a)	0.6	0.6	0.6
Gulf Oil Corporation	7.0	5.8	5.8	5.3
All Other Companies	26.8	24.2	24.6	25.9

(a) Share of market unknown included with "all other companies."

Source: McLean and Haigh, 104.

Table M-6A
Changes in Ownership of Turkish Petroleum Company[1]

	1914[2]	1919[3]	July 31, 1928[4]
Anglo-Persian	47.5%	47.5%	23.75%
Royal Dutch/Shell	22.5%	22.5%	23.75%
Deutsche Bank	25%	—	—
C.S. Gulbenkian	5%	5%	5%
Compagnie Française des Petroles		25%	23.75%
Near East Development Corporation			23.75%

1. Re-named Iraq Petroleum Company in 1929
2. Organized by Gulbenkian
3. Result of the Treaty of San Remo in 1919
4. Gulbenkian drew the famous Red Line around the former Ottoman Empire (Sampson, 67)

Table M-6B
Ownership of Near East Development Corporation

Standard Oil (New Jersey)—25% (SONJ)
Standard Oil (New York)—25% (SONY)
Atlantic Refining—16.67% (sold to Socony-Vacuum in 1931)
Gulf Oil—16.67% (sold half to SONJ and half to SONY in 1934)
Pan American Petroleum and Transport—16.67% (sold to SONJ in 1930)

By 1934 Near East Development Corporation was owned 50% by SONJ and 50% by Socony-Vacuum.

Source: Wilkins, 119

Table M-7A
Petroleum Consumption by Area
(TBD)

	1932[1]	1939[2]	1950[3] API	1950[3] BP	1955[4] API	1955[4] BP	1960[5]	1965[6]	1970[7]	1973[8]	1976[9]
World	3638	5534	9412	11040	13520	15690	21640	31060	46330	56425	58790
U.S.	2296	3384	5781	6500	7216	8460	9688	11300	14350	16815	16980
Canada	88	139	323	360	548	630	841	1142	1525	1755	1790
Other W.H.	186	345	721	880	970	1170	1592	2101	2820	3490	3840
Argentina	52	85	155		192		244	381			
Brazil	13	24	86		177		273	329			
Mexico	40	51	139		180		299	340			
Venezuela	16	26	56		110		155	178			
W. Europe	440	731	1049	1300	1959	2320	3855	7526	12660	15155	14075
France	92	124	200	90	345	400	559	1100	1920	2555	2385
Germany	59	148	67	90	222	250	627	1620	2655	3070	2885
Italy	28	60	88	90	208	240	444	1030	1740	2100	1970
U.K.	162	242	340	380	510	590	942	1495	2060	2285	1870
Asia	167	257	375		715		1490	3055	NA	NA	8440
Japan	38	70	42		189	200	663	1742	4000	5425	5195
Australia	22	48		120		180	260	351	600	670	765
USSR, E. Europe, and China	345	540		900		1660	2950	4455	6840	8775	10850

Notes: NA—not available

1. American Petroleum Institute, *Petroleum Facts and Figures*, 5th ed., 1937, p.8

2. Ibid., 7th ed., pp. 18-19

3. Ibid., 10th ed., p. 244; BP *Statistical Review*, 1959, p. 21. For 1955 and 1960 there is an unaccountable discrepancy between data from API and BP sources.

4. API, PF&F, 1959, p. 451; BP *Statistical Review*, 1959, p. 21

5. API, PF&F, 1963, pp. 276-7

6. BP *Statistical Review* 1973, p.21

7. BP *Statistical Review* 1976, p. 8

Table M-7B
Annual Growth Rates in Petroleum Consumption by Area

	1932-39	1939-50 API	1939-50 BP	1950-55 API	1950-55 BP	1955-60 API	1955-60 BP	1960-65	1965-70	1970-73	1973-76
World	6.2	5.0	6.5	7.5	7.3	9.9	6.6	7.5	8.3	6.8	1.3
U.S.	5.8	5.0	6.1	4.5	5.6	6.1	2.8	3.1	4.9	5.5	0.3
Canada	6.8	8.0	9.1	11.2	11.9	8.9	5.9	6.3	5.9	4.8	0.7
Other W.H.	9.2	6.9	8.9	6.1	5.9	10.4	6.4	5.7	6.1	7.4	3.2
Argentina	7.3	12.8		4.4		4.9		9.3			
Brazil	9.2	12.3		15.5		9.1		3.8			
Mexico	3.6	9.5		5.3		10.7		2.6			
Venezuela	7.2	7.3		14.6		7.1		2.8			
W. Europe	7.5	3.4	5.4	13.3	12.3	14.5	10.7	14.3	11.0	3.7	(1.5)
France	4.4	4.5	4.5	11.5	14.9	10.1	6.9	14.5	11.8	5.9	(1.4)
Germany	14.1	NC	NC	27.2	22.7	23.1	20.2	21.7	9.7	3.0	(1.4)
Italy	11.5	3.6	3.8	18.8	21.7	16.4	13.1	18.3	11.0	3.8	(1.3)
U.K.	5.9	3.2	4.2	8.5	9.2	13.1	9.8	9.7	6.4	2.1	(4.1)
Asia	6.4	3.7	NA	13.8	13.8	15.8	15.8	15.6	NA	NA	NA
Japan	9.1	4.8		35.1	36.8	16.4	13.1	21.3	18.1	10.7	(1.5)
Australia	11.8		8.3		8.5		7.7	6.2	11.3	3.8	4.5
USSR, E. Europe, and China	6.6	NA	4.8	NA	5.2	NA	12.2	8.6	9.0	8.7	7.4

Notes: NA—not available
       NC—not comparable

7

The Strategy of Standard Oil of California

7.1 Overview

The Standard Oil Company of California (Socal), as its name implies, was the California and Pacific coast organization of Standard Oil before the divestiture in 1911. (It changed its name to Chevron on July 1, 1984.) It was a miniature Standard Oil: strong in refining, marketing, pipelines, and purchasing (PL1-9) with small production and a western or Pacific basin view of the world far removed from the financial markets, companies, and government on the U.S. East coast.

In 1976, as shown by the strategic measures in Table 1-2 (Chapter 1), Socal was a worldwide company with about 55% of its refining and marketing volume in the Western Hemisphere and 45% in the Eastern Hemisphere (1976 figures). However, 85% of its crude supplies came from the Eastern Hemisphere, 65% from Saudi Arabia. Though one of the three smallest Majors (with Gulf and BP) in financial size, it was one of the larger producers, having the largest market share in both the Middle East and Far East. With the smallest petrochemical operation among the seven Majors and 97% of its revenues from oil and gas, it was one of the least diversified Majors (PL16) (Table 1-2, Chapter 1).

It will be apparent in the narrative that follows that the strategy which took Socal from 1911 to 1976 is both similar to and different from the strategies of the other two Standard Oil heirs among the Majors, Mobil and Exxon. In its U.S. refining and marketing operations, its main business before 1911, policies of the Standard Oil strategy are still in force (PL1-4,7-9), testifying to the longevity of strategy. But upstream and abroad Socal has followed its own path. It became a great explorer with discoveries in the western U.S., the Middle East, the U.S. Gulf coast, and the Far East (PL10-12). It will be known forever as the discoverer of Saudi Arabia. In the Eastern Hemisphere it joined Texaco to refine and market its products through Caltex and also became one of the largest crude sellers (PL11,14). The great explorer, conservative, Pacific basin oriented Socal (PL6,7,10) evolved differently than Exxon, the worldwide refiner-bulk marketer with weakness in exploration, or scrappy Mobil, the gasoline and lubricant marketer "long on brains, but short on crude." These comparisons show that homogeneity is not always maintained even when competitors were once parts of the same organization. Events, resources, personalities, and values bring about such differences that merger, which Exxon and Socal examined closely in the 1930's, becomes impossible.

The three most important strategic decisions from 1911 to 1976 for Socal were the decision to become, first of all, an upstream or exploring (PL10) company, the decision to refine and market (PL13,14) abroad through Caltex, and its geographic expansion in the U.S. over the barrier of the Rocky Mountains to the Gulf and East coasts to become a national rather than regional U.S. oil company (PL12). The first two of these decisions were departures from the strategy of its forbear, Standard Oil.

7.2 The Shift to Production; Domestic Success; Foreign Futility

This section covers Socal's domestic operations from 1911 until approximately World War II and foreign operations from 1911 to 1928. Before World War II Socal was a regional rather than a national company in the U.S. with operations only in the western states (PL6). Before 1928 it searched unsuccessfully for oil abroad and had small product exports to the Pacific basin (Hawaii, the west coast of South America, Australia, Philippines, Far East). But a radical and permanent shift in its strategy, approximately coincident with the 1911 divestiture from Standard Oil, took place as Socal became an exploring (PL10) rather than refining-marketing company.

Prior to 1911 the California subsidiary of Standard Oil had been primarily a refining and marketing company which purchased most of its crude, a pattern typical of most Standard Oil companies (PL1-3). Standard's participation in the Far West had begun with sales through commission merchants in about 1873, followed by a branch office of Standard of Ohio in 1878, and a marketing company, Standard Oil Company (Iowa) in 1885 (PL2). Anxious to gain a share of the growing western oil traffic which had hitherto gone by sea, the railroads were willing to grant favorable rates, allowing Standard to supply the West out of their Cleveland refineries (PL1).[1] By 1880 Standard already handled nearly 9/10 of the western oil traffic.[2] Between 1885 and 1900 Iowa Standard overcame its major marketing rivals, chiefly by obtaining preferential rail rates, and cornered about 96% of the market in the light oils, kerosene and naphtha, with a smaller share in lubes and painter's supplies (PL2, R11a).[3]

But it had little share in the rapidly growing market for fuel and gas oil, supplied mainly from asphaltic California crude.[4] Small amounts of crude had been produced in California since the 1860's but after the 1895 discovery of the Los Angeles City field by E.L. Doheny, his first discovery in a legendary career, California crude production soared (EC1). From 1.4% of the U.S. in 1894, by 1900 it was 6%; 35% in 1910; 23% in 1920; 25% in 1930; and 17% in 1940. From

1903 to 1928, when Texas surpassed it, California was usually the largest producing state.[5] In 1895 Iowa Standard made its first purchases of California crude for sale as fuel oil. In order to facilitate entry into the fuel oil market, Iowa Standard purchased the Pacific Coast Oil Company (PCO) in 1900. PCO was one of the three largest producers, transporters, and refiners in California with about 4% (429 b/d) of state production (R11b).[6]

Production in California also contributed to Standard's global strategy. In the 1880's, in order to maintain its dominance in growing markets, Standard reversed its earlier policy of leaving production to others and maintaining control of crude supplies by purchasing crude and by providing pipelines and storage facilities to move and store it.[7] It began to buy producing properties (PL3) in 1885 with the discovery of the Lima, Ohio field. By 1888 it produced 35% of the U.S. total.[8] The change in policy was typical of the bold, large-scale strategic moves Standard was capable of making. Between 1895 and 1900 it entered new producing areas in Kansas and Texas. To challenge Shell's grip in the Pacific basin it looked for promising producing properties in Alaska, Peru, the Dutch East Indies, China, and Japan. As California production rose during the 1890's so did Standard's interest. California crude fitted naturally into Standard's Pacific strategy with the added attraction of a booming domestic market. The main problem was that heavy California crude did not make a good kerosene which could be exported to the Far East. In 1900 Standard began to build storage facilities and buy the heavy crude to process and sell as fuel oil. After its acquisition PCO became Iowa Standard's principal supplier (mainly from purchases of local production) (PL3).[9]

Standard poured money into refining, pipelines, tankers, and storage but production atrophied to 167 b/d (= barrels per day) in 1906 (PL1-4).[10] A 3-TBD (= thousand barrels per day) refinery was completed on San Francisco Bay at Richmond in 1902.[11] It was the largest on the West coast at the time but by 1906 Standard had raised its capacity to 28 TBD, third largest in Standard's system (PL1), to supply Iowa Standard's western market and the Far East. However producers at the Coalinga and Santa Maria fields refused to sign crude sale contracts at the current price of 40¢/bbl., recently down from 95¢/bbl.[12] The threat of being unable to supply Richmond caused Standard to establish a producing department in California on January 1, 1907 (H1). It sent Col. J.J. Carter, its premier producing man, to California. But his recommendations to spend large sums on purchasing and developing promising producing properties rather than buying crude were rejected by J.M. Tilford, Standard's New York marketing man in charge of the West coast, in favor of a cheaper but indifferently pursued wildcatting program. Carter quit, leaving Standard in California without any experienced producing man.[13] Some valuable properties in the Coalinaga area were purchased and discoveries were made in the Midway field, but in 1911 the company's production of 8695 b/d, though much larger than in 1900, was still only about

4% of California's total.[14] Standard's California unit had definitely become a producer but was not yet much of an explorer (PL3,5).

Standard Oil Company (California)[15] was the fourth largest piece split off from Standard Oil in 1911. It had revenues of $23.6 million, earnings of $3.1 million, and assets of $49 million (total) ($39 million net). In addition to 8695 b/d of production, it purchased 52,928 b/d (23% of California production), piped 61,622 b/d (27%), refined 31,654 b/d (2/3 or more of California refining capacity), and marketed 32,589 b/d (21% of products by volume in its territory). Its home territory was California, Oregon, Washington, Arizona, Nevada, Alaska, and Hawaii. Standard was dominant in refined products with 83% (5417 b/d) of the kerosene market, 71% (1184 b/d) in gasoline and naphtha, 63% (910 b/d) in distillates, and over 85% (653 b/d) in lubes. But these light products were only 7% of market volume, the rest being fuel oil in which Standard had only a 17% (24,415 b/d) share. Fuel oil, sold to railroads and ships, was often no more than raw crude. In 1909 refinery runs in California were only 37 TBD (24%) of the 156 TBD produced. In 1911 2/3 of California's production was consumed in the West and 1/3 was exported. From 1911 to 1919 approximately 3/4 of Socal's products were consumed in the West and 1/4 exported (R11,PL1-5).[16]

Socal inherited from Standard (R11) a dominant regional company with an able management, a Pacific orientation (PL6), and a Standard Oil strategy (PL1-5). It was strongest in pipelines and tankage (PL4); it was the largest refiner (PL1) and principal refined product marketer (PL2) with a dominant market share (PL2). It was the price leader (PL8) and stressed high quality rather than price (PL9). It was a principal purchaser in the state (PL3) and had acquired some production (PL5), though its production ranked only sixth largest in the state.[17] Within California its operations were centered in San Francisco, where its Richmond refinery was located and to which its principal pipelines led, and which is the site of its headquarters today, rather than Los Angeles, which was the center of a very competitive producing area. San Francisco was also the main West coast port for shipments abroad, including the Orient (PL6).

But the company that emerged from the divestiture immediately departed from the Standard Oil strategy. From the first days of its independence it became an exploring company both at home and abroad (PL10). It made several important purchases of producing properties between 1913 and 1917 (PL5), but it also explored vigorously in California (PL10). By 1919 about 40% of its production came from properties it had purchased and developed and 60% from the results of its wildcatting activity since 1911.[18] What shows the shift in vertical strategic (PL10) emphasis more clearly is that it had found enough oil by 1919 to become the largest producer in California and the nation as well. In 1919 it produced 70 TBD, piped 112 TBD, refined 96 TBD, and marketed 82 TBD.[19]

Divestiture (EC2) alone was not responsible for this shift in strategy, but it was crucial to creating the opportunity. Socal shared in the boom in California production (EC1). Standard's strategy of fiscal conservatism (maintaining large cash balances for opportunities) (PL7) and size gave Socal the resources to make the shift. Standard's (1) growing interest in production after 1885, (2) its Pacific basin strategy initiated in 1895 against Shell, and (3) its establishment of a producing department and wildcatting program in California in 1907 all contributed to the development of production there. Without divestiture Standard would have become a large producer in California, but it is not likely it would have become an explorer. The crucial factor to the shift in strategy made possible by divestiture was the belief of an able executive in exploration (PV1).

In January 1911 before the divestiture decision Standard sent out another of its outstanding producing executives to fill the void Carter left. At 48 F.H. Hillman was a 32-year veteran with Standard companies.[20] Stranded in California by the divestiture, it was Hillman who largely effected the change in Socal's strategy. With a strong and gifted producer in charge Socal's production began to grow. He became a vice-president on June 2, 1914 and headed Socal's producing department until April 5, 1927.[21]

With abundant opportunities in the undeveloped West (EC1) and Hillman's expertise (R11g) Socal began to shape a new vertical strategy (PL10,11) based first of all on finding and producing crude oil (PV1, PL10,11). Until the end of World War I Socal expanded production in California. Afterwards Hillman pushed Socal into other western states and into foreign exploration (PL11). After a noncommercial find at Rangely, Colorado in 1917, Socal explored in Washington (1919 or earlier), Montana (discovery in 1921), west Texas (discovery 1923), Alaska (1922), New Mexico and Wyoming (1924), Idaho (1929), and Utah (1930). It was also one of the most vigorous international explorers. In 1917 Socal looked at the west coast of Mexico (as much to protect its California markets as to find oil) and the island of Timor in the Indian Ocean. In 1919 it looked at Central America and the Philippines, acquiring exploratory rights on the vast holdings of the United Fruit Company, which stretched from Guatemala to Colombia. Concessions in Colombia (1921-50%), Argentina and Ecuador (1922), and Venezuela (1924) followed (PL10,11).[22]

For all its effort Socal found very little from 1917 to 1932 either abroad or in the U.S. outside California.[23] Large discoveries had been made in Mexico and Venezuela but Socal was left out. Despite its lack of success Socal continued to search outside California and outside the U.S. and spent several million dollars doing so.[24] Why it did so is an important strategic question because Socal's quest eventually led it into Saudi Arabia. But conventional thinking requires strategy to be justified by economic analysis or competitive necessity. Neither of these would seem to apply to this search. Economic analysis would not justify it since early in the 1920's crude surpluses appeared in California and after 1927 in Texas that lasted until World War II and conditions for producers steadily deteriorated. Nor did Socal have the refineries or markets to profit from the crude if it found it. Of its competitors only Exxon and Shell, whose interests were quite different, engaged on such a course, so it did not come from competitive necessity. But its effort shows the very long range nature of effective strategy and its link to identity rather than justification by analysis.

An explanation was given by R.G. Follis (chairman 1950-1966), who first joined Socal in 1919.[25] Like other companies after World War I Socal went abroad with the idea of discovering crude for import. It developed the organization and expertise to handle foreign production. The number of geologists of Socal's payroll grew from 8 in 1919 to 22 a year later.[26] A "Foreign Crude Oil Production Department" was established in 1921 even though Socal as yet produced no crude abroad.[27] As surplus production developed, the organization to find foreign crude was in place but the goal changed. Instead of finding crude to import Socal would find crude to sell and profit from that business. It was an organizational commitment to foreign production that sustained the wildcatting effort, even though no oil was found. The idea of producing crude for sale thereafter became an important policy (PL11) but one far removed from the Standard Oil strategy of its origin.

Events (EC1,3-6) aided the shift in strategy. In contrast to the feared postwar shortages east of the Rockies and abroad, soon after World War I Socal had found more crude in California than the West coast market could absorb. Though it had sold products abroad since Standard had entered the Pacific (1890's) it sold crude (PL11) outside California for the first time in late 1922, a 20-million-barrel crude and fuel oil contract with Jersey.[28] The opening of the Panama Canal (August 1914) (EC3), the postwar tanker surplus (EC4), and the feared world crude shortage (EC5) made California crude competitive on the East coast for the first time.[29] Socal has been a seller of crude (PL11) ever since, even before its discoveries in the Middle and Far East.

The rise of scientific geology (EC6) also encouraged the shift to a producing strategy (PL10-12) by producing a flood of new discoveries where no surface indications were found. Kenneth Crandall, former Vice-President for Exploration at Socal, recalled that when he was preparing to study geology at Stanford in 1920 reserves had been declining, surface geologists had been all over California, and it was thought that most of the existing oil had been found.[30]

The methods which produced these discoveries were the result of technical progress in measurement and instrumentation as well as scientific geology.[31] The torsion balance outlined the perimeters of salt domes on which oil was found by measuring the rate of change in gravitational attraction between lighter salt domes and surrounding sediments. In the late 1920's the refraction seismograph came into use to map subsurface structures. It was followed by the reflection seismograph in 1933-1934. They worked like underground radar. The magnetometer was used to indicate the depth of basement rocks and therefore the thickness of the overlying sediments. The explorer's principal tools today are refinements of these technological breakthroughs.

The downstream policy accompanying this shift in vertical strategy (PL10) toward emphasis on exploration and production was to expand refining and marketing geographically where Socal had production (PL12). Beginning with its large refinery at Richmond on San Francisco Bay, late in 1911 a small refinery was completed at El Segundo near Los Angeles. Plans for it had begun in 1910 before the divestiture as the demand for gasoline and the volume of light refinable crude in the area had expanded steadily since 1907. A third refinery was placed in operation in 1913 at Bakersfield to "top" the light crude from the Southern Pacific Railroad's oil subsidiary. The lightest fractions were refined into gasoline and distillate for the local market. The rest was piped to Richmond as "crude oil equivalent" in exchange for fuel oil for the railroad. The timely acquisition of this subsidiary from the Southern Pacific in 1926 boosted Socal's production by nearly 50% (R26). In 1927 Socal built a 14-TBD refinery at El Paso, Texas to be supplied with crude from Socal's production in the Yates field in west Texas (R27a,b). Another majority-owned refinery at Colorado, Texas was placed in operation not later than 1929 also to handle west Texas crude. In 1935 Socal acquired its distributor in Vancouver, British Columbia and built a 2-TBD refinery there. In 1948 it opened a 25-TBD refinery in Salt Lake City, Utah supplied from production it had developed during World War II in the Rangely field in Colorado.[32]

Thus after 1911 Socal became an exploring-producing company, departing from its Standard Oil vertical strategy, due chiefly to the presence of F.H. Hillman and favorable circumstances. It found oil in California and the West, and it expanded downstream to refine and market it (PL10,12). It became a crude seller (PL11). It did not explore east of the Rocky Mountains and it did not find oil abroad and, consequently, did not expand there (PL12,H2).

7.3 1928-1939: Bahrain, Caltex, and Saudi Arabia

This brings the narrative to the decade of the 1930's when Socal found the pot of gold at the end of the rainbow. Hillman's organization survived the 1920's without finding oil, even though K.R. Kingsbury, the chief executive, was a marketer by background, having been liaison with Standard in New York before 1911. The commitment to foreign production, the absolute determination to find oil (PL11,12), can only be shown by the events of the 1930's. It illustrates, perhaps as well as any incident in the history of the industry, the triumph of strategy over conventional rationality.

In the first set of circumstances that led to the opportunity in the Middle East, the most important fact was the British domination of the area, both in oil, where it dominated the producing areas in Iran and Iraq, and politics, where it conducted the foreign policy of the entire area.[33] After years of jockeying, five American companies finally gained a toehold through the purchase of 23.75% of Turkish Petroleum Company on July 31, 1928.[34]

Having been forced to admit the Americans to the area, the Europeans contained them with the Red Line agreement. Its restrictive provisions were based on a preceding 1912 agreement among the British, Dutch, and German partners who formed Turkish Petroleum Company Limited, in October 1912, agreeing to develop the Ottoman Empire concession only as a group and to produce or refine crude oil in the Empire only through Turkish Petroleum.[35]

The Red Line agreement (see Chapter 4) was signed on the same day as the purchase. This agreement, viewed by industry critics as the most blatantly collusive agreement in the history of the industry and the main support of the theory that the industry before World War II was a company cartel, is perhaps more accurately viewed as a foreign policy maneuver by the British to box in the Americans. The American companies gained nothing by it, and in time lost a great deal from the stigma. Furthermore, among the Americans, no sooner had it been signed that it began to break up.[36]

The constraints of the Red Line agreement produced Socal's opportunity in the Middle East. The concessions Gulf had purchased from Frank Holmes on November 30, 1927 all lay within the Red Line area except Kuwait.[37] It was bound by that agreement to offer them to Turkish Petroleum. But the British geologists at Turkish Petroleum were not interested. Reluctantly, Gulf offered its concessions in Bahrain, the Neutral Zone, and the province of al-Hasa in eastern Saudi Arabia to Socal, a newcomer to the international scene not obligated by the Red Line agreement. Socal bought the Bahrain concession (R28) for $50,000 on December 21, 1928, but because of arrears in protection fees which might have rendered the options invalid, it did not take the Neutral Zone or al-Hasa. Bound by the Red Line, Gulf let these options formally lapse on April 1, 1932. Socal took the opportunity because its strategy inclined it to do so (PL10,11). It had not previously looked in the Middle East and had not sought it out. Others had been offered the concessions and had turned them down.[38]

The British government insisted that any concession on Bahrain must be operated by a British company, with a London office, British directors, and a local government representative in Bahrain who must be approved by the British government and work through the British Political Agent in his dealings with the Sheikh of Bahrain.[39] It took until August 1, 1930 to work out these details by setting up a Canadian corporation, Bahrain Petroleum Company (Bapco) (R30) to hold the Bahrain concession, with Frank Holmes as the local representative.

In the meantime the world had entered the Great Depression, and the east Texas field and cheap Russian oil[40] (EC29,30) flooded the world market. All during the time of negotiation with the British government there were questions about whether to go on with the venture. Socal had no foreign markets or refineries and all it had to show for millions of dollars poured into foreign exploration was a dozen years of dry holes. In a depression with crude selling for 10¢ a barrel in Texas, even if all the undiscovered crude in the world lay beneath the Middle East (as it nearly did), what could it be worth?

But Socal persevered, carried on by the enthusiasm of two directors, Maurice Lombardi and William H. Berg (whose career at Socal began as stenographer and ended as president in 1937-1939), Reginald C. Stoner, manager of the producing department, and the two chief geologists (PV2-6), Clark Gester and "Doc" Nomland.[41] On June 1, 1932, they had discovered (R32) oil in commerical quantities on Bahrain,[42] though what they could do with it was much less clear. Having tasted the nectar of success, they began to gaze at the similar hills 20 miles across the water on the eastern shore of Saudi Arabia (PL10,11).

Unknown to Socal, King Ibn Saud was in desperate need of money and was well aware of the good luck of his friend, the Sheikh of Bahrain. His advisor, Harry St. John Philby, a former British civil servant converted to Islam, suggested development of Saudi Arabia's mineral resources.[43] Contact was arranged through an American philanthropist for a survey by Karl Twitchell, an American mining engineer. Twitchell was encouraged by his findings and tried to hawk Saudi concessions to the American oil companies, first Texaco, then Jersey, then Gulf. The first were not interested and Gulf pleaded its Red Line commitments.[44]

In the summer of 1932 Socal finally made contact with King Saud, having tried fruitlessly to arrange a contact through Holmes since May 1930. BP was also interested in the concession—at least they were interested in keeping the Americans out—and sent a negotiator, Stephen Longrigg, but were prepared to offer only £10,000 in rupees when gold was demanded.[45] Socal put up a loan of £50,000 plus a yearly rent of £5,000, all in gold. The loans were to be repaid from the four gold shillings/ton royalty on any oil found.[46] The king was pleased and a concession was granted in August 1933 (R33).

Socal had first become a seller of crude (PL11) in 1922 and has remained so every since. Throughout much of this time, contrary to the present, there was a surfeit of crude and selling it was a very difficult job, perhaps a harder experience, shared by many Socal executives since the 1920's, even than finding it (H3). A former chairman related his experience in Europe in trying to sell Socal's recently discovered crude from Bahrain in the 1930's during the depths of the Great Depression (EC29,30).[47] Rounds of the major purchasers, including Jersey Standard and Shell, proved fruitless. Smaller marketers, including some government owned companies, would gladly have bought refined products, but they had no facilities to refine the crude nor could they afford to pay in a convertible currency. Socal tried without success to buy a refinery in Europe. Prospects for moving crude from far-off Bahrain were very discouraging. He finally achieved limited success by barter. Socal sent Bahrain crude to some small German refiners in exchange for refinery equipment it could install in Bahrain and then use to supply products to others. The Germans, hit as hard as were the Americans by the Depression (EC30), were glad to exchange a badly needed raw material for capital equipment, even though they could not pay in hard currency. The (R36a) 10-TBD Bahrain refinery opened in 1936, built with German equipment swapped for crude. These experiences convinced Socal executives that their need for markets was as great as others' need for crude (H3) and surely inclined them toward taking partners with markets, as they did in 1936 with the formation of Caltex and with the 1947 sale of 40% of Aramco to Jersey and Mobil.

The most important organizational event in Socal's history was one which did not happen, namely the near merger of Socal with Jersey in the late 1920's and early 1930's. In 1929 the president of Socal, Kenneth R. Kingsbury, approached Walter Teagle, his counterpart at Jersey, about a possible merger. Teagle was concerned about the legality of combining the former Standard subsidiaries and sought legal advice from outside counsel who consulted the Justice Department and advised against it. The two companies then decided to postpone further discussion until the outcome of the suit brought against the Socony merger with Vacuum Oil Company was known.[48] When that merger was permitted in 1931 the discussions proceeded in earnest (EC31).[49]

On August 25, 1931 the two companies issued a press release stating the reasons for the merger:

> The California Company has important domestic crude oil reserves but no important foreign reserves[50] and only limited foreign distributing facilities. The Jersey Company has limited domestic crude oil reserves but has large foreign reserves and an important foreign system of distribution. The Jersey Company operates on the Atlantic Seaboard and in the Gulf Coast states, the California Company in the states and territories west of the Rocky Mountains. Therefore the operations of the two companies are in effect complementary and the merger would make possible the most economic use of their reserves and facilities.[51]

In other words Jersey would gain U.S. crude reserves and markets in the West which suited its strategic desires to build up crude production and to re-establish nationwide distribution while Socal would conquer the hurdle of the Rockies.[52]

However, difficulties soon appeared. On Jersey's board there was disagreement between the producers and the marketers. Everit J. Sadler, Jersey's head of production, was opposed, arguing that Socal's West coast crude would be a high-cost source of supply in Jersey's eastern, southern, and Euorpean markets.[53] For the marketers F.H. Bedford, Jr. wrote, "We undoubtedly will ultimately have to meet our competitors by becoming nationwide in our activities and there is no way we can accomplish this so well and so cheaply and so quickly as by combination with a company having a Standard name and brand," and Heinrich Riedeman said, "my heart is set on this merger with California strongly."[54]

There were disputes over the valuation of crude reserves in Humble's holdings in the new east Texas field and Socal's heavy crude in the new North Kettleman Hills field. Humble executives finally accepted a significant reduction in the original estimates of their east Texas reserves, though the original estimates later proved to be correct.[55]

And there were personality clashes. "King Rex" Kingsbury got along with neither the prickly Sadler, nor Humble's president, William S. Farish (who succeeded Teagle at Jersey), nor Teagle himself. There was a meeting in 1933 at Lake Tahoe between Kingsbury and Teagle where it became apparent that, aside from the issues, a merger would not accommodate the colossal egos involved.[56] The negotiations were discontinued late in 1933.[57]

Even so, discussions between Jersey, Socony-Vacuum, and Socal continued with regard to Socal's oil in Bahrain to which the other two former Standard companies very much wanted access but were limited by the Red Line agreement and the lack of interest of other IPC partners. These talks dragged on fruitlessly for three years until Socal found another partner (PL13).[58]

On July 1, 1936 Texaco acquired a half interest in the Bahrain Petroleum Company Limited (Bapco) in exchange for Socal's receiving half interest in Texaco's marketing subsidiaries east of Suez (R36b).[59] The Bahrain property included production potential of 30 TBD and a 10-TBD refinery.[60] Texaco had been marketing abroad since 1905[61] and had about 10% of the market east of Suez.[62] The book value of these marketing properties at the time of merger was about $27 million.[63] Bapco set up a separate subsidiary, the California Texas Oil Company Limited (Caltex), to administer these marketing properties with each company being represented by two directors on the board. A refinery expansion to 25 TBD begun to supply Caltex and Bapco was in business.[64]

In a second agreement in December 1936 Texaco acquired half interests in Socal's concessions in Arabia, the Netherlands East Indies, and New Guinea for $3 million in cash and $18 million in payments from production that might be developed.[65] A Socal subsidiary, the N.V. Nederlandsche Pacific Petroleum Maatschappij (Dutch Pacific Petroleum Company) had concessions and exploration permits in the Netherlands East Indies and a 20% interest in a New Guinea concession.

In a third agreement Texaco granted Socal an option until July 1, 1939 to acquire a half interest in nine marketing subsidiaries in Europe, one in the Canary Islands, and consignment agencies on the west coast of Africa and in Madeira (Spain). The book value of these properties on April 30, 1936 was

$24.5 million. The option could be exercised only if reserves in the Arabian concession and 20% of reserves in the New Guinea concession totaled at least 72 million barrels. Payment by Socal was to be cancellation of the $18 million production payment for 50% ownership of those properties.[66] This option was not exercised.

It took almost five years before commercial oil[67] was discovered in Saudi Arabia on March 4, 1938 (R38). Surveys were conducted in 1934. The first six wells were drilled in the Bahrain zone, approximately 2300 feet deep, from 1935 to 1938. The first was a noncommercial discovery in 1935. The second well flowed spectacularly (3840 b/d) but then became a water well, producing 225 b/d oil and 1,965 b/d water by the end of 1936. Enthusiasm for the Saudi Arabian adventure rose and fell with the results of drilling. The third well in the Bahrain zone produced less than 100 b/d. The fourth was a dry hole. Five and six had not struck anything by the end of 1936.

The seventh well, begun in July 1936, was the first test well in the Arab zone, about 5000 feet deep, which had produced only gas on Bahrain. By November 1937, the board in San Francisco was getting restive and instructed suspension of work on the other wells until the results from No. 7 were in. Additional equipment for Arabia was getting hard to squeeze out of U.S. department heads. Reg Stoner, general manager of production at the time and one of the true believers in Saudi Arabia, was quietly "borrowing" equipment from California operations and shipping it to Arabia, hoping that something good might happen before the end of the year when the equipment had to be accounted for.[68]

It is easy to underestimate the difficulties in 1937 in light of today's perspective. It was difficult for San Francisco to maintain contact or control over events halfway around the world. The geology was unfamiliar. There were few surface indications in the rugged desert and almost no geophysical or structure data. Hollow limestone formations that reverberated too much for intelligible results prevented collection of seismographic information in some of the most important areas. As Stegner relates, "Stoner had heard enough reverberations of discontent from the limestone members of the board of directors to be uneasy. The expenses of Saudi Arabia had already run to a good many million dollars, and in late 1937 and early 1938 dollars did not grow on trees. The stock market was nearly as low as it had been at the bottom of the depression. . . . The company had pulled out of foreign wildcats before; it could pull out of this one too."[69] But once again Socal persevered, as it had through the dry years before Bahrain and through the intrigues surrounding the concessions. On March 4, 1938, it brought in Dammam No. 7 and, with it, commercial production.[70] It was the twentieth year of Socal's search for oil abroad (PL10,11).

As one looks back on this period in Socal's history, it is apparent that they were doing the right thing. They were taking risks others would not take. They were exploring new areas. They did not give up. Their actions maximized long-run profit. But it is also clear that they acted from strategy, not from economics. For no set of cost-benefit calculations would ever have justified taking such a course of action or sustaining it for so long in times of crude oil surplus.

7.4 The Aramco Sale

It took as much determination to hold onto the Saudi Arabian discovery as it had to make it (PL11). After nearly losing control of the concession to the U.S. government in 1943 (EC43) Socal and Texaco brought in Jersey and Mobil to help develop the concession and thereby strengthen their hold.

As it had during World War I, the U.S. in World War II supplied huge amounts of oil to the Allies, approximately 6 of the 7 billion barrels required.[71] This 2-billion-barrel-per-year drain, when U.S. reserves were only 20 billion barrels,[72] was a cause of great concern by 1943 to U.S. officials. In that year a mission from the Petroleum Reserves Corporation, one of Interior Secretary Harold Ickes'[73] agencies for supplying the war effort, visited Saudi Arabia to plan for supplying the offensive in the Pacific. The Japanese fleet, preparing to invade Hawaii, had been mauled at Midway in June 1942, a battle which proved to be the turning point in the Pacific war. Capacity at the Bahrain refinery was in the process of being doubled again. By 1943 the reserves in Saudi Arabia were estimated at possibly 5 billion barrels.[74] The head of the mission, Everett L. DeGolyer, the preeminent geologist, predicted, "The center of gravity of the world oil production is shifting away from the Gulf-Caribbean areas to the Middle East, to the Persian Gulf area, and is likely to continue to shift until it is firmly established in that area."[75] A refinery was planned at Ras Tanura to produce aviation fuel and other products for the Pacific theater.

The U.S. government came, it saw, and it wanted to buy in, as the British had bought into BP in 1914 for the Navy when Churchill was First Lord of the Admiralty. It would finance the refinery and a pipeline to the Mediterranean. Socal and Texaco proposed an agreement calling for government finance of the pipeline in exchange for a naval reserve of the lesser of 1 billion barrels or 20% of their Saudi Arabian reserves to be sold to the Navy at a discount of 25% from market prices.[76] Not only would it be a huge crude sale, but the pipeline would also give them a great competitive advantage in Europe, a market where they were weak. But selling the Arabian concession was not exactly what Socal and Texaco had in mind. The uproar that followed from the British, fearing American intrusion into the Middle East, and from U.S. producers, fearing a government owned competitor for U.S. exports, and from Exxon, BP, and Shell killed the idea but set the stage for the postwar events in Saudi Arabia.[77]

Socal and Texaco were mostly responsible for this near loss of their concession. In February 1943 Harry Collier and Star Rodgers, the presidents of Socal and Texaco, paid a visit to Harold Ickes to plead for lend-lease aid for Saudi Arabia, aid offered to any country whose defense was deemed vital to U.S. defense. They were seeking the aid to counter the $20 million in aid advanced to the king by the British during the war when his revenues from pilgrimages to Mecca and oil payments by Casoc, Socal's Arabian subsidiary, had been severely reduced (EC40).[78] They began to fear that the aid to King Saud by the British might be used to deprive them of their concession, particularly after they found geologists mysteriously present in a British expedition sent to deal with locusts.

In 1940 when his revenues from wartime cutbacks in oil operations and pilgrimages fell, Ibn Saud demanded a doubling of the royalty from 21¢ to 42¢ per barrel,[79] a dispute not settled until 1950.[80] In 1943, after three years of bickering and strained relations, plus an obvious British attempt to reassert its influence in the area, Socal and Texaco felt like small fish in a very big sea.

Like U.S. citizens in distress abroad everywhere, they turned to their government for help, stressing the value of the concession to Ickes in that February 1943 meeting. He was receptive. American presence in the area should be maintained. Oil was needed for the Pacific war. Lend-lease was authorized on February 18 as the two presidents requested. The DeGolyer report was exciting. By June 1943 the Navy, Ickes, and Abe Fortas wanted to buy the place. As Herbert Feis, Economic Adviser to the State Department put it, "they had gone fishing for a cod, and caught a whale (EC43)."[81]

This time they got away from all three predatory governments (the British, Ibn Saud, and the U.S.), vowing to do things differently in the future. The 75-TBD Ras Tanura refinery was built with Aramco funds. World War II was won. Still needing help, they turned in May 1946 to a slightly less predatory competitor: Jersey Standard (PL13).

There were political and economic reasons for wanting to sell part of the concession. By 1946 it was apparent that the reserves in Saudi Arabia were very large. There were no public estimates available, but estimates of ultimate reserves from internal memoranda[82] within Socal were 20-50 billion barrels, a number as large as total U.S. proved reserves at the time.[83] Politically there would be intense pressure from the Saudis, as there always is from leaders of developing countries, to develop the concession in line with its share of world reserves. The British would take advantage of any opportunity to push the Americans out. The prewar crude glut was expected to continue, only made worse by huge discoveries throughout the Middle East (H3). Postwar competition for markets would be, therefore, even more intense. The Russians were thrashing about in the Middle East, looking to Balkanize it.[84] Greece, Turkey, and the whole area looked vulnerable to Communist takeover unless it were strengthened with western aid. The king was old and the identity and disposition of his successor were unknown. The regime was not stable. If they did not develop the concession rapidly enough, stabilizing the area in the process, they felt they would lose it. And they might lose it anyway. With it they would lose a $92 million investment[85] and the only foreign venture they had that was likely to be profitable. From 27 years and $50 million of foreign exploration expenditure stockholders had not yet received a dime in dividends. What profits there were (from Bahrain) had been reinvested to develop the properties.[86]

Between May and August 1946 there were divergent views about the sale within Socal and among and within each of the companies involved. Reg Stoner, vice-president of production, one of the true believers (PV4) in Saudi Arabia, was against it. T.L. Lenzen, later a vice-president, was the main spokesman within Socal for the sale.[87] R.G. Follis, vice-chairman at the time, thought $250 million a reasonable price for a one-third interest.[88] Texaco thought $650 million in cash was closer to the present value of a one-third interest.[89] Jersey thought it should be allowed to participate for a nominal sum, say $1, Socal and Texaco profits from increased volume being adequate compensation.[90] With Saudi Arabia as the key to the Middle Eastern supply patterns, there then began an intricate dance among the Majors of adjustment to the postwar realities.

Unknown to Socal and Texaco, Jersey had consulted with Mobil, its American partner in Iraq Petroleum and the Red Line in July 1946, and found Mobil interested in taking part.[91] Jersey and Mobil representatives Oliver Harden and Harold Sheets went to London to negotiate with the British partners in Iraq Petroleum.[92] BP, worried that Arabian oil would undercut their Iranian production, was appeased with a contract for 50 million barrels over 20 years at cost, plus a shilling per barrel, plus assistance with a pipeline to the Mediterranean.[93] Desperate for crude at having lost its Indonesian production, Shell settled for crude purchase contracts with Mobil,[94] in addition to its major purchase of half of Gulf's production from Kuwait under long-term contract. This left the French (CFP) and Gulbenkian to appease.

In September or October, Socal and Texaco learned that both Jersey and Mobil wanted into Aramco.[95] In November they learned about the BP deal and were worried about its effect on Aramco, fearing it would frustrate their attempt to develop the concession.[96]

The next issue was minimum liftings. Socal and Texaco wanted to bind Jersey to minimum liftings to collect profits through Aramco from Jersey and to prevent its holding Saudi Arabia as a reserve. Otherwise, their whole purpose in making the sale to develop the concession and gain a market for the crude would be frustrated.[97] Jersey would only bind itself if Socal and Texaco did likewise. Texaco, with markets in Europe and the U.S. East coast, agreed, but Socal with foreign outlets only east of Suez through Caltex could not.[98] Mobil was willing to take half interest if one-third of the concession was offered to Jersey, but the thought of half of a 50% interest would be more than it wanted.[99] The Mobil board also had doubts about the size of the reserves and the antitrust aspects,[100] and they thought the price was too high.[101]

Socal and Texaco wanted to retain at least a majority interest.[102] The outcome of these issues, as Sampson tells it, was that "Exxon and Mobil were first offered 20 per cent each, but Mobil, after bitter arguments on the board, decided that the cost was too high for the dubious benefits, and took only 10 per cent—a misjudgment so huge that it has been the subject of recriminations ever since." Thus Aramco was created with three equal partners, Socal, Texaco, and Jersey, with 30 percent each, and a fourth, Socony-Vacuum, less than equal and jealous ever since.[103]

In order to get Socal to agree to minimum liftings, Texaco suggested that Socal exercise the lapsed 1936 option to buy half of its downstream interests in Europe. The price was $7.5 million in cash and $2 million per year for ten years at 2% interest paid to Texaco by Caltex (R47a).

Agreement in principle was reached by the four parties in December 1946, and was signed on March 12, 1947. The price for 40% of Aramco was some $240 million,[104] the largest acquisition up to that time (R47b).[105] Terms are given in Exhibit SC-7.

On the same day, March 12, 1947, President Truman granted military and economic aid to Greece and Turkey, assuring an American commitment to contain communism in the area. Socal and Texaco may have moved too fast. The timing was, as William S.S. Rodgers, chairman of Texaco, told a Senate Committee, "...absolutely coincidental...We were amazed...when we heard it. If I had known about it, I think we would have probably held up our arrangements with Standard Oil of New Jersey and Socony-Vacuum."[106] R.G. Follis, retired chairman of Socal, confirmed the fact, suggested by Rodgers above, that political fears were the deciding factor in making the sale. As he put it, "The question was whether Socal and Texaco could build a market big enough to live with the Arabs. If not, they might take some [unspecified] action." With the sale they could face the Arabs with the assurance that even if crude sales were less than the Arabs would like, "the Arabs had four great companies working for them, and that was about as good as you could do (PL13)."[107]

A consequence of the sale was the death of the Red Line agreement in November 1948, after litigation by the French and Gulbenkian against the other IPC partners, Jersey and Mobil. The old order was replaced by the new. Jersey and Mobil argued that (1) the Red Line agreement was now illegal under U.S. antitrust law, and (2) the shares of the French and Gulbenkian were forfeited when both were under German occupation. The French had wanted into the Arabian concession, but finally settled for a sixfold expansion of IPC production and another pipeline to the Mediterranean from Iraq.[108]

It is interesting that it was with this sale, not with their discoveries, that Socal and Texaco achieved recognition with Jersey and Shell as major international oil companies.

7.5 East of the Rockies

After World War II Socal overcame the physical and psychological (PL6) barrier of the Rocky Mountains and entered eastern U.S. markets (PL15). It did so in typical fashion, exploring first (PL10) and expanding downstream where it had production (PL12), continuing its cautious, conservative style of management (PL7).

Though the impact of Aramco on the postwar industry was enormous, it did not end Socal's exploratory activities. Socal and Magnolia (a Socony-Vacuum subsidiary) were the first to make discoveries off the shores of Texas and Louisiana (PL10). These discoveries were triggered by a technological advance in exploration (EC9), the use of the torsion balance over water. The torsion balance measures differences in specific gravity. Salt domes, the dominant oil bearing structure in Louisiana, have lower specific gravity than surrounding areas. The problem was to eliminate the disturbance of waves on the balance during the 24-hour measurement period during which the instrument had to be held motionless. Karl Hasselman, who had once worked for Socal in Texas, had gone to Germany to use modern oil exploration methods there, but came back to Texas before the war and solved the problem by placing the torsion balance on a tripod fixed into the sea bottom in the inside of a well cut into the center of a barge. The first discoveries were made in the late 1930's but accelerated in 1943-1944 as exploration in the industry revived after the turning point in mid-1942 for the Allies in both the Pacific and Europe. Socal was the first Major to develop (R43a,b) expertise in offshore exploration. Its efforts were led by Kenneth H. Crandall who became Vice-President for Exploration from 1949 to 1969, picking up the reins that had passed from R.H. Hillman to R.C. Stoner. For several years Socal had 75% of offshore production in the U.S.[109]

It then began to move downstream east of the Rockies. Much of the credit for this move should go to Harry D. Collier, Socal's chief executive from 1940 to 1949.[110] In late 1945 or 1946 it acquired a 20-TBD refinery, tanks, and marine terminal at Perth Amboy, New Jersey, on New York harbor.[111] The Perth Amboy refinery could be supplied from Saudi Arabia or the U.S. Gulf where its offshore production was rapidly expanding. Socal had finally established a toehold east of the Rockies. It marketed products in eight eastern states under the brand name Calso.[112] In 1951 Perth Amboy was expanded to 60 TBD and designed to run on Middle Eastern crude.[113] In 1957 Socal penetrated eastern Canada by a joint venture with the Irving Oil Company Limited, distributors in the maritime provinces and Quebec, which had formerly been supplied by Jersey Standard and Shell from Caribbean refineries. Socal would supply the crude; Irving would market the products, and a new 45-TBD, $50 million refinery at St. Johns, New Brunswick would be jointly owned.[114] The joint venture resulted because Irving did not want to be acquired by Socal.[115] In the U.S. Midwest, Canada, Central and South America it has never developed a strong market position, in part because when it finally made a discovery in Venezuela in 1948, it found heavy Boscan crude rather than light crude, and sold rather than refined it (PL10,11).[116]

Socal took from Jersey again in 1961, this time as part of a consent decree between Jersey Standard and the Justice Department[117] which forced Jersey to terminate its 1956 contract to supply approximately 80% of the products required by Standard Oil of Kentucky (Kyso), the principal marketer in the southeastern U.S. with 8700 retail outlets. Kyso was dirt poor and had neither refineries nor crude supplies, aside from small production in Mississippi.[118] To gain clearance for the acquisition from the Justice Department, which was interested in bringing more competitors into the Southeast, Socal agreed to build a refinery and to continue to buy products from independent refiners in the area for a number of years (R61).[119] The 100-TBD Pascagoula, Mississippi refinery would be supplied with offshore Louisiana crude and Middle Eastern crude (PL12,R63).[120]

With the Kyso acquisition Socal attained a position in all the major U.S. coastal markets east of the Rockies. Today, in fact, its production east of the Rockies is larger than in the West.[121] But not until at least 1965 did refinery runs exceed its domestic production east of the Rockies,[122] and ever since then its refineries continue to be supplied by its Middle Eastern crude (PL12). This is not to imply that Socal's East coast refineries were supplied solely or were limited by its offshore Texas and Louisiana production. To the extent that import controls permitted, they were supplied with cheaper Middle Eastern crude. The Texas and Louisiana crude was sold, or swapped, principally to Exxon and Gulf. One of the factors in the Kyso acquisition was that Jersey had discovered oil in California and wanted to swap it for Louisiana crude, rather than continue to buy, facing Socal with the prospect of losing the market for its Louisiana crude if it chose not to swap, which it did not (PL11).[123]

Other Socal downstream policies have more closely followed its Standard heritage. Its refinery scale and location policies have been to build large refineries close to its markets (PL1). Socal has never been a price marketer. Its gasoline was usually priced 1-2¢ above the trade. Due to loss of market share during the last 5-6 years, it is no longer the price leader in most of its markets (PL8). It follows a premium price policy (PL9), often being the last to cut its price and the first to raise it. It has stressed high quality products and attempts to be the best full-service marketer in its areas (PL2,9)[124] These policies came not only from its early leadership in refined products in the West but also as responses to local price wars from small price-cutting independents near crude sources, especially in the Los Angeles area over the years.[125] It has maintained margins, a strong dealer network, and has steadily conceded market share. One exception to this policy is its strong position in the jet fuel market, an extremely price-competitive market, where Socal's market share is nearly double its overall 6% U.S. market share.

Penetrating eastern U.S. markets in some respects had a greater impact on Socal than the discovery in Saudi Arabia. Before these two events it was a regional U.S. company, isolated from other oil companies, the media, the government, and the financial centers in New York and London by its geographic location on the West coast of the United States. Both its production and natural markets in 1911 were in the western states. From its San Francisco headquarters one looks west to the Pacific, rather than east over the formidable barrier of the Rocky Mountains. Today, with its largest assets in California, Indonesia, and Saudi Arabia, it is still a Pacific basin company. Entering eastern markets made it a national company and, with its Middle Eastern discoveries, brought it into contact with a larger world (PL12,15).

That is not to say that the strategy which grew from its western origin (PL6) changed. It remained a large, low profile, technically (PL17) oriented, financially conservative, independent Pacific basin company (PL7) . Its low profile came from its western isolation, making it almost unknown in the eastern U.S. and Europe. It was treated as an outsider by other companies in the Middle East and was not a party to the Red Line or Achnacarry or other market sharing agreements before World War II, which have tarred the public image of other oil companies. It pursued a low key public image. Among the public it is probably the least recognized of any of the seven Majors. It is the only one which does not market its products under its corporate name. Few people outside the local area would associate Chevron or Standard Oil of Kentucky with Standard Oil of California. Neither was it well known to the media. It did not hold its first press conference in San Francisco until 1963.[126] Before then its representatives in New York held occasional press conferences, supplemented from time to time by top management flown in from California for the occasion. They then flew back and had no regular contact with the New York based media. Top management earned a reputation for aloofness and inaccessibility.

Another policy which flows from its western origin is conservatism, both in finance and management (PL7). Among the major oil companies, which are surely among the most prudent organizations on earth, Socal has the reputation for being the most conservative. It has rarely moved rapidly or radically in new directions. This policy is partly due to its inheritance of Standard Oil's conservative financial policies (no debt, long or short term; large liquid assets, slow response). Its western location gave it little access to eastern capital markets and financial sophistication. It grew up in its own conservative, independent way. The personal conservatism and style of long reigning chief executives reinforced its reputation (PV7-9). In addition to its downstream policy (PL12), it has been the slowest of the Majors to diversify outside the oil industry. It did not get on the bandwagon in the 1960's to refine and market in all 50 states. It did not borrow large amounts of hard currencies to expand in the 1970's. It does still have some male secretaries at the officer level (like longer hemlines they may come back into style).

A third characteristic of its identity is its emphasis on engineering accomplishments (PL17). Technical expertise has always been held in high regard within the company. It has had a central board of engineers since 1921 to standardize and coordinate engineering efforts, while most other companies relied on departmental engineering. Its chief engineering position is one level below a vice-president. It views itself as being managed by engineers—methodical, conservative, careful, technical people. Many of Socal's greatest accomplishments in Saudi Arabia were feats of engineering, often peripheral to the oil business in bringing or adapting the benefits of technological civilization to the climate and culture of Saudi Arabia (PL17,SR1).

Two rather interesting illustrations occurred during the building of TAPLINE, the Trans Arabian Pipeline from Saudi Arabia to the Mediterranean over 1000 miles of desert. Pumping stations were required to boost pipeline pressure. Men were required to operate and maintain the stations and to inspect the line. Maintaining the men required the construction of living quarters, communications, and the drilling of a water well. The water attracted camels. Soon there was a miniature civilization radiating around the pipeline pumping stations for which the king obliged the companies to provide sanitation, then medical services, hospitals, housing, schools, and roads. The company town was born again. The greatest impact of companies like Socal on the average Arab has come not from oil but from the peripheral engineering feats of bringing western technological civilization to the desert (SR1).

The other illustration concerns specifications of the pipe for TAPLINE. One of the expenses in constructing a pipeline is the cost of transporting the pipe, particularly to a remote location like Saudi Arabia.[127] The huge 300-TBD projected capacity of the line required a pipe 30 inches in diameter. The larger the pipe diameter, the more wasted space there is in transporting the pipe and the higher is the cost. Socal engineers working on TAPLINE made a small change in specifications that cut the transportation cost in half. They ordered 500 miles of 30-inch line and 500 miles of 31-inch line. Then they fitted sections of the 30-inch line snugly inside the 31-inch pipe and shipped them to Saudi Arabia together at half the cost.

7.6 Caltex

In the Eastern Hemisphere Socal's downstream policies have been quite different. Aside from its exports to the Pacific basin, Socal had no downstream activities before the discoveries in Bahrain and Saudi Arabia. Through Caltex it shared until after World War II a small part of the market east of Suez, which was supplied by small source refineries in Bahrain and Saudi Arabia. Since then not only did it develop enormous production in Saudi Arabia, but using its lead in offshore drilling technology (R43) developed in the Gulf of Mexico by Socal, Caltex developed the largest production in Indonesia in the 1960's (R57,PL15).

Its downstream problem was how to move more crude (H3). In 1951 Caltex scored an important victory when it was the first oil company to reenter the Japanese market when it signed a 50-50 joint venture with Japan's leading refiner and marketer, Nippon Oil Company, providing for three shared refineries supplied with crude by Caltex.[128] After the war Socal took the option in 1947 to join Texaco west of Suez as part of the sale of Aramco recounted above. Caltex refineries went on stream in France (Bec d'Ambes, 13 TBD, 1950), Holland (Pernis, 20 TBD, 1950), Spain (Cartagena, 12% purchased 1949-50), and Italy (San Martino, 1952, 25% interest) (R50).[129]

Socal also wanted dividends from downstream operations for exploration and development in Saudi Arabia, offshore Louisiana, and Venezuela. This desire and its lack of downstream outlets led it to minimize investment in refining and marketing facilities (PL14). To move crude with minimum investment Caltex entered joint ventures all over the world (PL14) where it supplied the crude. In many developing countries the local market would not support more than one refinery of efficient size. Often it would support less. When the government insisted on having a refinery built locally, local marketers often split ownership according to their market shares. In 1975 in the Eastern Hemisphere Socal had a median equity of 25% in 41 refineries, compared to its median equity of 100% in 17 refineries in the Western Hemisphere (R50).[130]

The most recent change of importance in Socal's foreign operations took place in 1967 when most of Caltex was split up in Europe. Socal acquired most Caltex operations in Belgium, Denmark, Italy, Luxembourg, Switzerland, and the Netherlands. Assets and operations were split equally in Germany. France, Spain, and Turkey continued to be owned jointly, and Socal got small holdings in Sweden and the United Kingdom. Texaco got the U.K., Ireland, half of Germany, Greece, and Scandinavia.[131] With the split Socal began marketing directly in Europe for the first time.

The Caltex difficulties in Europe had begun in 1956 in the U.K., where Caltex had a joint marketing venture with Regent-Trinidad Leaseholds. Texaco acquired Trinidad Leaseholds and Regent, giving it a 75% share in the U.K. Differences arose over the value of the U.K. market and Socal was placed in the position of having to buy in or continue the friction in Britain. The solution in 1967 was to split the European operation in half, giving Texaco the U.K. market and leaving Socal freer to pursue its exploration interests in Africa. The move affected about 25% of Caltex's volume.[132]

7.7 Diversification

Socal is one of the less diversified of the major oil companies. It did not go into coal as did other companies, led by Conoco, in the 1960's. Its petrochemical business is perhaps the smallest of the seven Majors. It has not tried conglomerate diversification. As J.D. Bonney, Vice-President for Corporate Planning, explained its policy today,[133] Socal views itself not as an energy company (i.e., oil, gas, coal, nuclear energy suppliers), but perhaps someday as a natural resources company. This posture (PL16) on diversification has been a matter of conscious choice by Socal's management, reflecting the independent, cautious conservatism (PL7) of the company and its preoccupation with Saudi Arabia. To become a natural resources rather than an energy company reflects its feeling of expertise in upstream activities (PL10)—finding and developing resources.

Among Socal's non-oil and oil-related activities are asphalt, agricultural chemicals, gilsonite, some petrochemicals, geothermal energy (liquids underground), a uranium plant in Texas, and 20% of Amax. Socal became the leading marketer of asphalt in the U.S. in the 1960's. Only Shell is larger worldwide. Blessed with surplus heavy, asphaltic California crude and a demand for cheap all-weather roads needed to cover long distances in the West, Socal purchased Crown Emulsions, a firm with British asphalt emulsion patents, in the late 1920's and licensed the emulsion technology worldwide between 1928 and 1932 (75 countries, colonies, and protectorates in 1930). After its discovery of heavy Boscan crude in Venezuela in 1948 it built asphalt refineries east of the Rockies as well.

In July 1931 Socal acquired a large interest in California Spray-Chemical, a maker of oil-based orchard sprays, in order to protect its accounts receivable. By 1935 it marketed the products worldwide.[134] In the 1960's it added fertilizer to its agricultural lines. It is the dominant company in the home and garden market today.[135]

Socal built one of the earliest petrochemical plants, a polymerization plant at Richmond in 1936.[136] To support the war effort it built toluene (1942) and butadiene (1943) plants.[137] In 1943 it formed Oronite Chemical Company to sell industrial chemicals, most of which were developed from its own research. It developed paraxylene, the base chemical for dacron, before World War II and sold it in bulk to DuPont.[138] Socal developed the patents for alkane (alkylated benzene), the base chemical used to make detergents, when fats and oils used in soap making were in short supply at the end of World War II.[139] It nearly cornered the market for this detergent ingredient after the war. Both Colgate-Palmolive and Procter and Gamble once bought nearly all their alkane supplies from Socal. It is still a major business today.[140] It also patented isophthalic, used in plastics and surface coatings, and built the first plant to manufacture it in 1954-1958.[141] Oronite also developed a number of lubricating oil additives.

Socal began making some of the more conventional petrochemicals with a phenol plant in 1953 and an ammonia facility in 1955, both at Richmond.[142] In 1961 it established a polymers division in California to sell polypropylene.[143] It also began to build chemical facilities abroad in the mid-1950's. From 1968 to

1974 Socal began moving downstream in the fibers industry by building or acquiring three synthetic fiber plants.[144]

On May 30, 1975, nearly a decade after some of the other major oil companies, Socal made its first large move outside the oil and petrochemical industries with the purchase of 20% of Amax for $333 million.[145] Amax is a billion dollar (sales) metals producer with interests in coal, molybdenum, potash, iron ore, copper, nickel, and aluminum. In one stroke Socal got a lot of diversification of its earnings base while increasing its U.S. asset and earning base. Amax got cash for mining projects.[146]

An important reason for Socal's minimal diversification cited above has been its position in Saudi Arabia and its continued success in finding oil. In the 1960's Caltex became the leading producer in Indonesia. Most recently Socal participated in the Ninian field in the North Sea. In Africa it had significant positions in Libya and Nigeria. Being crude long in the Eastern Hemisphere and crude short in the U.S., it has doubled its refinery capacity in the Western Hemisphere since 1960 from 500 TBD to over 1000 TBD and added large refineries (200 TBD+) at Pascagoula, Mississippi and Freeport, Grand Bahama Island. In 1970 it produced 2558 TBD worldwide (474 TBD U.S.), refined 1741 TBD (707 TBD U.S.), and marketed 1919 TBD (935 TBD U.S.). It chose to channel its capital funds into downstream expansions rather than into diversification, so that it could move more crude.[147] When the OPEC aftermath cut back that crude flow it found its West coast refineries providentially equipped to process North Slope rather than Arabian heavy crude, both of which require a special desulfurization processing, of which it has the lion's share on the West coast.[148]

Exhibit SC-1
Socal Policies

PL 1	To be a large-scale refiner: (a) S, V; (b) 1906-present.
PL 2	To be a dominant marketer in the West of refined products: (a) G,H,V,P; (b) 1880-present.
PL 3	To be a large purchaser and trader of crude oil: (a) V,H,P; (b) 1900-present.
PL 4	Maintain a strong pipeline system in the West to facilitate PL 3; maintain sufficient marine transportation to serve Socal and affiliates: (a) G,V,P: (b) 1900-present.
PL 5	Crude supply by purchase and discovery of producing properties: (a) V; (b) 1906-1919.
PL 6	Independent Western location and outlook: (a) G,A; (b) 1911-present.
PL 7	Low profile and conservative style in finance and management: (a) A; (b) 1911-present.
PL 8	Price leader: (a) P,H; (b) 1880-World War II.
PL 9	Stress quality rather than price; charge premium price, offer full service; last to cut and first to raise prices: (a) P; (b) 1880-present.
PL 10	Become an upstream (exploring and producing company): (a) V; (b) 1911-present.
PL 11	Find crude for sale outside California and abroad: (a) G,V,P; (b) 1921-present.
PL 12	Geographical expansion downstream where supplied by own production: (a) V; (b) 1920's-present.
PL 13	Take partners abroad to reduce risks and spread development costs: (a) G,V,A; (b) 1936-present.
PL 14	Caltex uses joint ventures downstream to minimize investment: (a) G,V,A ; (b) 1950-present.
PL 15	Explore offshore and east of Rocky Mountains: (a) G,V; (b) 1943-present.
PL 16	To continue to be an oil company; little emphasis on diversification: (a) P; (b) 1968-present.
PL 17	Appreciation of engineering skills: (a) A; (b) 1918-present.

Notes:
(a) Type of strategic choice

S—scale	PL1
G—geography	PL2,4,6,11,13,14,15
H—horizontal concentration	PL2,3,8
V—vertical integration	PL1-5,10-15
P—product diversification	PL2-4,8,9,11,16
A—administration	PL6,7,13,14,17

(b) origin and duration

Exhibit SC-2
Socal Environmental Conditions

EC 1	California becomes a major producing area with discovery in Los Angeles—1895.
EC 2	Breakup of Standard Oil—1911.
EC 3	Opening of the Panama Canal—August 1914.
EC 4	World War I and postwar tanker surplus.
EC 5	Post World War I geared crude shortage and price rises— 1918.
EC 6	Rise of scientific geology—1920's.
EC 9	Use of torsion balance over water—late 1930's.
EC 29	Great Depression—1929.
EC 30	Crude glut—1930.
EC 31a	Socony-Vacuum merger permitted—1931.
EC 31b	Talks with Jersey Standard about merger—1929-33.
EC 40	World War II—1939-45.
EC 43	U.S. government tries to buy Saudi Arabian concession.

Exhibit SC-3
Socal Resources

R 11	Inheritance from Standard Oil.
R 11a	Approximately 90% market share in California by Standard of Iowa 1885-1900.
R 11b	Standard's acquisition of Pacific Coast Oil Company in 1900. Entry into production, refining, and pipeline transportation in California.
R 11c	Marketing in California, Oregon, Washington, Arizona, Nevada, Alaska, and New Mexico.
R 11d	Dominant market share in refined products (60-90%) in the West— 1900-11.
R 11e	Refinery at Richmond, Cal. 1902.
R 11f	17% share in Western fuel oil market—1911.
R 11g	F.H. Hillman's expertise in exploration and production—1911.
R 19a	Purchases of producing property 1911-19.
R 19b	Discoveries of crude oil in California 1911-19.
R 26	Acquisition of the oil subsidiary of the Southern Pacific Railroad.
R 27a	Discovery of production in west Texas.
R 27b	Refinery at El Paso.
R 28	Concession in Bahrain.
R 29	Refinery at Colorado, Texas.
R 30	Formation of Bapco.
R 32	Discovery in Bahrain.
R 33	Concession in Saudi Arabia.
R 35	Refinery at Vancouver, British Columbia.
R 36a	Refinery on Bahrain.
R 36b	Formation of Caltex.
R 38	Discovery in Saudi Arabia.

R 43a	Developed expertise in offshore exploration.
R 43b	Offshore discoveries in the Gulf of Mexico.
R 47a	Extension of Caltex to Europe.
R 47b	Sale of 40% of Aramco.
R 48	Refinery at Salt Lake City.
R 50	Caltex joint venture refineries in Eastern Hemisphere.
R 57	Beginning of large discoveries in Indonesia.
R 61	Acquisition of Standard Oil of Kentucky.
R 63	Refinery at Pascagoula, Mississippi.
R 75	Purchase of 20% of Amax.

Exhibit SC-4
Socal Handicaps

H 1	Shortage of controlled crude in California—before 1919.
H 2	Lack of operations east of the Rocky Mountains.
H 3	Lack of downstream outlets for crude abroad.

Exhibit SC-5
Socal Personal Values

PV 1	F.H. Hillman—belief in production and discovery of oil.
PV 2	Maurice Lombardi—belief that oil would be found in Saudi Arabia.
PV 3	William H. Berg—belief that oil would be found in Saudi Arabia.
PV 4	Reginald C. Stoner—belief that oil would be found in Saudi Arabia.
PV 5	Clark Gester—belief that oil would be found in Saudi Arabia.
PV 6	"Doc" Nomland—belief that oil would be found in Saudi Arabia.
PV 7	K.R. Kingsbury—autocractic, conservative personal style.
PV 8	R.G. Follis—conservative personal style.
PV 9	O.N. Miller—autocractic personal style.

Exhibit SC-6
Socal Social Responsibilities

SR 1	Bringing western technology and civilization to Saudi Arabia.

Exhibit SC-7
Aramco Sale Terms

**STANDARD OIL COMPANY (NEW JERSEY)
AND SOCONY-VACUUM OIL COMPANY INC.
PURCHASE OF ARAMCO STOCK DECEMBER 2, 1948**

Total Purchase Price	$240,357,845.30
Detail of Cash Flow:	
Purchase of 40% of Aramco stock by Jersey/Socony (additional issue)	466,666.67
Capital contribution to Aramco by Jersey/Socony	101,891,167.43
Total cash paid into Aramco by Jersey/Socony	102,357,834.10

Preferential dividends to Socal/Texas out of Aramco
  earnings subsequent to Jersey/Socony purchase:

1948	$15,000,020.00	
1949	29,999,970.00	44,999,990.00

10¢ per barrel to Socal/Texas beginning
  1950 on first 3 billion barrels production

	300,000,038.00
	345,000,028.00

Less Socal/Texas equity in earnings from which
  preferential dividends paid (60% of
  $345,000,028.00)   207,000,016.80

      Cost to Jersey/Socony in preferential dividends   138,000,011.20

      Total cost to Jersey/Socony $240,357,845.30

Purchase of December 2, 1948 was retroactive to date of basic agreement, March 12, 1947, between Jersey/Socony and Socal/Texas. Aramco borrowed $102,000,000.00 from banks in 1947 which was guaranteed by Jersey/Socony. Out of this sum, $79,765,241.94 of loans was repaid to Socal/Texas and $22,234,730.00 in dividends was paid to Socal/Texas in 1947, a total of $101,999,971.94. In 1948 Jersey/Socony made a capital contribution of $101,891,167.43 to Aramco to repay the bank loans.

Source: Socal records

8

The Strategy of Texaco

This chapter identifies the policies of Texaco's strategy from the events of its history.

8.1 1902-1913: "Buy It in Texas, Sell It up North"

The event (EC2) which triggered the formation of the Texas Company, or Texaco as it was called by its cable address like Socony (R2g), was the discovery of oil in large quantities at Spindletop in 1901, near Beaumont, Texas, by the J.M. Guffey Company, which became Gulf. Joseph S. Cullinan (R1a), a practical oil man who had worked for Standard in Pennsylvania, had struck out on his own and gone to Texas in 1895 where, with the backing of two Standard officials, he formed J.S. Cullinan & Company to develop a small field at Corsicana (EC1). It was the first oil field discovered in Texas and Cullinan built a $150,000 refinery there.[1] A few weeks after the Spindletop strike he formed on paper the Texas Fuel Company with the strategy of buying oil cheaply in Texas and selling it to Standard and other northern refiners (PL1). By comparison with the large-scale expansion of the Guffey company, which had quickly leased 1 million acres and begun building a pipeline and a refinery, it was a strategy for limited capital.

Cullinan's financial backers were not from Standard Oil, due no doubt to its antitrust difficulties in Texas (EC3), but a former Texas governor, James Stephen Hogg, whose Hogg-Swayne Syndicate contributed Spindletop leases (R1b) for $25,000 in stock and a New York financier, Arnold Schlaet, the agent of John J. and Lewis H. Lapham, leather merchants and investors in New York, who supplied $25,000 cash (R1c).

Texas Fuel went into business on January 2, 1902. The initial capital went toward a pipeline, tanks, and a refinery site at Port Arthur (PL2,R2a,b), but not for producing properties. Texas Fuel was not chartered to produce oil—i.e. drill wells—and with a glut on the market Schlaet, whose interests were finance (PV8) and sales (PV9), considered integration backward into production unappealing or at least premature (PV10). Two weeks later, without Schlaet's approval, Cullinan organized an affiliate to produce oil (PL3, R2c), the

Producer's Oil Company, with the backing of the Chicago promotor and industrialist, John W. "Bet-a-Million" Gates (R2d).

The action was to be typical of the abrasive relationship between Cullinan and Schlaet and the style of autocratic leadership which would rule Texaco thereafter (PL4). "Buckskin Joe" Cullinan, whose grandfather had escaped from the Irish potato famine, was a bold, independent field commander who was interested in production and vertical integration (PVl-5) and who had left Standard to run his own show.[2]

He was a straightforward buccaneer, a tall, powerful, bushy eyebrowed leader and a financier of Irish causes who in later years flew a black skull-and-crossbones flag over his offices "as a warning to privilege and oppression."[3] The austere, German-born Schlaet (R2e) was a cautious, financially conservative New York merchant (PV6-l0) whose contributions were East coast and foreign sales organizations (PLll,12) and a tradition of parsimonious financial management of the company for his investors (PLl4,15).[4] Eventually the sparks flew between them.

The ambitions of Cullinan and Schlaet had quickly grown beyond the $50,000 capital of the Texas Fuel Company. The Texas Company was organized on April 7, 1902, to take over the assets of Texas Fuel, including Producer's Oil Company. Though its assets had cost less, the books were valued at $1,250,000.[5] Thus almost from the beginning Cullinan, a one-man show (PL4), and the investors betting on his expertise (Rla), attempted in addition to purchasing and selling crude to build a vertically integrated organization (PL5), integrating upstream into production and downstream into transportation and refining.

Cullinan's judgment about the importance of entry into producing (PL3) was vindicated when Spindletop ceased flowing in August 1902, placing the company in peril. Producer's first great discovery (R3a) at Sour Lake in January 1903 restored the company's fortunes, along with participation in discoveries at Saratoga and Batson in Texas and another bonanza at Jennings in Louisiana, all in 1903 (R3b-d). Producer's, under the genius of Walter Bedford (R2f) until his death at 42 in 1912, participated in all the great new southwestern strikes—Humble (1905), Glenn Pool, Oklahoma (1905), and Goose Creek and Markham fields (1908) near Houston (R5a,b, 8b,c). By 1915 Texaco accounted for 7% of U.S. production.[6]

Texaco's sales during its first two years were mainly crude oil to northern refiners (PL1), crude and later fuel oil (PL6) to railroads (R3f) and Mississippi River sugar planters, and asphalt for roofing (R3g) (PL6), the first product to have a Texaco brand name (R2g) in 1902 (PL7). Since it could not make good kerosene from Texas and Louisiana crude and since Standard was dominant in that market, when its refinery (R3b) at Port Arthur began operations late in 1903 (PL2) it adopted the refining policy of making all of every barrel into profitable products (PL8), rather than the standard practice of refining what salesmen could sell and discarding unwanted byproducts like gasoline and asphalt. Refining would control sales (PL9) and nothing would be wasted.[7] By 1909 its products were sold under the Texaco trademark: a red star with a green *T* (R9b, PL7).[8]

Port Arthur's principal refined products were gas oil and naphtha rather than kerosene (PL10). Gas oil was used to enrich bottled gas for lighting and cooking in the U.S. and Europe. Naphtha, a low grade gasoline, was used in stoves, torches, industrial engines, as a paint solvent, and as fuel for a tiny new market, the 23,000 passenger cars on the roads in 1902.[9]

Both products proved to be of enormous strategic importance in Texaco's growth. Gas oil was the first product exported and led to the development of an international sales organization which attracted Texaco to Socal in Saudi Arabia in 1936. Gasoline was the product which developed Texaco's U.S. market position as the leading retail marketer among the oil companies. By having to develop the markets for these products, Texaco was forced to develop a strong selling organization (PL11).

Establishing strength in transportation (PL2), a policy all oil companies learned from Standard, benefitted other levels of the organization. By 1905, with four tankers and barge lines to New Orleans, Texaco was in a position to buy surplus crude in eastern Texas and move it to market (PL1). By 1910 it had terminals at Bayonne (N.J.), Baltimore, Providence (R.I.), and New Orleans (R10e-h) in addition to Port Arthur. With tankers to provide transport (R5g) in 1905 Schlaet went to Europe and established Texaco's first foreign sales agencies (R5e), aimed at developing the foreign retail market (PL12). By 1913 the company had agencies in Europe, Latin America, Africa, Australia, China, and elsewhere in Asia (R13a-f) selling kerosene, gas oil, fuel oil, and lubricating oil (R3c,f,7b,13g).[10]

A strong pipeline network (PL2) aided vertical integration (PL5). Texaco had extended pipelines to its Texas fields from Port Arthur (R5d). In 1907 when its assets were just under $8 million and income was $1 million Texaco took a huge risk in building a $5.6 million pipeline (R7a) from Tulsa (Glenn Pool) to the Gulf coast, but the line carrying Oklahoma crude for the company and competitors (PL13) was profitable from the start. Oklahoma crude yielded twice as much kerosene of a much higher quality and five times as much gasoline as Texas crude. It allowed Texaco to introduce its first high quality kerosene in 1907 and in 1909 a product called No. 4 Motor Gasoline (R7b, 9a). By 1910 the company also had 1010 rail tank cars and refineries in West Dallas, West Tulsa, and Port Neches (R10a-d). Its refinery runs exceeded its production, a sign of the ascending importance of refining (PL9) at the company. In 1911 a fifth refinery was built at Lockport, Illinois near Chicago.[11] Strong transportation (PL2) and Oklahoma crude (R5b) were important factors in Texaco's rapid growth both at home and abroad.

No company was more vigorous in developing its marketing organization (PL11). At first Cullinan handled sales in the South and West; Schlaet, in the East. In 1908 a sales organization with seven divisions covered all but the five western states, the largest market area covered by any company except Standard (PL16, R82).[12] This policy of almost nationwide marketing (PL16) made it less vulnerable to local price cutting by Standard and was clearly a response to Standard (EC4). Nor can the importance of its reliance on gasoline be omitted (PL10). In the U.S., more than any other oil company, Texaco rose with the gasoline market (EC6) as Standard had with kerosene a generation before (and had remained dependent on it). By 1912 there were 902,000 cars and 41,000 trucks. Gasoline consumption had spurted to 55.6 TBD (thousand barrels per day) from 15.9 TBD in 1902 nationwide. In 1912 Texaco had 5.6% of the market. Sometime after 1935 it became the leading U.S. gasoline retailer until Shell overtook it in 1977.[13]

In 1913 Cullinan resigned and the center of power of the company shifted to New York. He had lost out in a power struggle with Schlaet and the Laphams after the death of Gates and his son, both directors. In 1913 he lost control of the Executive Committee of the Board, then and now the reins of power at Texaco (PL17), which spelled the end of his one-man rule (PL4).[14] Schlaet became chairman of the Executive Committee until 1919. A diplomatic administrator, Elgood C. Lufkin, became president.[15]

The differences between Schlaet and Cullinan were functional and geographical as well as personal (PV1-10). Cullinan, the autocratic leader, founder, and producer in Texas, expanded too rapidly and took on too many activities for the cautious New York financier and marketer. Also the New York sales office had grown to considerable size, a fact Cullinan ignored while he continued to run the company from the field in Texas. Schlaet complained that Cullinan treated New York as the "tail of the dog."[16]

8.2 1914-1933: Refining and the Gasoline Era

During this period Texaco rose with the gasoline market (PL10, EC6), greatly expanding its refining and marketing organizations with the (R2O) Holmes-Manley refining patents for the first continuous (rather than batch) thermal cracking process (PL9) and became the first company to market gasoline in every state (PL16).

One consequence of Cullinan's resignation was a decline in the importance of production at Texaco (PV10, PL13). Its percentage of U.S. production peaked at 7.03% in 1915 and then fell to a low of 2.48% in 1924. C.N. Scott, vice-president of production, did start a successful program to train college graduates to drill wells, a forerunner of the field of petroleum engineering, but without the backing of the board in New York sufficient funds were not available to maintain the company's position in production.[17] In 1916 it did expand producing activities geographically by taking leases in Wyoming (where it found oil) and Montana (where it did not) (R16a). But in Mexico it developed only a small production (R12) and in Venezuela it missed out (PL18).

During World War I (EC5) and possibly because of it, Texaco's drive to be vertically integrated (PL5) began to develop into a policy of self-sufficiency and reluctance to rely on and deal with outside firms (PL19). When it became impossible to buy tankers during the war, Texaco leased a shipyard in Maine to build ships for its fleet. It discovered its own clay and fuller's earth beds for its refineries. It made its own oil cans. It acquired timberlands and a sawmill to make packing cases. For a time it even made its own tank wagons (R17a-e).[18]

When the Burton process, the first commercial thermal cracking process, was placed in operation by Standard Oil of Indiana in 1913 (EC7), it greatly increased the gasoline yield from crude and thereby threatened Texaco's gasoline business. Texaco would either have to purchase an expensive license from Indiana Standard or develop a better process of its own. The disadvantage of the Burton process was that it was a batch rather than continuous process.

Ralph C. Holmes, head of the Refining Department since 1902, and Fred T. Manley developed and put into operation in 1920 a process (R20) named after them to convert the Burton pressure still (EC7) to continuous operation. It was one of four basic patents on continuous thermal cracking (see Chapter 4). Lawsuits over patent claims among the four processes were not settled until 1923 (EC8) when a truce was declared and cross licensing arrangements were made. After having lagged behind the industry average gasoline yield since 1914, with the Holmes-Manley process Texaco gained a lead over the competition after 1924, reinforcing its gasoline product strategy (PL10). Holmes became president in 1926.[19]

From 1920 to 1926 refining and marketing of gasoline and lubricating oils for the motorist were Texaco's principal activities (PL9,10,11,16,21). It extended the geographical spread of its marketing operations to the Rocky Mountains and then nationwide to 46 states by 1926 (R26) although 56% of its sales were in eight states: Texas, New York, Florida, North Carolina, Illinois, Virginia, New Jersey, and Pennsylvania.[20] Still, nationwide marketing (PL16) was a policy no other oil company pursued. It was an extension of the earlier desire to avoid vulnerability to local price cutting by Standard Oil (EC4) in any region. Until the 1920's the former Standard affiliates remained closely linked and there was considerable scepticism as to whether the 1911 divestiture had accomplished anything.

Meanwhile Texaco's production lagged. The surge in gasoline demand from 1910 to 1926 pulled the organization out of vertical balance. Between

1918 and 1924 Texaco bought a substantial part of its gasoline supplies from other refiners and it produced, at the 1924 low point, only 44% of the crude it refined.[21]

There was concern about production and belief in the central importance of refining (PL9) and marketing was not universal among the company's top executive ranks. Amos L. Beaty, a Texas lawyer (PV11), had guided important bills expanding the charter of Texas oil companies to allow interstate commerce through the Texas legislature in 1915 and 1917, and served as president from 1920 to 1926 as Lufkin became chairman. He was the first General Counsel to head one of the Majors and his rise marks the beginning of a long time policy at Texaco of unusual attention to and appreciation of legal matters (PL20). His resignation in 1927, after a year as chairman, was partly due to his dissatisfaction with Texaco's producing efforts (PL18).[22]

Holmes, his successor, believed that the long-run success of the company depended on the efficiency of its refineries (PV12), a narrower view which characterized his administration (PL9). From 1926 to 1933 the company almost doubled its refinery runs.[23] In 1928-1929 it purchased a new refinery at Amarillo (Texas), doubled the size ot its Lockport (Illinois) facility, and built refineries at Cody (Wyoming), El Paso and San Antonio (Texas) (R28a-e). In 1928 Texaco acquired the Galena-Signal Oil Company, a Cullinan company, which included a 20-TBD refinery at Houston, four foreign marketing subsidiaries and other marketing properties (R28f). Also in 1928 it acquired the California Petroleum Company with three refineries in California (Los Angeles, Fillmore, and Coalinga) and one in Montana (Sunburst) with a total crude capacity of 42 TBD. California Petroleum also had gross production of 43 TBD, pipelines, tankers, marine terminals, and extensive marketing properties in the western states (R28g). A third acquisition in 1931 of the Indian Refining Company (R31a), a victim of the depression, added a 16-TBD refinery at Lawrenceville, Indiana, 20% of the gasoline business in Indiana and marketing facilities in five other states. Refineries of unknown origin at Louisville, Kentucky and Craig, Colorado complete the list (R31b-c). Even before the acquisition of California Petroleum Texaco operated more refineries than any other American oil company.[24] What Holmes had done was to provide a large number of relatively small-scale local refineries to supply a strengthened nationwide marketing organization (PL22). Refinery runs began to exceed sales in 1928.[25]

8.3 1933-1953: Rebuilding Production

When Texaco lost money in 1931, 1932, and 1933, there was dissatisfaction with Holmes' emphasis on refining at the expense of production. Though production from 1926 to 1932 had increased from 100 TBD to 165 TBD, due mainly to the acquisition of California Petroleum and a purchase of west Texas leases yielding 12 TBD net production (R28g,h),[26] in 1928 it still amounted to only about half the company's refinery runs (H1). Texaco stood fourth or fifth in crude reserves, second or third in refinery runs, and second in gasoline production. Better vertical balance was desired (PL5).[27]

After bitter and prolonged controversy over these issues Holmes was replaced in April 1933 by William Starling Sullivant "Star" Rodgers as president and Judge Charles B. Ames, a former General Counsel to Texaco (PL20), as chairman. They took steps to boost domestic production (PL25), reversing the trend prevailing since 1913 (PL18). Capital expenditures for the Domestic Producing Department were increased.[28] The department's technical organization was enlarged. From 1933 to 1951 it added 2.8 billion barrels to its U.S. reserves. By 1939 it was second in U.S. reserves (behind Jersey Standard).[29] Refining (PL9) took a back seat, leaving Texaco for the next four decades with the system of a large number of small-scale, local market U.S. refineries which Holmes had put together (R28-31).

New products and services bolstered Texaco's already strong marketing position in the 1930's. During the depression (EC9) refining but not marketing volume had been maintained but low prices produced losses.[30] Texaco had begun marketing aviation (R28i) gasoline (PL24) in 1928.[31] Acquisition of Indian Refining (R31a) had brought important dewaxing patents for making lubricating oil and Indian's established premium product, Havoline Waxfree Motor Oil (R31a, PL21). It was improved by a solvent extraction process developed at Texaco (R34) and marketed in 1934. Texaco's dewaxing and solvent extraction methods were adopted by the industry, even by refiners of Pennsylvania oils. To boost gasoline sales in 1932 Texaco introduced a new product, Texaco Fire Chief gasoline (R32), with a nationwide comedy radio program starring Ed Wynn. It was the first of many advertising compaigns launched on radio and television (PL26) including the Texaco Star Theatre starring Milton Berle after World War II and many Bob Hope specials.[32] A premium lubricant was introduced in 1938 (R38).[33] Advertisement of clean restrooms "Clean Across the Country" at Texaco's 40,000 rental outlets nationwide boosted its marketing business.[34] In 1935 it introduced a Texaco credit card (R35a) honored at all its stations in the U.S. and Canada.[35]

By 1935 domestic operations had recovered from the depression (EC9) and management began to look for production abroad (PL27). One may reason that this step was taken to supplement domestic production, which for years had been less than downstream needs (H1), not to mention those of a foreign network as well, to reduce purchases and increase self-sufficiency (PL19), and to achieve better vertical balance in foreign activities (PL5). These motives appear to have been internal needs rather than any external stimulus.

The only foreign production Texaco had developed before this time was in Mexico and that had declined quickly after peaking in 1920 at 27 TBD and ceased after 1935. It had been developed when Texaco began looking for foreign crude in 1911-1912 during the Cullinan years. Production began in 1912 (R12). In 1928 with the California Petroleum acquisition Texaco had obtained a 50% interest in some Venezuelan leases but no production. Also by 1928 annual reports record some concession acreage in Colombia. No published sources indicate any other foreign exploring or producing activity before 1935 (PL18).[36] Texaco also purchased crude from Venezuelan and Mexican sources and had a topping plant at Tampico.[37]

The following year it implemented its new policy (PL27) to seek foreign production. The quickest way was by purchase. In 1936 Texaco purchased a 50% interest with Socony-Vacuum in the Barco concession in Colombia from Gulf (R36a). In 1940 through a joint venture with Caracas Petroleum it obtained Venezuelan concessions on which it obtained commercial production in 1944 and export production in 1948 (R40,44,48).[38]

But the most important foreign crude acquisition was its 1936 joint venture with Socal to supply Texaco's markets east of Suez (R5c) with Socal's crude from Bahrain. The formation of Caltex (R36b), as the joint venture was called, produced a complete union of their activities east of Suez. As a result Texaco also acquired a 50% interest in Socal's then undeveloped concessions in Saudi Arabia (R36c) and what is now Indonesia (R36d). Enormous discoveries have been made in both areas, though as with Venezuela and Colombia (where exports [R39b] began in 1939) there was little development until after World War II.

All Texaco's foreign producing interests before World War II were acquired (PL27). Though its domestic crude was acquired through exploration, it had not been successful in exploring abroad, except in Colombia and Venezuela to a modest degree.[39]

The joint venture with Socal which made Texaco an international producer as well as a marketer was possible because of the foreign marketing organization Texaco had built up since 1905 (PL12). Its foreign strategy was similar to its domestic strategy (PL10,11). It specialized in products like gas oil and lubricants (PL10, 21). Its share of foreign markets was not large, about 10% east of Suez, for example, but its geographic coverage was wide (PL16). It marketed nearly everywhere. Texaco also acquired production in Canada through its marketing subsidiary. In 1935 it had begun buying stock in McColl-Frontenac (R35b), acquiring a majority by 1948, when production was developed on McColl-Frontenac leases in Alberta. Through a subsidiary of McColl-Frontenac, Texaco also acquired production in Trinidad (R35c). McColl-Frontenac became the third largest marketer in Canada with refineries at Montreal and Toronto.[40]

Despite the central importance of refining at Texaco in the U.S. (PL9), Texaco did very little refining abroad before World War II even though it had widespread marketing interests (PL12). Before the war it refined abroad only in Bahrain and possibly Canada and France.

Another major event before the war concerned Texaco's chairman. Judge Ames had died in 1935 and Torkild Rieber was named chairman while Rodgers remained president. "Cap" Rieber was a Norwegian ship captain, an associate of Cullinan's cut from the same autocratic buccaneering mold (PV13-15, PL4). He rejoined Texaco in 1928 with the Galena Signal acquisition and took charge of Texaco's foreign operations and marine department. A man of great daring, he was responsible for putting Caltex together and, therefore, for Texaco's entry into Saudi Arabia. He had punched a 260-mile Colombian pipeline across the Andes for Texaco. But, as Sampson aptly puts it, "He had a sailor's internationalism, but without any real political instinct: the world was a market with no barriers or taboos."[41]

He was the third chief executive of the Majors to be disgraced by Hitler. He had first gotten into trouble by supplying oil to Spain in 1937 during the civil war, an important contribution to Franco in violation of the U.S. neutrality law. Through Spain he met leading Nazis and agreed to supply oil from Colombia to Germany in barter for 3 tankers. He also gave them his diplomatic support, presenting Göring's plan for the surrender of Britain to Roosevelt. Sensing sympathy, German intelligence agents infiltrated Texaco. Their discovery by British Intelligence disgraced the company. Rieber was compelled to resign, the fourth of six Texaco chief executives to be fired or leave in disagreement over policy. The position of chairman was eliminated and Harry T. Klein, General Counsel (PL20) became Executive Vice-President. Texaco began sponsoring weekly radio broadcasts of the Metropolitan Opera without advertising at this time. It was a rare exception to the company's policies of acting only for profit (PL14,15).[42]

Like other American multinational companies Texaco worked with special diligence to reaffirm its loyalty during the war (EC10). It was the second leading producer of aviation fuel (PL24). In 1942, after tankers were sunk by German submarines within sight of the Florida coast, Harold Ickes, Roosevelt's Secretary of the Interior and Petroleum Administrator for War, decided to build a 24-inch pipeline to carry crude from east Texas to Pennsylvania and later a 20-inch line to carry products from Texas to New York. The president of Texaco's pipeline subsidiary, Burt E. Hull, supervised construction of the "Big Inch" and "Little Big Inch" pipelines (R43a,b). He had helped survey Texaco's first line from Tulsa to the Gulf coast in 1906 and his reputation is evidence of Texaco's strength in pipelines (PL2). The company lost nine of its 40 plus tankers to enemy action including two of its ten foreign flag (Norwegian, from Rieber) fleet. Almost 2/3 of all tonnage shipped overseas during the war consisted of oil products. Texaco also produced toluene for TNT at its Lockport, Illinois refinery. The Atlantic Refining Company, Gulf, Pure Oil, Socony-Vacuum, and Texaco formed the non-profit Neches Butane Products Company (R43c) to manufacture butadiene for synthetic rubber from butylene at a government plant at Port Neches (where Texaco had a refinery) during the war.[43]

The years after World War II for Texaco were times of smooth expansion. It had rebuilt its domestic crude ratio (R39a) to almost 3/4 of refinery runs. Its U.S. refining and marketing organization was in place. Abroad it had new sources of production in Canada (1948),[44] Trinidad, Venezuela (1943), Indonesia (1939), New Guinea and, above all, Saudi Arabia. Its strategic deficiencies, which had divided the organization since Cullinan resigned in 1913, were remedied. Star Rodgers and Harry Klein moved up to chairman (recreated) and president in April 1944. They worked together more harmoniously than any of their predecessors with Rodgers handling operations and Klein, legal, tax, administrative, and service matters. They retired together early in 1953.[45]

In contrast to their counterparts at Socal, BP, Gulf, and possibly others who thought the crude glut of the 1930's would continue in the expected postwar recession, Rodgers and Klein were optimistic. As a consequence they were far less interested than Socal in sharing Saudi Arabia with Jersey and Socony-Vacuum.[46] Political uncertainties in the area (EC11-12), where huge capital expenditures would be required, rather than lack of outlets for Saudi crude, were apparently the factor which persuaded them to agree to the Aramco sale. When President Truman announced the policy of granting military and economic aid to Greece and Turkey on March 12, 1947, thus committing the U.S. to containing communism in the Middle East, the same day the Aramco sale papers were signed, Rodgers told a Senate Committee, it was "absolutely coincidental . . . we were amazed . . . when we heard it. If I had known about it, I think we would probably have held up our arrangements with Standard Oil of New Jersey and Socony-Vacuum."[47] In a transaction related to that sale Texaco sold its European and west African marketing organizations to Caltex (R47), leaving it only limited foreign marketing networks in Canada (McColl-Frontenac), the Caribbean, and South America. The entire Eastern Hemisphere was administered by Caltex.

Their optimism proved correct. There was a postwar boom worldwide. Texaco built a new refinery near Camden, New Jersey in 1949 (R49), its first on the Atlantic coast, expanded capacity elsewhere, and closed or converted four refineries at Craig, Colorado, Cody, Wyoming, West Dallas, and San Antonio, Texas to sales terminals.[48] Abroad they participated through Caltex or other partners in refineries in Puerto La Cruz, Venezuela (1949-30 TBD), Rotterdam, Netherlands (20 TBD-1950), Cartagena, Spain (1950-24% Caltex), Bordeaux, France (13 TBD-1950), and Milan, Italy (1952-10 TBD-50% Caltex with Fiat).[49] These refineries followed the U.S. pattern of small scale and geographical dispersion (PL22). But the emphasis on refining abroad was clearly different from that in the United States and small scale (PL28). They existed not as a principal activity (PL9) but to move crude from the producing to the marketing organization with minimum investment (PL29).

8.4 1953-1976: The Long Years

Augustus C. Long became president of Texaco in 1953.[50] Had his middle initial come first it would surely have stood for Caesar, for no chief executive in any industry has ever dominated a large company so completely for so long as did Gus Long.

He did not build Texaco. Its resources and strategy were already in place when he took over. Rodgers and Klein had left it in good shape. Its crude reserves had been rebuilt since the 1930's (PL25, R39a). The system of numerous local U.S. refineries (R28-31) was Holmes' contribution (PL22). Texaco had pursued nationwide retail marketing (PL16) since at least 1908 (R8c) and had achieved it in 1926 (R26). In 1952 refinery runs (527 TBD Western Hemisphere, 303 TBD U.S.) and marketing volumes (541 TBD) in the Western Hemisphere were nearly equal, while net production (334 TBD Western Hemisphere, 303 TBD U.S.) supplied 63% and gross production (395 TBD Western Hemisphere, 359 TBD U.S.), 75% of crude requirements. In the Eastern Hemisphere in 1952 Texaco's share of the production (R36b) of Aramco (247 TBD) and Caltex (Bahrain, 15 TBD and Indonesia, 16 TBD) was soaring, while its shares (PL28) in the refinery runs of Aramco (51 TBD) and the Caltex Group (126 TBD) struggled to move the crude (PL29). As in the U.S. marketing volumes were approximately equal to refinery runs.[51] Overall Texaco was well balanced (PL5).

What Gus Long did was to operate it brilliantly. During his administration, as *Forbes* noted, "No oil company management had a greater talent for doing all the things that oil men most prize, whether it's sipping bourbon, spudding a well, outfoxing the competition, or telling the government where to shove its latest bright idea."[52] From 1959 to 1971 Texaco was the most profitable Major, exceeding the growth rates and rates of return on assets and equity of the other six.[53]

It was not by chance. He squeezed every possible nickel of profit from the organization. In addition to a well-balanced system (PL5) that could move volumes of low cost crude (PL16,22,23,29), parsimony was the chief financial policy (PL14). Cost control and profit improvement were company watchwords. Texaco executives worked longer and harder for less pay than their counterparts at other companies. Texaco negotiated tough contracts with suppliers and partners. It was said to have sent an important telegram involving millions of dollars to a host government by the cheap night-letter rate. It refused to send its experts to industry meetings. It did without layers of staff who supplied the planning, controls, and forecasts and other accoutrements of modern management at other companies. Recently when it moved its headquarters to a new executive office building outside New York City it neglected to tell its employees that the workday would be 45 minutes longer for the same pay. Pennypinching became a cult within the company.[54]

Autocracy under Gus Long returned to a degree even Cullinan would have found hard to believe (PL4). He was a 1926 Annapolis graduate who joined Texaco in 1930 after deciding he would never become an admiral. A giant, exuberant Southerner, Gus Long was a granite-faced salesman who made his reputation as a "no" man and a hard-driving competitor (PV16-19). He attracted other Naval Academy graduates. The quasi-military style which resulted reinforced his authority. Deference to the Commander (his World War II rank), secretiveness, and uniformity of opinion blinded the company to events in the outside world. "Gus Long says there's no oil in Alaska" was a company dictum long before the North Slope discovery at Prudhoe Bay. So, for Texaco, there wasn't.[55]

He was Mr. Texaco in more than name. Every capital expenditure of more than $10,000 required his personal approval.[56] Any of Texaco's 40,000 stations could not be built or equipped without his okay. Negotiations involving Texaco were delayed by the constant referral of decisions to headquarters for approval.[57] The degree of centralization of authority under Gus Long was unrivaled at any major company. Unencumbered by staff or dissent, he was the organization (PV20).

Texaco achieved greater devotion from its employees than any other Major in spite of their treatment. It had the strongest retail network of franchised dealers (PL11), nationwide distribution (PL16), a large credit card operation (R35a), engaged heavily in radio and television advertising (PL30), and projected the image of a winner. The uninformed may be forgiven for mistaking Bob Hope for the president of Texaco and the highway patrol for its dealers. Its contracts were reputed to be unbreakable. And it backed them to the hilt. A competitor recalled with awe that Texaco once took a dispute over a single dealer to the U.S. Supreme Court. It had more vice-presidents than any other Major (almost all in New York, of course) (PL4).[58]

Texaco was geniunely hated by its competitors. "We all hate Texaco," an Exxon man said.[59] This was partly earned for its habits above: cutthroat competition, secrecy, parsimony, pettiness, short sightedness, and financial success. But mostly it was for its go-it-alone attitude, its smug self-sufficiency, its reluctance to exchange anything with other members of the industry (i.e., product swaps, industry associations) (PL19), and its unabashed independence from any consideration of anything that did not favor the narrowest definition of its short-term, profit-maximizing self-interest (PL14). As Gus Long used to say, "We have never run with the herd."[60]

Specific incidents are illustrative. When the rest of the industry was prepared to unite in resistance to the wage demands of the Oil, Chemical, and Atomic workers, Texaco broke ranks and worked out a settlement that the others were forced to accept. It refused to support an American Petroleum Institute campaign advertising that the U.S. was running out of oil. But its most important break with the industry came in 1970 when with Socal it was the first of the Majors to cave in to Libya, paving the way for the spread of similar terms to the Middle East.[61]

In the U.S. during Long's administration Texaco exploited its 50-state marketing system (PL16, R26), its strength in gasoline and lubricants (PL10,21), its strong dealer network (PL11), and the new interstate highway system begun in the mid-1950's to move huge volumes of its low cost domestic crude (R39a). It put the Texaco Star (R9b) all along the 40,000 mile system, provided full service at each station (PL21), linked it all together with its credit card operation (R35), and reinforced it with national radio and TV advertising.[62]

Its refineries were relatively inefficient, designed to serve local markets and use local crudes (PL22) and therefore inherently more expensive to operate. They lacked the economies of scale which lowered costs and the flexibility to handle different types of crude and vary product mix as widely as the 100 to 200-TBD regional refineries being built by competitors. But with adequate supplies of low cost crude they were unnecessary. Texaco's refineries were inexpensive to build and produced a full range of products (PL8).[63]

Gus Long explained Texaco's success in terms of its balance. This meant three things at Texaco. First it meant vertical balance: producing about as much as it refined and refining about as much as it sold (PL5). It did not have to buy large amounts of high-priced U.S. crude as Mobil did, nor cut prices to move crude in a glutted market as Gulf did in the U.S.[64] and the crude heavy companies (Gulf, Socal, BP) did abroad. Texaco was not a price cutter. Second, it meant geographical balance (PL11,12,16,27,29). Texaco was the leading marketer hardly anywhere. But it aimed for and generally achieved the second or third largest position, perhaps 10% of the market, everywhere (PL23). Anywhere it had crude or products it could market them with minimum interaction with competitors (PL19). Third, balance meant marketing every product produced from its crudes (PL8): fuel oil, distillates, aviation fuel, gasoline, lubricants, asphalt, and so on right down to, as Senior Vice-President Alfred C. DeCrane put it, "the squeal of the rig."[65]

What Texaco did so well at home, Long sought to do abroad. Thanks to Socal's discoveries and its partnership in Caltex, Texaco had access to almost limitless supplies of low cost Eastern Hemisphere crude in Saudi Arabia, Iran, Bahrain, Indonesia, and later Libya (R36,54,63). Caltex built small-scale, local-market refineries (R60) everywhere on the Texaco model: minimum investment to move the maximum volume of crude (PL29). In the Western Hemisphere it became the third largest marketer in Canada. It became the largest producer in Trinidad and Colombia and had production in Venezuela as well. It developed a special affinity for the oil bearing possibilities in the Amazon basin straddling Colombia, Peru, and Ecuador (on the west side of the Andes), eventually developing a sizable find with Gulf in Ecuador (R68). It refined locally and marketed everywhere, balancing production, markets, and products (PL8,12, 27,28).

Maintaining balance required imagination. Gus Long, the marketer (PV19), matched crude with markets in a series of brilliant tactical moves. In 1956 Texaco acquired Trinidad Oil (R56) and expanded its refinery to 350 TBD to process large volumes of crude from Saudi Arabia. Texaco then had its offshore refinery like those of Exxon and Shell in Aruba and Curacao. It shipped the gasoline to Trinidad Oil's marketing outlets in Britain and the fuel oil to the U.S. It acquired Paragon Oil Company in 1959 (R59) to distribute fuel oil, making Texaco nearly as large a factor in home heating oil as in gasoline. Acquisition of Seaboard Oil Company (1958) added reserves in the U.S., Canada, and Venezuela; the TXL Oil Corporation (1962) brought 2 million promising acres in west Texas; and the Superior Oil Company of Venezuela added substantial production in Lake Maracaibo (R58,62,64).[66]

The Trinidad Oil acquisition gave Texaco a larger market share in Britain, but in so doing it destabilized the Caltex partnership (R47) in Europe, which had been put together in 1947. With Socal and Texaco equal partners in Europe, transfer prices for Middle Eastern crude affected both equally. But when the Trinidad Oil acquisition changed the important U.K. market to 75% Texaco/25% Socal, disputes arose. Socal did not want to buy into more marketing, desiring instead to fund exploration activities in Africa, while Texaco desired a larger market share in Europe. After 10 years of friction Caltex broke up in Europe in 1967. Texaco acquired Deutsche Erdöl in Germany (R66) in 1966 and several other European marketers. With its share of Caltex's European marketing it ended up with about 7% of the European market including strong positions in Germany and Britain. It had established a position similar to that in the U.S. As in the U.S. it did not make money in marketing, but its system moved a lot of Middle Eastern crude (775 TBD Eastern Hemisphere in 1967) on which profits were high. It had balanced its operations again. Worldwide in 1967 Texaco produced 2318 TBD, refined 2155 TBD, and sold 2369 TBD.[67]

Gus Long gets the highest marks for profitable financial performance and tactical or operational virtuousity, two of three areas by which managers may be evaluated on the scale of history. These two are the short run areas. In the third area, long run strategic planning, the measure of which is the state of the organization he leaves to his successors, he fares less well.

Under Gus Long Texaco did not build for the future as well as did competitors. The company was run for short-term profit (PL15). Profit improvement, a Texaco watchword, meant cutting costs and doing without rather than investing for higher returns.[68] Lulled by success into believing the world in which it was the best competitor was eternal, it did not plan or invest for other possibilities.

It would continue to find crude where it had always found crude in the U.S. (PL3,25). Abroad, except for the world wars, crude had not been in short supply since the beginning of the industry or since Russian discoveries led to price wars with Standard in the 1880's. Consequently, Texaco's exploration efforts abroad were probably the least vigorous of the Majors during the 1960's. It did not begin to look for higher cost alternatives to low cost domestic and Middle Eastern crude in the North Sea, North Slope, or off the U.S. Gulf coast as did BP after the Iranian nationalization, Gulf or Exxon because they worried about the security of Middle Eastern supplies, Shell and Mobil because they were crude short, or Socal because it was interested in exploration. As Maurice F. Granville, Long's successor in 1972 acknowledged, "We didn't feel the need to make major capital investments in exploration outside the U.S. to the degree that some of our competitors did, or even in some of the early offshore activities."[69] In those areas, offshore California, offshore Texas, the Gulf of Alaska, and the Amazon basin, where it made major efforts, results were disappointing. Texaco was caught unprepared when its U.S. production peaked out at 940 TBD in 1971 (R39a) (94% of refinery runs) and declined thereafter (65% of refinery runs in 1976) and its foreign sources were nationalized beginning with Libya and continuing to the present (H1).[70]

Neither did it upgrade its refining and marketing facilities to make a profit. When it began to run short of domestic crude it was forced to pay a $1 or $2 a barrel premium for low sulfur oil because its refineries lacked the flexibility (H2) to use lower priced high sulfur crudes. The losses from its fifty-state marketing network of full service stations, which numbered 40,000 at the peak, were tolerable when crude was cheap, but no longer are. Nearly 10,000 have been closed in the last decade, 5,000 since 1975. In 1975 Texaco gave up its commitment to fifty-state marketing (PL16), pulling out of Montana, North Dakota, and parts of Idaho and Washington and soon Minnesota and four other midwestern states. Its three small 20-TBD refineries in Casper, Wyoming, Amarillo and El Paso, Texas are in trouble, and their troubles may foreshadow a withdrawal from west Texas and the Rocky Mountain region as well. Texaco has a long way to go. It served 7.5% of the U.S. gasoline market with 30,000 stations while Shell did the same job with only 18,000.[71]

Nor did Texaco diversify very much (PL30). Its primary petrochemical investments were two joint ventures it shared equally with partners. The Jefferson Chemical Company with American Cyanamid as partner began operation in 1948 at Port Neches to produce permanent antifreeze. The Texas-U.S. Chemical Company was a joint venture with the U.S. Rubber Company organized in 1954 to produced synthetic rubber for tires. Both of these diversifications seem related to Texaco's full-service product policy (PL21,8) and its previous history of self-sufficiency (PL19). It has made large investments in petrochemicals beginning only in 1974. It has recently acquired interests in coal gasification, coal lands, tar sands, oil shale, and agriculture. It sold its fertilizer operation in 1976. It is nearly the least diversified of the Majors measured by non-oil income and assets.[72] Without large-scale refineries it could not generate sufficient volumes of petrochemical feedstocks (which in total are 5% or less of refinery volumes) to justify the large fixed investment required at most locations (H2). Nor was it the policy under Gus Long (PL14) to make such investments.

The low cost crude supply to which Texaco's system was so finely tuned by Gus Long came to an end after 1970. As it receded, the inefficiencies of the refining and marketing system, which had been submerged beneath the flood of crude for so long, became visible. Without vertical balance (PL5), geographical and product balance (PL8,16) were not always assets. Texaco fell from first to sixth in profitability.[73] After the Long years Texaco had a lot of rebuilding to do.

It is a measure of the Gus Long's total dominance of Texaco (PL4) that even after his retirement, many only half believe that he is no longer running the company. In 1964 he retired as chairman and was succeeeded by J. Howard Rambin. But he remained chairman of the executive committee and retained control as Schlaet had done (PL17). Rambin was granted early retirement on September 10, 1970, continuing the tradition of Cullinan, Beaty, Holmes, and Rieber. Long formally regained the title of Chief Executive Officer in addition to Chairman of the Executive Committee. On December 31, 1971 he retired again and Maurice F. Granville became Chairman of the Board and Chief Executive Officer, but Long remained chairman of the Executive Committee until he retired as a director in April 1977. Even though corporate headquarters have been moved from the Chrysler building on 42nd street to Westchester County, at 73 Gus Long remains a $50,000 a year consultant with offices in the old building just in case he should be needed. Perhaps the sceptical are right. Giant men cast long shadows. Gus Long's influence on Texaco was so overpowering that it may appear that he is controlling policy for decades to come.

The rebuilding job has begun. It is required in every function. The marketing operation as noted above is being slimmed down and the policy of geographical coverage is changing. Texaco has invested $200 million to convert its refineries at Port Arthur, Texas and Convent, Louisiana to run on high sulfur crude. It took a 27% interest in the $350 million Louisiana Offshore Oil Port (LOOP), which would have allowed the largest tankers (VLCC's) to dock in the U.S. for the first time in 1980. It has stepped up exploration in the U.S. Gulf, the North Sea, and Canada. It has arranged participation agreements in Nigeria, Ecuador, and Angola and service contracts with Venezuela and Saudi Arabia. But the profits on these contracts will be only 15 to 21¢ per barrel, one-third to one-half the level in the heydays of the 1960's. But in production these are stopgap efforts. As Senior Vice-President Richard Palmer explained, "At some level we hope to stabilize our production. We're not sure what the level is."[74] Texaco clearly has a long way to go.

By now, the reader may agree with the conclusion that Texaco's current status and strategy, if not unique, are at least not identical with those of the other six major oil companies in this study and are not unrelated to its past. There would appear to be important determinants of its behavior and performance present which it would be difficult to derive from the simple phrase "profit maximization," even in this most profit maximizing company. These policies, set forth in the exhibits that follow, show Texaco to be a legalistic (PL20), autocratically and centrally administered company (PL4), long concerned with vertical balance (PL5), wide rather than deep geographic coverage (PL16,23), full product line service and brand name advertising (PL8,11,21, 26), the U.S. gasoline market (PL10), parsimonious financial management (PL14,15), and self-sufficiency to the point of aversion to any transactions with outsiders. Only its customers and shareholders would say, "Vive Texaco."

Exhibit TX-1
Texaco Policies

PL 1	Buy crude in texas and sell it to northern refiners: (a) V,G; (b) 1902-6.
PL 2	Transport (pipelines and tankers) and refine Texas crude: (a) V; (b) 1902-present.
PL 3	Discover and produce oil in Texas: (a) V; (b) 1902-present.
PL 4	Autocractic, centralized executive leadership: (a) A; (b) 1901-present.
PL 5	Vertically integrate to balance activities at all levels: (a) V; (b) 1903-present.
PL 6	Produce crude, fuel oil, and asphalt: (a) P ; (b) 1901-approximately 1910-20; lingering effects to present.
PL 7	Use Texaco as a brand name and a red star with a green *T* as a trademark: (a) A; (b) 1901 (TEXACO brand name), 1909 (trademark)—present.
PL 8	Make profitable products out of all parts of every barrel of crude, rather than refining what salesmen could sell and discarding the unwanted byproducts: (a) P; (b) 1903-present.
PL 9	Refining as principal activity: (a) V; (b) 1909-1933.
PL 10	Product specialization in gas oil and naphtha (gasoline) rather than kerosene: (a) P; (b) 1903-present.
PL 11	Develop a strong sales organization: (a) V; (b) 1905-present.
PL 12	Develop an international retail marketing organization: (a) G,V; (b) 1905-present.
PL 13	Transport competitors' crude as well as its own: (a) V,P; (b) 1907-early years.
PL 14	Parsimonious financial management: (a) A; (b) 1902-present.
PL 15	Maximization of investor wealth: (a) A; (b) 1902-present.
PL 16	Nationwide domestic retail marketing: (a) G,V; (b) 1908-1976.
PL 17	Control through the executive committee of the board: (a) A; (b) at least 1913-19 and 1965-77.
PL 18	Decreased emphasis on production: (a) V; (b) 1913-1933.
PL 19	To be self-sufficient: reluctance to deal with or rely on outsiders: (a) V,A; (b) 1915-present.
PL 20	Appreciation of the importance of legal matters: (a) A; (b) 1915-present.
PL 21	Produce lubricants and provide full-service marketing: (a) P; (b) 1913-present.
PL 22	Large number of small-scale, local-market U.S. refineries: (a) S,G: (b) 1928-present.
PL 23	Substantial (second or third) but rarely leading market share in geographical markets: (a) H,G, (b) 1902-present.
PL 24	Produce aviation gasoline: (a) P; (b) 1928-present.
PL 25	Boost domestic production: (a) V; (b) 1933-present; really a return to PL3.
PL 26	Nationwide radio and television advertising: (a) A; (b) 1932-present.
PL 27	Seek foreign production through acquisition: (a) G,V; (b) 1935-46.

PL 28	Small-scale, partial interests, joint ventures in foreign refining: (a) G,V,A; (b) 1946-present.
PL 29	Design foreign refineries to move the maximum volume of crude with minimum investment: (a) A,V,P: (b) 1946-present.
PL 30	Little diversification: (a) P; (b) 1946-present.

Notes:

(a) type of strategic choice

S—scale	PL22
G—geography	PL1,12,16,22,23,27,28
H—horizontal concentration	PL23
V—vertical integration	PL1-3,5,9,11-13,16-19,25,27-29
P—product diversification	PL6,8,10,13,21,24,29,30
A—administration	PL4,7,14,15,17,19,20,26,28,29

(b) duration

Exhibit TX-2
Texaco Environmental Conditions

EC 1	1895: Discovery of oil at Corsicana, Texas.
EC 2	1901: Discovery at Spindletop.
EC 3	1900-9: Standard Oil's antitrust difficulties in Texas.
EC 4	1870-1911: Standard Oil.
EC 5	1914-18: World War I.
EC 6	1908-24: Rise of the automobile: the gasoline market.
EC 7	1913: Burton process placed in operation by Indiana Standard.
EC 8	1923: Settlement of lawsuits over continuous refining patents.
EC 9	1930's: Great Depression.
EC 10	1939-45: World War II.
EC 11	1946: Political uncertainty in Saudi Arabia.
EC 12	1946: Russian influence in Middle East.

Exhibit TX-3
Texaco Resources*

R 1a	Joseph H. Cullinan (1901-13).
R 1b	Hogg-Swayne leases at Spindletop for $25,000 stock (probably 1901).
R 1c	$25,000 cash from Schlaet and the Laphams (late 1901).
R 2a	Pipeline from Spindletop to Port Arthur (1902).
R 2b	Tanks and refinery site at Port Arthur (1902).
R 2c	Producer's Oil Company (1902).
R 2d	Financial backing of John W. Gates (1902-13).
R 2e	Arnold Schlaet (1902).
R 2f	Walter Bedford Sharp—head of Producer's Oil Company (1902-12).

* Note: Resource numbers (R##) are generally the years 19## when the resource became significant to Texaco or when it was acquired.

R 2g	Brand name TEXACO (1902).
R 3a-d	Discoveries at Sour Lake, Saratoga (Tx.), Batson (Tx.) and Jennings (La.) (1903).
R 3e	Refinery at Port Arthur (1903).
R 5a	Production at Humble field (1903).
R 5b	Production at Glenn Pool, Okla. (1905).
R 5c,d	Tanker fleet and pipelines.
R 5e	Foreign sales organization (1905).
R 3f-i	Product strengths in fuel oil, asphalt, gasoline, and gas oil, the products of the Port Arthur refinery (1903).
R 7a	Pipeline from Tulsa to the Gulf coast (1907).
R 7b	Familylite—Texaco's first high quality kerosene (from Oklahoma crude) (1907).
R 8a	Strong domestic sales organization marketing in all but 5 western states (1908).
R 8b,c	Production at Goose Creek and Markham fields (1908).
R 9a	No. 4 Motor Gasoline (1909).
R 9b	Texaco trademark: red star with green *T* (1909).
R 10a	Rail tank cars.
R 10b-d	Refineries in West Dallas, West Tulsa, and Port Neches (Tx.) (1910).
R 10e-h	Marine terminals at Bayonne, N.J., Baltimore, Providence, and New Orleans (1910).
R 12	Mexican production (late 1912).
R 13a-f	Sales agencies in Europe, Latin America, Africa, Australia, China, and elsewhere in Asia (1913).
R 13g	Texaco produces lubricating oil.
R 16	Leases in Montana and Wyoming; production in Wyoming (1916).
R 17a-e	Shipyard in Maine, clay and fuller's earth beds, oilcan manufacturing facilities, timberlands and sawmill to make own packing cases, tank wagon manufacturing facilities (1914-8).
R 20	Patents on Holmes-Manley continuous thermal cracking process (1920).
R 26	Marketing operations in 46 states (nationwide) (1926).
R 28a-e	Refinery at Amarillo, Tx. purchased, size of Lockport, Ill. refinery doubled, refineries built at Cody, Wyo., El Paso, Tx., and San Antonio, Tx. (1928-9).
R 28f	Acquisition of Galena-Signal Oil Company (1928).
R 28g	Acquisition of California Petroleum Company (1928).
R 28h	Purchase of west Texas leases yielding 12 TBD.
R 28i	Texaco produces aviation gasoline (1928).
R 31a	Acquisition of the Indian Refining Company (1931).
R 31b,c	Refineries at Louisville, Ky. and Craig, Colo. (1931).
R 31a,d	Havoline Waxfree Motor Oil (from 31a).
R 32	Texaco Fire Chief gasoline (1932).
R 34	Texaco develops a solvent extraction process for refining lubricating oil (1934).
R 35a	Texaco credit card (1935).
R 35b	Interest acquired in McColl-Frontenac (1935).
R 35c	Production in Trinidad through a subsidiary of McColl-Frontenac (1935).

R 36a	50% of Barco concession in Colombia acquired from Gulf (1936).
R 36b	Joint venture with Socal—formation of Caltex (1936).
R 36c	Interest in Saudi Arabian concession acquired from Socal.
R 36d	Interest in Indonesian concessions acquired through Caltex.
R 36e	Production in Bahrain acquired through Caltex.
R 36f	Refinery in Bahrain acquired through Caltex.
R 38a	Texaco Sky Chief Gasoline (1938).
R 38b	Discovery in Saudi Arabia.
R 39a	U.S. crude reserves rebuilt.
R 39b	Export production from Colombia begins.
R 40	Venezuelan concessions acquired from Caracas Petroleum.
R 43a	Participation in Big Inch crude pipeline.
R 43b	Participation in Little Big Inch products pipeline.
R 43c	Joint venture to manufacture butadiene at Port Neches.
R 44	Commercial production in Venezuela.
R 47	Caltex in Europe.
R 48	Export production from Venezuela.
R 49	Camden, New Jersey refinery.
R 50a	Refinery at Bordeaux, France.
R 50b	Refinery at Cartagena, Spain.
R 51	Refinery at Edmonton, Alberta, Canada.
R 52	Refinery at Milan, Italy.
R 54	Production in Iran.
R 56	Trinidad Oil acquired: production in Trinidad, marketing in U.K.
R 58	Seaboard Oil Company acquired with production in U.S., Canada, and Venezuela.
R 59a	Paragon Oil Company acquired.
R 59b	Discovery in Libya.
R 60	Caltex refineries in E.H. (1950's, 1960's).
R 62	TXL Oil Corp. acquired—2 million acres in west Texas leases.
R 63	Export production from Libya.
R 64	Superior Oil Company of Venezuela acquired—production in Lake Maracaibo.
R 66	Deutsche Erdol acquired (German marketer).
R 68	Discovery with Gulf in Ecuador.

Exhibit TX-4
Texaco Handicaps

H 1	Crude short after 1908 in the U.S.
H 2	Small-scale, inflexible refining system.

Exhibit TX-5
Texaco Personal Values

PV 1	Cullinan—bold
PV 2	Cullinan—independent and autocratic
PV 3	Cullinan—field leader, frontiersman
PV 4	Cullinan—interest in production
PV 5	Cullinan—desire for vertical integration
PV 6	Schlaet—cautious, conservative
PV 7	Schlaet—Eastern background and outlook
PV 8	Schlaet—interest in finance
PV 9	Schlaet—interest in sales
PV 10	Schlaet—lack of interest in production
PV 11	Amos L. Beaty—Texas lawyer who believed in the importance of production
PV 12	Holmes—belief in the central importance of refining
PV 13	Rieber—Norwegian sea captain
PV 14	Rieber—internationalism, no barriers to trade
PV 15	Rieber—autocracy
PV 16	Long—hard-driving competitor
PV 17	Long—parsimonious
PV 18	Long—autocracy—naval background—military style
PV 19	Long—marketer
PV 20	Long—centralization—no staff, make all decisions personally.

Exhibit TX-6
Texaco Social Responsibilities

SR 1	Sponsorship of Metropolitan Opera broadcasts (1940- present)

9

The Strategy of Royal Dutch/Shell

9.1 Overview

The Royal Dutch/Shell Group (generally known as Shell after its seashell trademark) is the result of a 1907 joint venture between the Royal Dutch Petroleum Company (60% ownership) of the Netherlands and the Shell Trading and Transport Company, Limited (40% ownership), a British company. As a joint venture with dual nationality it has resolved the inevitable coordinative problems by decentralizing authority geographically to its subsidiaries to a greater degree than possibly any other Major (PL17)[1]. Decentralization was also a response to Sir Henri Deterding of Royal Dutch who, with Marcus Samuel of Shell, was responsible for the joint venture and who ruled it with brilliance and tyranny for nearly 30 years. Under Deterding, Shell was Standard's and later Jersey's nemesis, the second largest oil company in the world. Multinational and worldwide from its beginning when it shipped refined products in bulk from Russia and the Dutch East Indies to Europe and Asia, it served its markets by taking an interest in every major producing area until World War II—the Dutch East Indies, Russia, Romania, Mexico, the U.S., Venezuela, and Iraq—and matching low cost production with nearby markets. This policy (PL6) contrasted sharply with Standard's practice until the late 1920's of serving its foreign markets primarily with exports from the U.S. It was Shell's great misfortune (H3) not to share, except in Iraq, in the great discoveries in the Middle East. It had to be content with long-term contracts, principally with Gulf from Kuwait. Locked out of Saudi Arabia, Iran (until 1954), and Kuwait, Shell seemed to lose its expertise in finding oil after World War II and has gone to great lengths to attempt to regain it. This loss forced it to rely on its large-scale, worldwide refining and marketing organization to move purchased crude. Shell specialized in lubricants, jet fuel, gasoline, specialty products, and service in the attempt to add value to crude and, partly as a consequence, developed the largest position in petrochemicals among the oil companies (PL2-8,19,20).

In 1976, as shown in Table 1-2 (Chapter 1), Shell is still the second largest Major, nearly as large as Exxon, with the largest fleet and a 60% exposure downstream to the Eastern Hemisphere. Also shown in Table 1-2 are the policies described above: decentralization, crude shortage, position in chemicals, and product mix. Other than chemicals Shell's principal diversifications are metals through the acquisition of a Dutch multinational metal firm and a half interest in the losses of Gulf's General Atomic subsidiary.

The strategies of Exxon and Shell are perhaps more similar than those of any other two Majors in their global scope, vertical emphasis on refining and marketing, and degree of decentralized management. They show that some degree of homogeneity of strategies is possible among competitors from different origins and circumstances. But there are also differences, with Shell being weaker in production but stronger in retail marketing, lighter products, chemicals, tankers, and in the Far East and with a more locally autonomous, indigenous organization (PL3,4,5,7,10,11,19,20).

9.2 The "Group"

To understand Shell's strategy one must first understand its organization. As with BP there are significant differences from what American readers at least will regard as the norm for the top echelon of the organization. The usual pyramidal organization composed of layers of vice-presidents of varying degree heading the operating divisions and culminating with a chief executive and board of directors who meet monthly is absent. In its place are two parent companies, the Royal Dutch Petroleum Company and the "Shell" Trading and Transport Company, Limited, each with its own board of directors. The Dutch and British parents hold stock in a 60:40 ratio in two holding companies, one Dutch, the other British, both called Shell Petroleum more or less. These two holding companies hold stock in equal amounts in some 500 operating companies and ten service companies. Top management of the Group is composed (in 1976) of eight Managing Directors and below them a number of geographical and functional Coordinators. "Managing Director" is a nineteenth-century European title (EC2) for the chief executive who manages a business for his investors. The Committee of Managing Directors is headed by a chairman and a vice-chairman. This organization is partly European, partly post-Colonial, partly McKinsey, and partly expedient. It is distinguished from its American competitors by being a joint venture with unequal partners, two nationalities, and an unfamiliar management structure (EC1,2).

The parent and operating companies seem more familiar. The "Shell" Trading and Transport Company, Limited is a British company whose shares, traded publicly, are almost entirely held in the U.K. The Royal Dutch Petroleum Company is Dutch. Its shares are traded widely with a majority held outside the Netherlands. But by law its chairman and directors are Dutch. National identities are preserved by design (PL14). The operating companies are geographical and functional units, some vertically integrated within a geographical area. They are not all wholly owned. Many are joint ventures with other companies or governments. Some, like the Shell Oil Company in the U.S., have minority interests. Operating companies have their own subsidiaries and boards of directors.

The ten wholly owned service companies—oil (2), chemicals (2), marine (1), metals (1), nuclear (1), gas (1), coal (1), and research (1)—conduct the activities in the Group that are not decentralized (PL2). In oil these are exploration, group finance, oil purchase, sale, and transportation to and from operating subsidiaries. Coordinators orchestrate these movements.

The Royal Dutch/Shell Group is a group of companies, not a corporation. The Shell Centre complex in London and the offices in The Hague are its central offices, not corporate headquarters. The operating companies are coordinated, not managed, by the central offices. By the tax laws (EC3) of the U.K. and the Netherlands, if they were managed by the central offices, their profits would be taxable in the home countries of the owners. So for reasons of dual nationality and tax liability the decentralization of operations, most often geographically by market (PL17), is not a facade. It is a particularly important issue for executive compensation.

On the other hand Sampson's sly suggestion that if an observer went up to the "bridge" of the leviathan he would find no one there[2] clearly understates the role of the central office and the eight Managing Directors. The chairman of the Managing Directors behaves just as he would if he were the chief executive.[3] "Coordination," if it is less than management, where profit responsibility can be assigned and taxed, is much more than an advisory, technical service role. It is, perhaps, indirect management. Daily operations are not controlled from the central offices. The authority to turn the valves to change product mix at the Curaçao refinery does not come from the central office as it might in other companies. Neither does a central computer in Chicago make out all pay checks worldwide. Instead the tools of coordination are personnel assignment, planning, and internal competition.

Compared to Exxon where decentralization is more on a regional basis with tighter control within each region, Shell's decentralization (PL1) seems to be more on a country basis with more emphasis on local management to the extent that in many places it is viewed as a local company rather than a multinational giant. The British think it is British; the Dutch think it is Dutch; the Americans think it is an American oil company with foreign subsidiaries.[4] And while in some areas it is more decentralized than the other major oil companies, it is perhaps more centralized, as all oil companies are, than other European multinationals like Unilever and Nestle.[5]

Shell's policies of geographic decentralization (PL17) and emphasis on the preservation of national identities of all units, not just the British and Dutch, (PL14) are the result of its organization shaped by its joint venture status, unequal ownership, dual nationality, the tax laws, European models of management, and a reaction to and desire not to repeat Deterding's tyranny (PV14). The disadvantages—inefficiency and overstaffing—inherent in this organizational structure (H4) are greatest in upstream areas where, not incidentally, Shell today is weakest. They are smallest in marketing where Shell is strongest. As expected it has generally ranked before 1973 among the lowest of the Majors in profitability of its operations since World War II (though because of its size its net income has traditionally been second largest).[6]

However, on another increasingly important scale, political acceptability, it must be ranked at the top. The considerable effort at Shell in diplomacy required to coordinate and preserve national identities has yielded an advantage in political acceptability and therefore long-term viability. Shell has rarely drawn criticism or been confused with Standard Oil. It has avoided political controversy and concentrated on serving its customers, a consequence of its multinationality and a position that Churchill recognized as long ago as 1914.[7] Its diplomatic style of management (PL15) and the indigenousness of its local subsidiaries (PL17) are greater strategic investments in political acceptability, if lower in profitability, than other Majors have chosen to make. In these aspects of its organization and strategy Shell differs from the other Majors. It is the United Nations of oil companies, as that body wishes it could function. What follows is the story of its development.

9.3 Formation: Shell before 1907

The story of Shell in its earliest years is not told here for the sake of historical completeness but because, as is also true of the other Majors, many of its most important policies today date from these early years. These include not only its organization discussed in the previous section but also its worldwide multinational character (PL2), its geographical strength in the Eastern Hemisphere (PL7,8), its vertical emphasis on marketing and shipping (PL3,4,5), the policy of vertical integration by geographical area (PL6), and its brand name and trademark (PL1). The first link between the partners of the Royal Dutch/Shell Group was forged in 1903 with the formation of the Asiatic Petroleum Company (R03). The motive was to survive and prosper against the price cutting of Standard Oil in the Far East. The link was forged by Henri Deterding, the Managing Director of Royal Dutch and the possessor of the finest commercial mind ever produced by the Netherlands (R96a). The three equal partners were Marcus Samuel's Shell, Deterding's Royal Dutch, and the Paris Rothschilds, who supplied Russian kerosene to Shell and Royal Dutch.[8] Asiatic was to combine the marketing organizations of Shell and Royal Dutch and market the kerosene of its three partners all over the Eastern Hemisphere (PL3-5).[9]

The "Shell" Trading and Transport Company (PL1,3) was the creation of a Jewish trader named Marcus Samuel (PV1,2). His father, also Marcus

Samuel, had begun with a small antique, curio, and bric-a-brac shop in the East End of London in 1833, including geegaws (EC4) for the Victorian household made from oriental seashells. The trade in shells became so profitable that he arranged for regular shipments from the Far East, and developed it into a general import/export business. In 1878 his son took over and not long after began to carry consignments of cased kerosene for "the lamps of China" as a sideline to the shell and general merchandise trade.[10] From the (R78) seashell trade came Shell's name and trademark (PL1), its experience with worldwide multinational commerce (PL2), and its interests in shipping (PL3), trade and consumer products (retail marketing rather than discovery or manufacturing) (PL4). These policies were part of Shell's strategy before it entered the oil business, which it did through its principal businesses, trade and transport (R80a-d).

In 1890 on a trip to the Far East Samuel saw some of the first oil tankers being operated on the Black Sea by the Nobel brothers' interests. Realizing the economics of bulk transport (PL5) compared to Standard's method of shipping Pennsylvania kerosene in metal cases by clipper ship, he secured a ten-year supply contract from Rothschild interests in Russia (R92a) and ordered eight tankers (R92b) for shipping Russian kerosene in bulk eastward through the Suez Canal to the Far East (PL5). He had difficulty in securing persmission to send tankers through the canal, partly due to the Standard interests he had outmaneuvered and partly because of the genuine fear that the new tankers were floating bombs. In 1892 the *Murex*, followed by the *Conch*,[11] sailed through the Suez Canal to the Far East (EC6) and into competition with both Standard and the Dutch producers in the Dutch East Indies.[12] Bulk stations and a marketing network were built throughout the Far East (PL7), but the venture almost failed for lack of demand. Customers were not willing to use their own containers for bulk oil, desiring the blue Standard tins as much as the kerosene. Bright new red Shell tins locally manufactured were soon competing with battered blue Standard tins which had traveled halfway around the world.[13]

The precedents set by this entry all became important parts of Shell's strategy. Shell entered the oil business as a trader, shipper, and marketer of kerosene (PL3,4,5,7), purchased from low cost Russian sources (PL6), and transported in bulk to the Far East (PL5,7). This strategy was shaped by shipping and Far Eastern trade (R2,3), the availability of low cost Russian crude which made inexpensive, low quality kerosene (EC7,8), the invention of the tanker (EC5), use of the Suez Canal (EC6), and the competition of Standard Oil (EC9).

The Royal Dutch Company for the Working of Petroleum Wells in the Dutch East Indies was formed in 1890 to develop an oilfield (R90) in Sumatra. Under the management of J.B. August Kessler a pipeline and a refinery at Pankalan Brandan began operating in 1892 (R92c). The business of Royal

Dutch was producing and refining in the Dutch East Indies (PL8). But it had difficulty with marketing (H1). Kessler hired a young banker and accountant (PV3), Henri Deterding, in 1896 to get, as Deterding put it " . . . a man whose mind would be sufficiently free from the production bias to enable him to concentrate more readily on the selling end."[14] Deterding began building tankers, bulk storage points, and a sales organization in the Far East (PL9), getting his start in marketing (PV4).[15]

Samuel's reply was to finagle a concession from the Dutch (R96b) on Borneo in 1896. He struck oil and started a refinery at Balik Papan (R96c,d) (PL9).[16] Almost from the beginning both Shell and Royal Dutch found it imperative to integrate vertically from production through marketing (PL9), Shell to assure low cost sources and Royal Dutch to survive price cutting. By 1897 Samuel's oil business had become so extensive that he formed a separate company (R97), the "Shell" Trading and Transport Company (PL1,3,4) to operate it. He also contracted for supplies of a crude suitable for making gasoline (R98) from a small independent field in Sumatra, adding a second product (PL10).[17]

The power of the British Empire (EC11) at the turn of the century, the last year of Victoria's reign, was at its zenith in form if not in substance. Germany, with the Prussian army, dominated the Continent, but Britain with her navy dominated the world. Between 1895 and 1899 British merchant ships carried 70.8% of world sea trade.[18] In 1898 Marcus Samuel was knighted for services to the Empire for donating the services of a Shell ship to free a Navy warship that ran aground in the Suez Canal.[19] The following year he first formally tried to persuade the Navy to test oil as a fuel, the fuel his own fleet used. With Lord Fisher at the Admiralty, Marcus Samuel pioneered the use of oil as marine fuel (PL11). For the next 15 years they tried to get the Navy to convert to oil. It was Marcus Samuel's lifelong dream for Shell to fuel the British Navy, the backbone of the Empire, with oil from its bunkering stations worldwide (PV5,6,SR1,PL11) in its days of glory. He even offered control of Shell as Disraeli had taken control of the Suez Canal for Britain.[20]

He was fulfilling the wish in his father's will that his sons "be united, loving and considerate and keep the good name of Marcus Samuel from reproach" (PV6,7).[21] Of humble origin and Jewish (PV1,2) all his life he sought acceptance for himself and his family from aristocratic Anglo-Jewry and the Empire more than wealth (PV7,8). An alderman in London since 1892, he became the third Jewish Lord Mayor of London in 1902-1903, the pinnacle of his civic career.

From 1897 to 1901 his business empire also reached its peak. In 1900, bolstered by production in the Dutch East Indies (R96c), he renewed favorably the Russian purchase contract with the Rothschilds (R92b). Shell expanded everywhere and determined to market gasoline in Europe by purchasing a German company from the Deutsche Bank (R01a), cracking the armor of the

Standard, Nobel, and Rothschild interests which controlled the market there.[22] When oil was discovered at Spindletop in Texas in 1901, Samuel contracted with the J.M. Guffey Petroleum Company (forerunner of Gulf) to purchase, transport and distribute Texas oil as fuel for ships (PL6,11). Shell would take up to half Guffey's production for 21 years at 25¢/bbl. plus 50% of the net profits (R01b), terms quite similar to those between Gulf and Shell in Kuwait in 1947. Spindletop alone could produce as much oil as Pennsylvania—half U.S. output—and Shell had access to it.[23] Standard's monopoly could be broken. Shell ordered still more tankers. At the end of 1901 Samuel's Shell was Britain's largest oil company, second only to Standard worldwide, poised to enter Europe, and the only company with worldwide sources of crude (PL6). His oil business was barely ten years old.

But Samuel's moment of power passed as adverse circumstances accumulated.[24] By 1907 he had lost control and 60% ownership to Henri Deterding and Royal Dutch which was 1/10 Shell's size in 1901. The difficulties began in 1899 with a Standard-Nobel-Rothschild (EC10) alliance in Europe that threatened to spread worldwide and extinguish Shell. Samuel expanded feverishly and took on large inventories at high prices in the rising market. Talks to ally with Dutch producers in the East Indies, incuding Royal Dutch, were inconclusive, as they had been in 1896, partly due to the breakout in 1899 of the Boer war in South Africa between British troops and Dutch settlers (EC13). In 1900 a slump began (EC12). Shell was caught with high priced stocks as prices fell. The aftermath of European intervention in the Boxer Rebellion in 1900-1901 cost Shell its markets in China (EC14).[25] In 1902 the Navy tried fuel oil with obsolete burners with disastrous results (EC15) and stuck its head back in the sands of tradition and declined the use of oil.[26] Also that year the British government of India (then a colony) refused to grant Shell a concession in Burma or relief from tariff on products of non-British companies in the mistaken view that Shell was not wholly British, an insult Marcus Samuel never forgot (EC16).[27] This gave the Burmah Oil Company a monopoly in India as Shell, Standard, Royal Dutch and others lost their markets there. The idea that Shell was or might become "foreign" came after Marcus Samuel rejected Standard's second attempt to buy Shell in 1901 (EC17). To weaken Shell Standard then began the rumor that it secretly controlled Shell.[28] In 1902 Spindletop went dry (EC18) and half of Shell's fleet was idled. It began shipping kerosene from Romania to Germany to keep its four largest tankers employed.[29] Sir Marcus's civic duties as Lord Mayor left him little time for business in 1902-1903 and the small family management was unable to cope in his absence.[30]

Meanwhile Royal Dutch was emerging from rough times. In 1898 salt water appeared in its wells and production fell. Deterding became interested in production. As he relates it, there were then two methods of drilling and refining in existence—Russian and American—and the American method was far better. Royal Dutch's drillers and refiners had been imported from America. But they were unscientific, practical oil men, "often exceedingly dictatorial gentlemen of the roughneck breed, with their blue shirts, riveted trousers, and bowler hats—a costume so general among them that it practically amounted to a uniform. . . . "[31] Deterding decided they had to go and replaced them with geologists, chemists, and engineers trained in Europe, often in the Netherlands. Royal Dutch was the first Major to use scientists in production (PL12), though geologists had been used in Russia for 35 years and Standard had begun to employ chemists in refining as early as 1885.[32] The application of Dutch technical expertise (EC18), in its golden age like other European science late in the nineteenth century, gave Royal Dutch a competitive edge and gave the Dutch an identity as suppliers of expertise on the technical areas of production and refining that is very much alive at Royal Dutch/Shell today.[33] It was a brilliant move by Deterding. New production was found, including Sumatran crude with good gasoline content. With supplies of Russian kerosene Deterding had begun acquiring in 1898 Royal Dutch expanded and began undercutting Shell in the Dutch East Indies.

In 1900 Kessler died, leaving as his last wish that Deterding be given complete charge of Royal Dutch. The directors agreed. In that year he obtained his first "working agreement" among Dutch producers. He decided "whenever and wherever possible, a definite system of cooperation with smaller trade rivals must be made an essential part of our business policy—in fact, our main working plan" (PL13). "Quality and service" he stated, "are the only sure foundations on which competition can survive." Price competition was not competition at all; it was "annihilation" (PV11).[34] Considering Standard's price policy, this was not an exaggeration. Standard strove to maximize market share by eliminating or absorbing competitors. To do this it often lowered prices in the product or area most important to a competitor while keeping them high elsewhere, subsidizing a war in one area from profits in another. For example Standard sold gasoline in America at £7 17s. 6d. while charging £4 10s. in Europe and also paying the shipping cost to Shell on a charter.[35] Though Deterding (PV11) detested this kind of price competition, even he recognized its value: "If our now friendly competitor, Standard Oil, and one or two more had not gone hitting us by price cutting . . . the Royal Dutch might quite conceivably have been a small company, almost unheard of, today."[36]

"Working agreements" to keep prices and business steady were Deterding's response to Standard. The Asiatic Petroleum Company in 1903 and then the Royal Dutch/Shell combination described below were such agreements: "it was really close cooperation, and never absorption, at which we aimed," he said (PL14,SR2).[37] From this viewpoint one can see how Royal Dutch/Shell became a 60:40 joint venture. This policy stands in contrast to that of Standard of paying generously and decentralizing but never without control.[38]

The Asiatic Petroleum Company, with Marcus Samuel as Chairman and Deterding as Managing Director, was to be the eastern wing of Samuel's vision with the Guffey contract supplying his European distributing organization as the western wing.[39] With Spindletop dry (EC19) and Shell's Far Eastern facilities constrained by the Asiatic partnership with Deterding, Shell's fortunes suffered. A trade slump from 1903 to 1906, a price war by Standard initiated in Europe in 1904, the Navy's refusal to grant Shell fuel oil contracts, and Deterding's control of the Far East forced Samuel to the position of having to withdraw from Europe and forfeit the heart of his fleet to the Deutsche Bank. In desperation he opened the question of amalgamation with Royal Dutch, offering equal terms as in Asiatic. But Deterding insisted on nothing less than 60:40 and Samuel, in defeat, accepted in 1907. Now Shell was "foreign" and his dreams of supplying the Navy with fuel oil were shattered (PL11,PV5,6). He felt himself a failure but blamed no one save the government and the Admiralty, for their short-sightedness, and to whom he would gladly have given control of Shell to keep it British and be of service to the Empire (PV6).

Marcus Samuel, though Chairman of Royal Dutch/Shell, began a brief retirement. But it was not victory over or control of Shell that Deterding and the Royal Dutch wanted. As much as its assets and markets they desired headquarters in London, access to its capital markets and association with the British Empire—Shell's Britishness.[40] Deterding had moved to London in 1902 to set up the Asiatic and had become more British than the British. It was Shell which imparted its name to the group and its products (PL1).

The amalgamation worked out differently than Marcus Samuel expected.[41] In fact it had brought together a great management team, something neither Shell nor Royal Dutch had ever had separately. Besides Sir Marcus from Shell and his brother Sam, there was Robert Waley Cohen, a Cambridge educated chemist from the Anglo-Jewish aristocracy (PV8) who became Deterding's right hand man; Mark Abrahams, a nephew who had built Shell's organization in the Far East; Walter Levy, a son-in-law (PV7), and others. From the Royal Dutch there came Hugo Loudon, an aristocratic diplomat who had begun as Kessler's driller in 1894, for whom the respect of everyone was so great that he made the amalgamation function smoothly from the start[42] and whose personality set the style of the organization (PV9,PL15), a level of courtesy no visitor to Shell can fail to admire today. There were Dutch scientists and geologists (PL12). And as Managing Director there was a genius, Henri Deterding. With Deterding's deft hands at the control of operations Marcus Samuel was recalled to active duty as elder statesman, counselor on policy, and public spokesman for the organization.

An unexpected result of the system of management Loudon put together was a lack of tension between the British and Dutch.[43] The two parent companies became holding companies owning shares in all the subsidiaries in the proportion 60:40. A production-refining operating company was set up in The Hague and a transport-storage company in London, reflecting the respective national specialities (PL3,8). Management tended to collect in London because of its preeminent position in world trade. There was little Dutch consciousness of their controlling majority. Perhaps because they were conscious of their dual nationality they were careful to make ability, not nationality, the qualification for a post. If tension arose, it was between London and The Hague, but not between British and Dutch (PL14). This policy produced a remarkable degree of accommodation and respect for national and individual differences which is very evident today.

Royal Dutch and Shell shared an important policy, the "policy of the straight line," Deterding called it,[44] from the pre-1907 era (PL6). At Royal Dutch it evolved in response to Standard (EC9). The advantage of Royal Dutch over Standard, which purchased, refined, and exported its products from America to the Far East, was that it produced its own oil and its production was much closer to its markets. To combat price cutting it resolved to deliver on as straight a line as possible, serving only those markets nearest its production. It was the same simple commercial principle of seeking the nearest source of supply to match a local demand (PL6).[45] This principle was to guide Royal Dutch/Shell's worldwide expansion.

9.4 1907-1929: Worldwide Expansion

The years which followed the amalgamation were Shell's golden years under Deterding. Shell expanded worldwide geographically in all functions and became a Major. Important policies of openness (PL16) (in contrast to Standard's secrecy), subsidiaries organized by country as indigenous parts of the local economy (PL17), service to the British Empire (PL18), and in the 1920's interests in petrochemicals (PL19) and specialty products (PL20) were initiated. Large scale, volume and growth oriented marketing became more important (PL4,5).

Shell expanded vigorously worldwide after 1907. It explored and discovered oil in British Borneo (1910), Mexico (1913), and Venezuela (1914). It acquired production in Romania (1906), Russia (1919), Egypt (1911), Trinidad (1913), and California (1913) (R06-14).[46] A different source places Egypt in the discovery category, citing a personal interest by Sir Marcus in the area from 1907 because of its strategic location on the Suez Canal and because it was within the British Empire. He never gave up on his dream to serve the Empire (PV6,PL18). Had the Anglo-Egyptian Oil Company had large production it might have furthered his dream.[47]

Shell was the first oil company to seek worldwide production. One reason was the technological edge its geologists had over the practical oil men used by other companies (PL12). Integrating vertically by geographical area was an application of its policy of the straight line (PL6) to minimize costs. Having its own production also, as W.L. Mellon of Gulf concluded,[48] prevented Standard from squeezing it with high crude prices as Marcus Samuel had experienced in Russia in 1905 (EC2).[49] Geographical expansion also made Shell less vulnerable to local price cutting by Standard (EC2).

Entry into the United States was the most important phase of this expansion and one that was absolutely necessary to deny Standard a secure home base from which it could finance price wars elsewhere. This expansion reflected the view of Deterding, who saw Shell solely as a business engaged in selling oil (PV4,10) worldwide (PL2) against Standard, as compared to Sir Marcus who sought Shell's greatness by fueling the Royal Navy and appending Shell to the British Empire (PV5,6). A geological survey was carried out in Oklahoma in 1908, but the entry really began after a marketing truce (PL13) from 1907 to 1910 (EC20) broke down. During that truce Asiatic had sold Standard excellent Sumatran gasoline (R98a) which was marketed in California.[50]

About this time, a famous story goes, Deterding made his first sales in America by sending Shell tankers filled with gasoline steaming past the Statue of Liberty in New York harbor and dumping them at low prices on Standard's doorstep. Deterding confirms the facts but insists there were only two ships which had been diverted from Germany in search of better prices in the ordinary course of business and that Shell did not at that time intend to begin marketing in America. But neither he nor Standard were unaware of the competitive threat implied.[51]

Shell's actual entry began in 1912 with formation of a marketing (PL4) company, the American Gasoline Company, in Seattle (R12b) and secret purchase of producing properties in the mid-continent which became (PL9,R12b) Roxana Petroleum Company later that year.[52] These events occurred after negotiations with the Mellons of Gulf to form a U.S. marketing combine did not pan out. In 1912 Standard dropped the price of gasoline from 18¢ to 10¢ a gallon in California, while maintaining the higher figure elsewhere in the U.S. Unable to sell its high quality Sumatran gasoline profitably in California on its own, Shell began looking for a source of production which could supply gasoline more cheaply by saving shipping charges, the policy of the straight line (PL6). In 1913 Deterding purchased large production in the Coalinga field in California (R13a).

Due to its joint venture character (R07) and the diplomacy required to manage such an organization (PL15), its worldwide scope of operation (PL2), the policy of the straight line (PL6) to minimize long supply routes, and the aim of live and let live toward partners and competitors (PL14), the Royal Dutch/Shell Group took a different attitude toward the management of its subsidiaries and amalgamations than most companies and one certainly different than that of Standard whose subsidiaries, though they had some managerial autonomy, were viewed locally as tentacles of the worldwide octopus and over which there was little local control. The Shell policy set forth by Deterding in 1911, and for which he was responsible, was that the local Shell subsidiary for each country should be, though ultimately and distantly controlled by the Group, an integral part of that country's economy and an instrument of its social welfare (PL17,SR3).[53] It was, and is, a policy aimed at cooperation and acceptability rather than profit maximization and control.

Another of Deterding's policies in great contrast to Standard's secrecy, secret companies, and secret deals was that of openly stating Shell's intentions and policies (PL16). Compared to the perfunctory annual reports of most American companies before the 1930's, which were rarely more than a balance sheet and sometimes an income statement, Deterding's reports to his shareholders were models of lucidity and candor, compared even to the present, often running as long as 40 pages. Shell's intention to enter the U.S. was published in the *Oil and Gas Journal* on April 12, 1911, eighteen months before it occurred.[54]

Between 1911 and 1914 Lord Fisher and Marcus Samuel were making some headway in converting the Navy to oil with Churchill's assistance.[55] But the old suspicions, ever since the Burma incident in 1902 referred to above, prevented it from becoming the Navy's chief fuel oil supplier, though it was Britain's largest oil company. It was to lose that honor to BP.[56] An oil shortage and high prices for which Shell was blamed (EC21) were partly at fault, but the main reason was inability to guarantee price and secure supply in long-term contracts. Churchill expressed the government's policy when he charged Fisher in 1912 to "find the oil: to show how it can be stored cheaply: how it can be purchased, regularly and cheaply in peace; and with absolute certainty in war.... "[57] On the issue of price Churchill attacked Sir Marcus personally in the House of Commons in remarks that were politically shrewd but anti-Semitic, with the result that the government bought BP and Shell lost out (see Chapter 10 for details).

When war came, however, Sir Marcus had the opportunity to prove (PL18) Shell's loyalty and realize his dreams of fueling the Navy and appending Shell to the British Empire (PV5,6).[58] He declared that Shell would make no profit from the hostilities and turned over Shell's fleet of 75 ships to the government while having to charter neutral ships at four times the rate it received from the government. The Dutch, who were neutral, consented and Deterding executed it brilliantly. Twelve Shell ships were sunk, including the *Murex,* which had first traversed the Suez Canal with bulk oil for the Samuels in 1892. During the war Shell regarded itself as a government agency committed to winning the war rather than as a commercial concern. All this is by way of tracing the wartime origin of an important Shell policy, derived from Marcus Samuel's dreams (PV5,6), to be of service to the British Empire (PL18). It is a very important part of the identity of Shell today as Sampson wrote in

1975 of the recognition of Shell executives as a kind of "junior diplomatic service."[59] Sir Marcus's dreams came true.

After the war Shell continued its worldwide expansion (PL2), especially in production (PL8), under Deterding's management (R96a). In 1918 it was already the world's largest producer (Table 1-3, Chapter 1). In the Eastern Hemisphere its Romanian properties were destroyed willingly during the war (1916) to prevent German use and soon after (EC22) the Russian producing properties it had purchased from the Rothschilds in 1912 were confiscated. But production was increased in Sumatra and a refinery built at Sarawak. In the Western Hemisphere refineries were built at San Francisco (Martinez—1915), St. Louis (Wood River, Ill.—1918), and Curaçao (1918) in the Netherlands Antilles (R15,18). In 1919 Shell purchased Lord Cowdray's "El Aguila" (Mexican Eagle Oil Company), Mexico's largest producer, and in 1920 set up a joint venture in Britain with BP called Shell-Mex/BP to market the products (R19,20).[60]

Shell led the way into Venezuela, purchasing concessions from General Asphalt and others in 1913 and making the first discovery (R14) at Mene Grande in 1914.[61] It was the crowning achievement in the careers of Deterding and the Dutch geologists (PL8,12). In December 1922 the Los Barossos well in the Maracaibo Basin blew out at 100 TBD (= thousand barrels per day) and Venezuelan production began to climb, always led by Royal Dutch/Shell until nationalization in 1976. For nearly half a century Venezuela was Shell's largest source of production.

Shell was well balanced vertically as well as geographically (PL9). It refined a little less than the 83 TBD it produced in 1918 (some crude used for fuel oil was not refined) and marketed a little more (from supply contracts before World War I with Russian and Romanian producers and after World War I with BP—it was BP's first customer). Shell rose with the demand for gasoline (EC23,PL10) from Sumatran, Oklahoma, and Venezuelan crudes marketed in the U.S. and Europe and fuel oil (PL11) from Mexican, Californian, and Persian crudes sold to ships, railroads, and power generating plants everywhere. It had the largest fleet after the breakup of Standard in 1911 as it does today (PL3,11).[62] In 1976 it finally surpassed Texaco as the leading U.S. gasoline marketer (PL10).

Other events of the 1920's were less dramatic as Shell steadily expanded. It made a major discovery of the Signal Hill field near Long Beach, California in 1921. In 1922 it merged its organizations in the U.S. West and Midwest with Union of Delaware, overextended in the 1920-1921 recession, to form Shell-Union (R22c) owned 72% by Shell and 28% by Union. Union's main assets were not in the East but in Oklahoma and California, including an option on a large block of Union Oil of California stock. It also gave Shell access to American capital markets and placed owernship in Shell subsidiaries in local hands, an important way of making Shell indigenous to the local economy by giving its citizens a share of ownership (PL17). Also in 1922 concessions were acquired in Argentina and a company was formed to acquire oil rights in Brunei, British Borneo (R2a,b). Shell Company of Canada was formed in 1926 (R26a). Turkish Petroleum, in which Shell had had roughly a quarter interest since 1912 (R12a), struck oil in Iraq in 1927 (R27). In the U.S. Shell expanded rapidly, becoming the second Major (after Texaco) to market nationwide in 1929 (R29). While nearly half its business was now in the U.S. abroad it was the largest producer, refiner, and marketer, especially strong in Japan (64% market share in 1935) India, Eastern and Western Europe, with worldwide volume nearly equal to Jersey Standard's (see Table 1-3 1927, Chapter 1). In 1928 it formed joint marketing ventures with BP in the Near East and Africa and with Burmah Oil Company in India to market their products (R28a,b).[63] The Achnacarry agreement of 1928 (R28c), coming after the failure in 1927 to control overproduction in the U.S., between Walter Teagle of Jersey, John Cadman of BP, and Deterding of Shell attempted to cartelize foreign marketing. It was revised in 1930 and extended in 1932 to include Socony-Vacuum, Atlantic Refining, Gulf, Sinclair, and Texaco in the "Heads of Agreement for Distribution" agreement (R32).[64] With his joint ventures, geographical coverage everywhere (PL2), and these cartel agreeements, Deterding had, on paper at least since they never worked very well, nearly achieved his objective of limiting competition (PL13,PV11) and sewing up foreign markets with Shell as the world leader. As he had taken as the Group's motto upon the formation of Asiatic in 1903, "Our field is the world" (PL2).[65]

Its product line expanded in the 1920's as well. It set up an (R20a) aviation department to fuel K.L.M. Airlines in 1920 (PL20), eight years before Jersey did so. Diesel fuel, home heating oil, lubricants, and asphalts became significant businesses in the 1920's (PL20, R20b-e). With these additions came a marketing-oriented dedication (PL4) to a full product line and strength in specialty products (PL20), in contrast to Jersey even in those products whose volume was low. They also represented a desire by the frugal Dutch to minimize waste and by Dutch chemists (PL12) to apply science to upgrade products to their highest value. The Shell Development Company (R26b), a patent licensing and research company (PL12), was founded in 1926. In 1977 Shell was the world's leading lubricant marketer with 23% of the market.[66]

What would in time become the largest petrochemical company among the Majors began with the founding in 1929 of Shell Chemical Company (R29b) to produce ammonia from refinery gases. The reasons were not only Dutch frugality (PV12) and scientific interest (PL12) but also oligopolistic response to Jersey Standard, which had been trying to manufacture isopropyl alcohol since 1920, and I.G. Farben (EC24,25), the German chemical giant who was already coming close to producing a synthetic gasoline from coal (and would do so in 1934).[67] Deterding decided that if Farben could enter the motor fuels business with German scientists, Shell could enter chemicals with Dutch scientists (PL12).[68]

It was nearly at the beginning of the organic chemicals industry. DuPont, an explosives manufacturer, entered chemicals in 1917, Union Carbide and Allied Chemical in 1920, I.G. Farben in 1925, and Imperial Chemical Inc. in 1926. Feedstocks were initially coal and coal tar but the gas, coal and steel companies which produced them were uninterested in small volume chemical operations, thus permitting the chemical companies to grow on their byproducts. Petroleum feedstocks from refinery wastes were desirable because of their greater purity, but the oil companies still nearly missed the boat. Even Shell Chemical today, the 13th largest chemical company, is only half as large as the top five chemical companies.[69]

With the commitments to petrochemicals and specialty products (PL19,20), particularly asphalt, lubricants, and aviation fuel, and with the achievement of nationwide U.S. as well as global market coverage (PL2), Deterding's edifice was complete. Though they would grow, adapt, and sometimes suffer, all the major policy elements of Shell's strategy were in place by 1929. Their vertical relationships were also established as, despite its worldwide discoveries and position as largest producer, Shell's markets, especially in the U.S. were already growing faster than its production in the 1920's.

In the years since its origin the Royal Dutch/Shell Group managed the unlikely combination of achievements of attaining diplomatic status while selling gasoline, marine fuel, and a full line of specialty products, openly, worldwide under a single name and trademark (PL1,2,10, 11,15,20). An indigenous cog in each local economy, supplied from the nearest source of production worldwide while fueling and serving the British Empire, Shell became the largest producer and shipper, a pioneer in bulk transport, a large refiner and volume-oriented marketer, vertically integrated to local markets, acquiring a uniquely accommodative view of individual, national, and competitive differences from the circumstance of its two European parents having begun business in Asia (PL3-8,17,18). It was a commercial counterpart of the British Empire on the way to utopia with all competitors accommodated, none destroyed, with price wars limited and volume steadily expanding, and with the Dutch scientific contribution of expertise to upgrade crude to products of maximum value—first fuels, then lubricants and specialty products, and soon chemicals (PL12-14,19). In its policies there was a sense of public service (SR1) and responsibility to its employees (SR4), to the countries where it operated (SR3), and even to its competitors (SR2). It was pursuing a different social model than the hard-nosed, close-mouthed, laissez faire capitalism often practiced by Rockefeller's minions.

9.5 1930-Present: Coping with Growth and Change

Until the diversification and nationalization movements of the late 1960's and 1970's Shell's management problems have for the most part been those of coping with growth and change rather than striking out in new directions.

The rapid expansion of the 1920's, especially in the U.S., bought at high prices in a rising market, made Shell vulnerable to the events of the 1930's (EC30). Its U.S. affiliates taken together suffered more than those of any other Major, losing $37 million from 1930 to 1934. What was damaging, however, was Shell's response. Cash was hoarded and its large debt load reduced, but also its exploration staff was dismissed and valuable undrilled leases were surrendered even though the U.S. affiliates were crude short. Except in California and east Texas to a limited extent Shell missed the opportunity to buy crude reserves. After 1935 volume and market share as measures of marketing success were replaced by "competitiveness"—the idea (PL21) that Shell's costs in each major market should be in line with those of the lowest cost competitor or the market abandoned. As a result it withdrew from Colorado, the Dakotas, Nebraska, Oklahoma, Kansas, Arkansas, New York, and western Pennsylvania, its first withdrawal since Marcus Samuel's troubles at the turn of the century.[70]

Deterding's career also came to an end in his seventieth year in 1936. Always hot tempered, in recent years he had an increasingly erratic effect on the organization as his autocratic tendencies (PV14) approached megalomania. Worse still, he became pro-Nazi (PV13). The (R10b) confiscation of Shell's properties in Russia without compensation in 1917 was an affront he never forgave. He became the world's leading anticommunist, was burned in effigy in Russia, and until 1927-1928 persuaded other oil companies not to buy Russian oil. One rumor has it that Shell's appearance on the U.S. East coast was in retaliation for Socony's purchase of Russian kerosene in 1927-1928. Influenced by his Russian (aristocratic and exiled) second wife and German third wife he came to see the Nazis as the only solution to communism. He became an embarrassment to the diplomatic managers of Shell and was finally eased out and went to live in Germany where he became intimate with Nazi leaders. Hitler and Göring sent wreaths to his funeral in 1939 just before the outbreak of war. After the war the niche in the entrance to Shell headquarters in The Hague intended for Deterding stood empty. So that no one may again tyrannize Shell as Deterding did, the company adopted the present top management structure of a group of three to eight Managing Directors whose chairman is ruthlessly retired at sixty.[71]

Without Deterding's strong hands at the reins, and partly in reaction to him, there was a tendency for the members of the Royal Dutch/Shell group to become less well coordinated, to go their own separate ways. This tendency was reinforced by a variety of factors. The age of the group's first generation of senior managers was one. Deterding's successor, J.E.F. de Kok, died in office in 1939. The hard times of the depression forced each subsidiary to consider its own future (EC30). The policy of localization (PL17) and the appointment of native chief executives rather than expatriates from Britain or the Netherlands gave more weight to local issues. During World War II the Royal Dutch managers and directors, able to exercise their authority only from Dutch soil, evacuated not to London but Curaçao, a location which made coordination with London and the subsidiaries more difficult. The war itself (EC39), centered in Europe and the Far East, cut off communication with more of Shell's properties than with those of any other Major.

World War II (EC39) also cost Shell more than any other Major. It lost its Romanian production (R06) and 87 of its 180 ships (R92a); its Far Eastern properties were destroyed; and its markets were in disarray. It had cooperated closely with the Allies in the war (PL18) and its U.S. refineries had made especially notable contributions in the provision of aviation fuel (PL20), gasoline (PL10), and fuel oil (PL11). Shell and other large-scale refiners led the way in the (R42) production of 100 octane gasoline for aviation fuel and in catalytic cracking. General Jimmy Doolittle had been Aviation Manager of Shell before the war and was instrumental in the development of long range aircraft. Catalytic cracking and the production of high octane fuel required for high performance flying led Shell's research scientists into the synthesis of aromatic petrochemicals like cumene (in 1942) which were first used as enriching agents to raise the octane rating of gasolines for aviation (PL12,19,20).[72]

But as events unfolded after the war Shell became aware of an even greater loss: they were the only Major without production from the enormous new reserves in Saudi Arabia and Kuwait (H3). They had production only in Iraq in the Eastern Hemisphere after the war (before their Dutch East Indies production was rebuilt). They had fallen from the position of being the leader in the development of new reserves and production to being seriously crude short and at a competitive disadvantage to the new Middle Eastern producers with enormous low cost reserves (H3). As a partial remedy Shell was able to negotiate the largest long-term purchase contract (R47) in the history of the industry with old friend Gulf, who was as anxious for markets as Shell was for crude, in 1947. Shell would purchase half of Gulf's Kuwait production up to 750 TBD and split the profits with Gulf. Still this handicap was the most significant fact in the postwar period for Shell.

In the complicated sequence of events that led to this result Shell was a victim of its own policies of cooperation (PL13) and misplaced faith in the British government (PL18). Having bound its competitors with marketing agreements in the period from 1928 to 1932, it let itself be bound by the Red Line exploration and production agreement after 1928 as its competitors, one by one, freed themselves to obtain interests in Saudi Arabia and Kuwait. It also allowed itself to be outmaneuvered by the British government.

From 1913 until it could no longer do so British foreign policy was to cordon off the Middle East for BP, which was its majority owner.[73] The Netherlands had attempted much the same thing in the East Indies for Dutch companies including Shell. The aim of this effort was to control the trade routes to the Far East over which Britain and France, and later Britain and Germany, had struggled since before Napoleon. Britain chose the sea route and needed a secure supply of oil to fuel her Navy, the backbone of her Empire. Disraeli in 1875 had bought control from the Khedive of Egypt of the Suez Canal the French had built. Germany, the power of the Continent, took the land route and built the Turkish railway from Baghdad to Berlin. It wanted control of the oil fields for its railroads. This nineteenth-century power struggle explains much of the activity of the Deutsche Bank in the oil industries of Romania and Turkey after 1901. Churchill engineered a British victory in the Turco-British treaty of 1913, whose fine print brought BP 25% of the future Turkish Petroleum concession and extended Turkey's boundaries eastward into the Persian oil regions. The following year the British government bought control of BP. Shell was eventually replaced by BP as operator of the Turkish Petroleum concession.[74] The purchase of BP denied equal treatment by Britain to Shell, its largest oil company. By relying on agreements like the Red Line and on a mistaken view of its place in the British Empire (PL13,18), Shell missed out in the really big finds in the Middle East. This is not to downplay Shell's heavy financial commitments in the U.S. and Venezuela in the late 1920's and financial difficulties in the 1930's as crucial factors. But just a few years earlier Deterding and Shell were leading the industry into such promising new areas. In the late 1920's and early 30's they did not.

In the postwar surge of demand Shell rebuilt and recovered, aided for once by a British Treasury agreement in 1946 (EC46), freeing Shell from exchange controls and permitting oil payments in sterling rather than scarce dollars, a slight advantage over American companies.[75] The Shell story from 1946 to 1973 is one of growth of marketing volume (PL4) and construction and expansion of large-scale refineries (PL5), including the world's largest at Pernis (near Rotterdam). Near its refineries Shell built large-scale bulk petrochemical plants (PL5,19), to produce the products resulting from its research (PL12).

Shell enjoyed able leadership in the postwar years from men developed within its own ranks: George Leigh-Jones, Frederick Godber, Sir David Barran, F.J. Stephens, and Sir Frank McFadzean, all British, and J.B. August Kessler, B. Th. Van Hasselt, H. Bloemgarten, F.A.C. Guepin, and L.E.J. Brouwer from the Dutch. Special mention must go to J.H. Loudon, General Managing Director from 1952 to 1964, who inherited his father's diplomatic skills and healed many of the divisions which had developed in the 1930's and during the war.[76]

The level of coordination between geographical units of Shell was often still imperfect, as for example when the Curaçao refinery was dumping products in Rotterdam while European affiliates were trying to maintain prices.[77] Shell brought in the consulting firm of McKinsey and Company in 1957 to address these organizational inefficiencies which were affecting profits. The role of the Managing Director was redefined, more directors were added, and a number of new positions for Group Coordinators for various functions were created, all no doubt with salutary effect. The nature of this response to lack of coordination was not to centralize authority or diminish the policy of local autonomy (PL17) but to work a little harder at coordination.

Shell's greatest problem, however, continued to be its shortage (H2) of crude, stemming from lack of equity in low cost Middle East (H3) production.[78] The Gulf contract provided vertical balance after the war but also delayed needed exploration.[79] As Shell's markets grew, the gap between production and sales widened. Shell explored worldwide, participating nearly everywhere there was exploration.[80] It was often the first to enter new areas, especially those with complicated geological structures. In this its strategy was the opposite of that of BP, which sought large, simple structures. Shell was a leader in secondary and tertiary recovery to get more oil from existing fields and in deep water drilling technology.[81] There were two reasons for these tactics. First, being crude short it had to take more chances than companies with huge Middle Eastern reserves (H3). And second, its efforts reflected its interest in technical expertise (PL12). It made its share of discoveries, including positions in Colombia (1945), Canada (1953), Algeria (1956), Nigeria (1958— with BP), Abu Dhabi (1963), Oman (1964), and the North Sea (1971, 1972). It developed a strong U.S. offshore position in the Gulf of Mexico and made a major discovery of gas with Exxon in 1960 in the Netherlands.[82] But on the whole the fields it found were smaller and less profitable than the largest discoveries of the postwar period,[83] many made by BP, and Shell did not regain the position of leadership in production it once enjoyed.

Like other oil companies Shell became interested in diversification in the early 1970's and made two significant commitments. It joined Gulf in the disastrous HTGR nuclear venture (R70a), perhaps seeing in the nuclear industry a natural application for its technical expertise and political acceptability (PL12,17). It also acquired in 1970 an old Dutch mining and metals producer, Billiton N.V. (R70b), with tin interests in Thailand, aluminum in Surinam, and coal in Australia.[84] Two factors in this decision were to use Shell's shipping (PL3) and trading expertise (PL4) in dry cargo businesses like ores and coal.[85] After the tanker market became glutted Shell hauled grain to Russia in the early 1970's as it once had turned its tankers into cattle boats when Spindletop went dry in 1902.

After the OPEC Crisis in 1973 (EC73) Shell reassessed the profitability of a number of countries in which it marketed and in a repeat of the 1930's, exited from 20 or so including Italy and ones like Venezuela in which its operations were nationalized. As Shell's Vice-Chairman explained, "It was once part of Shell's philosophy to be seen everywhere (PL2) but no longer." Now it would disengage from those who cannot pay their way (PL21). It also cut back on marginal product sectors like supplying fuel oil to utilities.[86]

To summarize the development and present state of Shell's corporate strategy I have drawn upon the expressions of Shell executives in a series of interviews I conducted in 1976, which were the starting points for this research tracing the origins and development of the patterns of behavior I have chosen to call Shell's policies. Taken with the numerical data in Table 1-1 (Chapter 1) they form a picture of the Royal Dutch/Shell Group which has had remarkable stability and success.

Shell's products are still sold worldwide under a single name and trademark (PL1). What began as a multinational operation between Russia, Europe, and the Far East became worldwide after the 1907 amalgamation when Shell soon entered the Western Hemisphere (PL2) and the U.S. to "attack Standard at home."[87] It is "the only true multinational"[88] and is "not just an American company spread abroad."[89]

It began as a shipper, trader, and marketer (PL3,4) even before it entered the oil business. Today it is the largest shipper (Table 1-2). It has "at least equal knowhow in trading/marketing business as in the oil business."[90] It is the largest trader with Eastern Europe today.[91] At Shell "international trade is a way of life, not a novel experience."[92] It has long "bought oil by type rather than price."[93] As a retail marketer (PL4) "Shell always tended to go closer to the final consumer than the competition in all product lines."[94]

Shipping kerosene in bulk (PL5) rather than cases was an operation requiring large scale from the start. For many years its tankers "led the industry in size."[95] So have its refineries and its chemical operations. Shell's scientists (PL12) are "good at finding low cost processes"[96] which were often large scale. With large scale comes a volume rather than profit margin orientation. Its businesses were "growth oriented until 1973, and then profit oriented"[97] just as in the 1930's. Bulk chemicals, not low volume, high margin products, were Shell's specialty.[98]

Both Shell and Royal Dutch integrated into all vertical functions early in the Far East and then in every geographical area in which Royal Dutch/Shell operated (PL6-9). It was the first oil company to engage in every function worldwide. Though always well balanced vertically, in both volume and organizationally in number of subsidiaries marketing has long been its dominant function. But before World War II Shell was also the largest producer and the only Major with worldwide production. Since then its "markets grew faster than production," a choice that was "deliberate."[99] It assumed "the marginal barrel had to be bought,"[100] a situation which has prompted it to try to "whiten the barrel"[101] to upgrade its crude to higher value products[102] like gasoline (PL10), chemicals (PL19), and high value specialty products like lubricants (PL20), international aviation fuel (PL20), and asphalt (PL20). In these products and marine bunkers (PL11) and gas (since World War II it has found huge deposits of gas rather than oil in Algeria, Brunei [formerly British Borneo], the Netherlands and the North Sea), it has the leading position in the industry.[103] With many of its Dutch executives today having graduated from the Technical University at Delft, Shell is "proud of its technical excellence"[104] and spends "more on R&D than anyone," [105] an interest which accounts for much of its success in refining and chemicals (PL8,19).

The prevention of price competition was Deterding's philosophy and Shell's policy (PL13) in numerous cooperative market sharing agreements before World War II. They were intended to prevent the erosion of market share and profitability through price competition. The motto on Shell's stations, "service is our business," continues to express its view about price competition. Sometimes it has been "too slow to react downwards"[106] on price and while "market share was important"[107] after 1973 Shell was "willing to shed volume"[108] rather than cut price.

The two policies (PL14,17) which define its relationships to its subsidiaries are perhaps Shell's most distinctive features. Accommodating individual and national differences, Shell has a planning staff of nine people of nine different nationalities (PL14).[109] But while it will accommodate minority local partners, it will not take a minority interest with a local partner.[110] In contrast to the American Majors who believed in never having a close relationship with a government, Shell is "close to the needs of governments and consumers—serving the energy needs of the state."[111] Its policy is to "adapt not fight."[112] It "optimizes locally from each country's point of view."[113] Its subsidiaries are "more prepared to be part of the countries in which we operate"[114] than are others. It has not so much an overall strategy as "100 motherhood strategies"[115] with Shell and Royal Dutch acting as "doting parents."[116] It seems clear that Shell is more willing than others to make concessions in economic efficiency to respect local preferences, a strategy designed to increase its long-term political "acceptability."[117]

Diplomacy, openness, and public service (PL15,16,18) also aid the pursuit of acceptability. "Shell is the least disliked foreign company where it operates."[118] The expression of favorable judgment for people and ideas is that they are "considered sound."[119] Whereas many of Shell's Dutch managers have technical backgrounds, its English managers are typically hired from liberal arts backgrounds directly from university or private school, one important selection criterion being that they are "good mixers with social and political skills" (PL15).[120] They then usually spend one or two years in the central office learning the business and are then posted abroad to operating companies.[121] There were once large numbers of British and Dutch expatriates, some affiliates being traditionally British and others Dutch.[122] Today there is also a substantial flow of native managers from operating companies posted to headquarters to become "group minded"[123] and more importantly to establish personal relationships with managers in the central offices. The principal tool of management even in the usually centralized finance function is not control but to "know people well," making a point of seeing them whenever they visit the central offices.[124] Since continuity of personal relationships is important, "we haven't fired anyone."[125] As a result "all big companies are envious of Shell personnel loyalty, dedication, and cooperativeness."[126] The development of personnel has always been exceptionally important at Shell and was a particular interest of Sir Frederick Godber.[127] Shell's openness (PL16) continues to stand in stark contrast to the habits of most European companies who, until recently, "had to be pushed to admit that they were in business at all. One could read their annual reports without knowing whether their products were nails, furniture, or bordellos."[128] Shell has had an "always high code of conduct" and has not been involved in banking or bribery scandals.[129] It "would love to think that virtue shows up in profits" (PL18).[130] What began as a commitment to serving the British Empire became a policy of public service generally (PL18).

This completes this chapter on the policies and events which have made the Royal Dutch/Shell Group the second largest oil company in the world. By no means is this list of policies complete. But it is sufficient to support the comment of Shell's Chairman of the Committee of Managing Directors, Gerrit A. Wagner, "We are neither better nor worse than other major companies—just different."[131]

Exhibit SH-1
Royal Dutch/Shell Policies

PL 1	To be known by the "Shell" name and trademark: (a) A; (b) 1892-present.
PL 2	Engage in multinational operations (Far East to Europe), world-wide after 1907: (a) G; (b) 1892-present.
PL 3	Engage in shipping and transportation: (a) V,P: (b) 1892-present.
PL 4	Engage in trading and marketing: (a) V,P; (b) 1892-present.
PL 5	Ship kerosene in bulk—large-scale operations: (a) S,P; (b) 1892-present.
PL 6	Supply crude by purchase or production from low cost sources near to markets (Policy of the Straight Line): (a) G,V,P: (b) 1892-present.
PL 7	Market products in the Far East: (a) G,V; (b) 1892 (Shell), 1896 (Royal Dutch)-present.
PL 8	Produce and refine oil in the Far East: (worldwide after 1907): (a) G,V; (b) 1896-present.
PL 9	Integrate vertically (engage in production, refining, transport, and marketing): (a) V; (b) 1896-present.
PL 10	Refine and market gasoline (enter fuels business): (a) V,P; (b) 1897-present.
PL 11	Promote the use of and market oil as marine fuel: (a) P,V; (b) 1897-present.
PL 12	Employ scientists (geologists, chemists, engineers) in production and refining—apply science to the oil industry: (a) A; (b) 1898-present.
PL 13	Limit price competition through cooperation and agreement—compete on quality and service only: (a) H; (b) 1900-World War II (price agreements), present (quality and service competition).
PL 14	Close cooperation with, not absorption of, acquisitions—accommodation and respect for national and individual differences: (a) A; (b) 1900-present.
PL 15	Diplomatic style of management: (a) A; (b) 1907-present.
PL 16	Openness: (a) A; (b) 1907-present.
PL 17	Organize subsidiaries by country as indigenous parts of their local economies: (a) A; (b) 1911-present.
PL 18	Serve the British Empire—public service: (a) A; (b) 1897-present.
PL 19	Derive chemicals from petroleum: (a) P; (b) 1920's-present.
PL 20	Produce and market specialty products: aviation fuel, diesel fuel, heating oil, lubricants, and asphalt: (a) P; (b) 1920's-present.
PL 21	Achieve "competitiveness"—costs in each major market should be comparable to those of the lowest cost competitor or the market should be abandoned: (a) G,P,A; (b) 1937-43, 1973-6.

Notes:
(a) Type of strategic choice

S—scale	PL 5
G—geography	PL 2,6-8,21
H—horizontal concentration	PL 13
V—vertical integration	PL 3,4,6-10
P—product diversification	PL 3,5,6,10,11,19-21
A—administration	PL 1,12,15-18,21

(b) origin and duration

Exhibit SH-2
Royal Dutch/Shell Environmental Conditions

EC 1	National and cultural differences between Dutch and British.
EC 2	European traditions of management.
EC 3	Tax laws of U.K. and Netherlands which tax income where "mind and management meet."
EC 4	Victorian demand for oriental seashells—1840's.
EC 5	Invention of the tanker—1888.
EC 6	Permission to sail tankers through the Suez Canal—1892.
EC 7	Availability of low cost Russian crude—1880's.
EC 8	High volume but low quality kerosene content of Russian crude.
EC 9	Competition with Standard Oil—1880's.
EC 10	Standard-Nobel-Rothschild alliance in Europe—1899.
EC 11	British Empire.
EC 12	Boer War in South Africa—1899.
EC 13	Slump in demand and prices—1900-6.
EC 14	Boxer rebellion—1900.
EC 15	Navy refuses to convert to fuel oil following the disastrous trial of the *Hannibal*—1902.
EC 16	British government of India refuses to grant Shell a concession in Burma or relief from tariffs on products of non-British companies in India, giving Burmah Oil a monopoly there as Shell and others lose markets—1902.
EC 17	Standard's second attempt to buy Shell—1901.
EC 18	Dutch technical expertise.
EC 19	Spindletop goes dry—August 1902.
EC 20	Marketing truce with Standard—1907-10—ends.
EC 21	Oil shortage and price rise in Britain—1911-12.
EC 22	Russian revolution and confiscation of Shell's properties—1917.
EC 23	Post World War I rise in gasoline and fuel oil demand.
EC 24	Jersey attempts to manufacture isopropyl alcohol—1920's.
EC 25	I.G. Farben begins to synthesize gasoline—late 1920's.
EC 30	The Great Depression—1930's.
EC 39	World War II.
EC 73	OPEC Crisis and nationalizations—1973-76.

Exhibit SH-3
Royal Dutch/Shell Resources

R 01a	PPAG acquired from Deutsche Bank (Shell enters European marketing)—1901.
R 01b	Fuel oil added to product line by long-term contract for Texas crude with Gulf—1901.
R 03	Asiatic Petroleum formed—1903.
R 06	Production acquired in Romania—1906.
R 07	Royal Dutch/Shell formed—1907.
R 10a	Discovery in British Borneo (Brunei)—1910.

R 10b	Production acquired in Russia—1910.
R 11	Production acquired in Egypt—1911.
R 12a	Turkish Petroleum (23.75% interest acquired)—1912.
R 12b	American Gasoline Company formed in Seattle (U.S. entry)— 1912.
R 12c	Roxana Petroleum Company formed in Tulsa (first U.S. production) —1912.
R 13a	Discovery in Mexico—1913.
R 13b	Production acquired in Trinidad—1913.
R 13c	Production acquired in California and Oklahoma—1913.
R 14	Discovery in Venezuela—1914.
R 15	Martinez (San Francisco), California refinery built—1915.
R 18a	Wood River (St. Louis), Ill. refinery built—1918.
R 18b	Curaçao refinery completed—1918.
R 19	Mexican Eagle Oil Company purchased—1918.
R 20a-e	Specialty products added to product line: (a) aviation fuel, (b) diesel fuel, (c) home heating oil, (d) lubricants, (e) asphalt 1920's.
R 20f	Shell-Mex and BP Company formed to market in U.K. (60% Shell) —1920.
R 22a	Concession in Argentina—1922.
R 22b	Concession in British Borneo—1922.
R 22c	Shell-Union formed in U.S.—1922.
R 26a	Shell Company of Canada formed—1926.
R 26b	Shell Development Company formed—1926.
R 27	Discovery in Iraq by Turkish Petroleum (23.75% Shell)—1927.
R 28a	The Consolidated Petroleum Company Limited, a joint venture with BP, formed to market in Near East and east and south Africa—1928.
R 28b	Burmah-Shell Oil Storage and Distribution Company of India formed—1928.
R 28c	Achnacarry and Red Line agreements—July 31, 1928.
R 29a	Nationwide U.S. marketing—1929.
R 29b	Shell Chemical Company formed—1929.
R 32	"Heads of Agreement for Distribution" pact signed—1932.
R 42	100 octane gasoline produced in large volume—1942.
R 47	Long-term contract with Gulf for Kuwait crude—1947.
R 45	Discovery in Colombia—1945.
R 48a-g	Refinery construction at (a) Stanlow (U.K.), (b) Shell Haven (U.K.), (c) Pernis (Holland), (d) Rouen (France), (e) Cardon (Venezuela), (f) Geelong (Australia), and (g) Bombay (India).
R 53	Discovery in Canada—1953.
R 54	Iranian Consortium interest (14%) acquired—1954.
R 56	Discovery in Algeria—1956.
R 58	Discovery in Nigeria (with BP)—1958.
R 59	Discovery of gas fields near Slochteren (Netherlands) (with Exxon)—1959.
R 63	Discovery in Abu Dhabi—1963.
R 64	Discovery in Oman—1964.
R 70a	Half interest acquired from Gulf in the HTGR reactor—1970.
R 70b	Acquisition of Billiton (mining and metals company)—1970.
R 71	Discovery in North Sea—1971.

R 72	Discovery in North Sea—1972.
R 78	Marcus Samuel's seashell and general import/export trade—1878.
R 80a	Shell name and trademark—date of origin unknown—before 1897.
R 80b	Expertise in international trade—1840's on.
R 80c	Expertise in shipping—1840's on.
R 80d	Expertise in marketing consumer products—1840's on.
R 90	Production of crude in Sumatra (Royal Dutch)—1890.
R 92a	Ten-year contract to purchase Russian kerosene (Shell)—1892.
R 92b	Eight bulk tankers—1892.
R 92c	Pipeline and refinery in Sumatra (Royal Dutch)—1892.
R 96a	Skills of Henri Deterding—1896 (Royal Dutch).
R 96b	Concession in Borneo (Shell)—1896.
R 96c	Production in Borneo (Shell)—1896.
R 96d	Refinery at Balik Papan (Shell)—1896.
R 97	"Shell" Trading and Transport Company formed—1897.
R 98	Gasoline added to product line—1898.

Exhibit SH-4
Royal Dutch/Shell Handicaps

H 1	Difficulty with marketing (Royal Dutch): 1896-1903.
H 2	Weakness in production: 1896-1907, 1930-present.
H 3	Lack of equity interest in production in Saudi Arabia or Kuwait: post World War II—1976.
H 4	Organizational inefficiency, duplication, overstaffing, and lack of coordination from decentralization.

Exhibit SH-5
Royal Dutch/Shell Personal Values

PV 1	Marcus Samuel: Jewish.
PV 2	Marcus Samuel: merchant (middle class) not professional (aristocratic) social class, even among British Jewry.
PV 3	Henri Deterding: banker and accountant.
PV 4	Henri Deterding: marketing background.
PV 5	Marcus Samuel: promote oil as fuel for ships.
PV 6	Marcus Samuel: patriotism—serve the British Empire.
PV 7	Marcus Samuel: close family ties.
PV 8	Marcus Samuel: sought social acceptance from upper class Anglo-Jewry and British Empire more than wealth.
PV 9	Hugo Loudon: aristocrat with extraordinary diplomatic skills.
PV 10	Henri Deterding: viewed oil solely in business, not political, terms—"a Dutch merchant and nothing else" (Gerretson).

PV 11	Henri Deterding: did not believe in price competition.
PV 12	Dutch frugality.
PV 13	Henri Deterding: anticommunist.
PV 14	Henri Deterding: autocractic.

Exhibit SH-6
Royal Dutch/Shell Social Responsibilities

SR 1	Serve British Empire—public service in general.
SR 2	Accommodation and respect for local, cultural, and individual differences.
SR 3	Local Shell companies should be integral parts of local economies and instruments of social welfare.
SR 4	Generous personnel policies.

10

The Strategy of British Petroleum

The strategy of British Petroleum (BP) may be briefly summarized by four of its strategic characteristics. BP is a British and once colonial company, an instrument of the British Empire, the only Major with government ownership, and possesses a long and intimate relationship with the British government. It is the founder of the Middle East oil industry and until recently its largest producer. It is primarily an Eastern Hemisphere (Middle East/Europe) company and, until its 1970 acquisition of Sohio for its North Slope crude reserves, the only Major without a strong U.S. presence. It is also the largest producer in the North Sea. It is the most successful explorer for oil in the world, the largest crude seller, and the weakest of the Majors downstream. The environmental conditions (EC#) resources (R#), handicaps (H#), social responsibilities (SR#), and personal values (PV#) which produced its policies (PL#) are keyed to the exhibits at the end of the chapter.

10.1 Anglo-Persian

BP's first name, the Anglo-Persian Oil Company, Limited, communicates the two most important elements in its strategy: a British, and then colonial, company (PL1) founded to develop Persian oil (PL2). It was formed in April 1909 to develop commercial production at the discovery in southwest Persia in May 1908 (R8) on a huge concession covering 4/5 of Persia (now Iran) which had been granted to William Knox D'Arcy in 1901.

His was the third attempt to find oil in Persia. The first two, in 1872 and 1889, had been backed by Baron Julius de Reuter, founder of Reuters, a British news agency. On the first attempt the concession had been rescinded under pressure from the Russians (EC3). The second found no oil and lost £94,000. But a French government mission brought back favorable reports about oil prospects in Persia which were published in 1892. With thousands of seepages and tar pits, it was widely believed that there was oil in Persia as well as other parts of the Middle East (EC2).[1]

But not by all. C.S. Gulbenkian, who put together Turkish Petroleum in 1914, had been offered and declined an oil concession in Persia for £15,000, a decision that rankled him to the end of his days. In his memoirs he wrote, "Between 1895 and 1900 the Concession which afterwards came into the possession of the Anglo-Persian Oil Company was a drag on the market. I submitted this business to my friend Sir Frederick Lane—also, I believe, to Deterding. But we all thought it was a wild-cat scheme and it looked so speculative that it was a business for a gambler and not at all for our trio. It was on my refusal that Mr. D'Arcy, a great speculator in mining businesses in Australia and elsewhere, interested himself and succeeded in forming a syndicate...."[2]

No additional venture capital was found for oil exploration in Persia until 1901 when Spindletop (EC1) in Texas set off speculative fever of fortune seekers all over the world. One of them was D'Arcy (PV1) who, in a British version of Horatio Alger, had emigrated to Australia and made a fortune in gold mining in the 1880's and had retired to a life of luxury in England.[3]

He was susceptible when called upon by the British Minister to Teheran, a Persian general, and two French geologists from the French mission a decade earlier. He sent two representatives to Teheran and was quickly promised a concession but the Russians objected (EC3). The concession was written in Persian and slipped past their scrutiny on May 28, 1901 when the only Russian diplomat who could read Persian was away from Teheran.[4]

The 60-year D'Arcy concession (R1) covered 480,000 square miles, an area almost twice the size of Texas and with, as it turned out, twice the reserves of oil. To appease the Russians the five most northern provinces, which bordered Russia, were excluded. The Persian government was to receive £20,000 cash, £20,000 of shares, and 16% of the net profits of the company which would develop the concession. In addition the Shah and other notables received £50,000 worth of shares, increased to £65,540 by 1914 (PL3). It was understood that the British Minister to Teheran who arranged the negotiations was not to be forgotten, should the venture prove profitable. Thus the D'Arcy concession was the product of bribery, intrigue, and subterfuge—practices for which the British blamed the weakness and corruption of the Persian government (EC4) but to which they acquiesced (PL3). An analysis by Zuhayr Mikdashi in 1966 did not find the terms unfair to Persia.[5]

D'Arcy nearly lost his fortune in Persia (PV1). After an initial discovery in January 1904 went dry, he approached the Rothschilds in Paris and other groups to discuss the possible sale of the concession. Fearing that it might fall into non-British hands (PL1),[6] Lord Fisher (who would convert the Royal Navy from coal to oil) persuaded the Burmah Oil Company to bail D'Arcy out (EC5). Glasgow-based Burmah was Britain's oldest oil company and supplied the Navy from fields in Burma and Assam (a state in northeast India), even further from Europe than Persia. The Concessions Syndicate Limited (CS Ltd.), for which Burmah would provide the working capital (R5) to find oil in

Persia for the Navy (PL4), was formed in 1905.[7] It was the beginning of a close relationship between company and government (PL5).

Lt. Arnold Wilson, a political officer in Persia, expressed British fears of Burmah's motives: "What is to prevent CS Ltd. from selling D'Arcy's rights to an American or German company? The directors of CS Ltd. are Scotsmen of the hard-headed, short-sighted (PV3) sort who would not hesitate to do so if they saw a profit ..... the Burmah Oil Company took a financial interest in CS Ltd. because they did not want a rival company within reach of India, where they have something like a monopoly at present."[8]

D'Arcy's next hurdle was political and economic. The authority of the Shah was minimal in the southwest part of Persia where the company wanted to drill next. The territory was controlled by the Bakhtiari tribesmen who demanded and got £3,000 per year for protecting company property plus a 3% equity interest in any companies which exploited oil in their territory (PL3, H2). From December 1907 through September 1909 the British government provided 20 soldiers, led by Wilson, to protect the drillers (EC6).[9] From its beginning the company had the interest and support of the British government (EC5,6, R4, PL5) as a British controlled source of oil for its navy (PL2,4) and a counter to Russian influence in Persia (EC3).

While D'Arcy deserves much credit for persevering (PV1) through these difficulties from England (he never set foot in Persia), his investment might never have paid out had it not been for the inspiration, leadership, scientific ability, and single-mindedness of purpose (PV2) of the man he selected as his field manager, G.B. Reynolds. Reynolds at age 50 overcame every physical and political obstacle with natives, terrain, equipment, labor, and home office for seven frustrating and disappointing years in his determination to find oil. A self-trained geologist and civil engineer, he could not be restrained from drilling at Masjid-i-Sulaiman (Mosque of Solomon) in the Madan-i Naftun (Plain of Oil) where the first major discovery was made on May 28, 1908 (R5).[10]

The Anglo-Persian Oil Company (APOC) was formed April 14, 1909 with £2 million in capital, £1 million provided by Burmah Oil for common stock (R5) and the rest by sale of 6% preferred shares to the public (R9). D'Arcy's wife felt his contribution should have been recognized in the name of the company, but instead he received reimbursement for all he had spent plus £900,000 worth of stock in Burmah Oil, and the satisfaction of having organized not one, but two, of the greatest discoveries in the history of the Empire.[11]

The Scots of Burmah Oil, who had been in charge since 1905 of the Concessions Syndicate (CS Ltd.), formally took over Anglo-Persian in 1909. Lord Strathcona, the 88-year-old head of Burmah, became chairman and Sir Charles Greenway became Managing Director, succeeding Strathcona in 1914. Greenway had been in charge of marketing in India (PV4) for Burmah before becoming a director of CS Ltd. in 1905.[12] He left no personal imprint on the company, but as Sampson notes, the Burmah executives " . . . contributed part of their character to the new company. Scots engineers and accountants dominated the headquarters, with a mixture of adventurousness and meanness" (PV3).[13]

D'Arcy had been eased out by the Scots in 1909 and Reynolds was the next to go. Following a common practice of the day Burmah hired a firm of "managing agents" to oversee its Persian operations. Having found oil, Reynolds felt entitled to command. Conflicts arose. In February 1911 Greenway and James Hamilton became the first directors to visit Persia, investigated the discontent, and dismissed Reynolds.[14] Neither D'Arcy nor Reynolds ever received recognition for their service to the Empire. Reynolds went to Venezuela and was forgotten. BP named its exploration subsidiary after D'Arcy until 1957.[15]

Due to the distance from Europe to Persia (H2) and the logistical difficulties involved, a 130-mile pipeline to the coast (R11) was not completed until 1911 and a source refinery (PL7) on the island of Abadan (R13a) was not completed until 1913, at which point the company once again needed financial help.[16]

Greenway had been pressing the Foreign Office for a subsidy since 1912, arguing that Shell was a foreign company whose loyalty was questionable and a marketing monopoly. The Foreign Office was convinced but the admirals were wary of involvement and interested in price. After a public outcry against high oil prices in 1911 Churchill, who became First Lord of the Admiralty in October 1911, was suspicious of Shell and came out in favor of APOC (EC14a), "Against this (the dominance of Standard and Shell), . . . the Burmah Oil Company, with its offshoot the Anglo-Persian Oil Company, is almost the only noticeable feature." Of Shell he said, "It is their policy . . . to acquire control of the sources and means of supply and then to regulate the production and the market price . . . We have no quarrel with Shell. We have always found them courteous, considerate, ready to oblige, anxious to serve the Admiralty and to promote the interests of the British Navy and the British Empire—at a price . . . The only difficulty has been price. On that point of course we have been treated with the full rigor of the game . . . we shall not run any risk of getting into the hands of these very good people."[17] An independent blue ribbon commission, including Sir John Cadman, was sent to Persia and reported favorably on the oil fields. Cadman (PV6) then a professor of petroleum, civil servant, and Britain's leading oil expert, subsequently joined APOC in 1921 and succeeded Greenway in 1927 as chairman.[18]

In 1914 Churchill convinced Parliament to bail APOC out (PL5,8), investing £2 million for just over 50% of the company (EC14b, R14a).[19] His desire like Fisher's was to convert the Royal Navy to oil because oil-fired ships were much faster, more efficient, and had greater range than coal-fired ships.

But Britain had no quantities of oil (EC5) and, until discovery in Persia, the only oil in the Empire was in Burmah and Assam.[20]

APOC got capital in exchange for an oil supply contract (R14a,b, PL4) for the Navy at a cut-rate price.[21] Some of the oil companies' greatest difficulties have come from navies who, since this precedent, have seemed to feel entitled to a cut-rate price. Examples are Socal's difficulties after World War I over price on the West coast, the post-World War I fear of shortage which led to the establishment of the Elk Hills Naval Reserve in California and a court battle with Socal over its perimeters, and the U.S. Navy's attempt to buy into Saudi Arabia in 1943 (see Chapter 7).

The agreement stipulated that the company must remain an independent British company and every director must be a British subject (PL1).[22] To represent its ownership the government was to appoint two directors with veto power over issues of national strategic importance. But otherwise the government was not to interfere with the conduct of the business (PL8). If this distinction was clearly specified in the government's agreement with the company, elsewhere the distinction of being an autonomous company majority owned by the British government was blurred.[23] "Even in Persia," Lord Strathalmond, chairman from 1940 to 1954, said, "they were never quite convinced that I had not come straight from Whitehall."[24] In Iraq the company and the British government were thought to be one and the same.[25] As events show, this special business-government relationship was both a strength (R14c) and a weakness (H1).

Mikdashi analyzed the financial and political aspects of the agreement, points 1-5 below, concluding that the government in retrospect got a good deal, though at the time Marcus Samuel of Shell considered it a "gigantic subsidy" to APOC and pleaded for similar treatment.[26] (1) The government purchased stock at par at a time when market values were much higher and a large shareholder of APOC (probably Burmah) urged in 1914 that existing shareholders be given the option of providing at par the additional capital required.[27] (2) The lifelong fuel oil supply contract with the British Navy (R14b) called for prices which were much lower than those prevailing in the market and provided for an automatic reduction in price up to 25% of the excess of profits above dividend requirements. Cumulative savings from this provision were estimated by Churchill in 1923 at £7.5 million and by the Persians in 1951 at $500 million.[28] (3) The contract assured the company a profit regardless of market conditions, insulated it from all competition, and spared the company the expense of building a marketing organization in competition with the large international oil companies (PL9).[29] But it also directed its production technology to a specific product, fuel oil (PL19). The government favored and assisted (R4) the company in the sale of products to other buyers, like the Iraqi and Indian railroads. Churchill openly declared that the British government was "bound to help and bound to enrich APOC."[30] Its political assistance in limiting competition and influencing Near Eastern governments greatly benefitted the company. (5) APOC made some "public service" expenditures it might not have made were it not for government ownership (PL6,10).[31]

As the visible symbol not only of the British Empire (PL1) but also of western civilization the company recognized its social responsibility in bringing that civilization to Persia. It trained labor and built roads, houses, schools, and communications in the country (SR1, PL10). As at Jersey Standard, the medical department made some of the greatest contributions. Providing medical care in Persia not only to their own employees but to anyone in the area was a company policy (PL6, SR2) from its earliest days. An extraordinary doctor and mediator, M.Y. Young, known to all as the Little Doctor, led this effort from 1907 to 1936. So great was the respect for him that the neutrality of Persia during World War I and the absence of trouble in the oil fields during the war has been attributed to his personal influence.[32]

It is impossible to place a dollar value on the benefits (PL5) of the special relationship (R14c) to the company. The government's restriction of competition is surely responsible for the company's leading market share in the Middle East. Without it Jersey would have entered Persia in 1921; there might not have been any Red Line agreement in 1928, and it is unlikely that APOC would have gotten into Kuwait in 1934. With a guaranteed contract for large amounts of fuel oil to the British Navy (R14b) the government gave the company a different product and marketing strategy (PL4) than the kerosene/gasoline-sold-to-private consumers-and-industry strategies of the other Majors. The contract also gave APOC large scale (R14d) in production, refining, and bulk sale marketing (PL11).

The characteristics of Persian oil—high quality, unlimited quantity, and low cost (EC9, R13c) also shaped APOC's product and marketing policies (PL4,9,11). They overcame the handicap (EC7) of Persia's remoteness from markets. With or without the contract (R14b) and special relationship with the government (R14c, H1) APOC would still have been able to compete successfully with any other company. The government's military protection was not required again until World War II and in 1951 when needed was not used. The influence of the British government with the Persian government (EC6) permitted the company to adopt courses of action and attitudes (PL13) in its relationship with the Persian government that ultimately cost it a great deal (H1).

In early 1915 as World War I (EC15) moved into full force and tanker rates soared, the company made another important strategic commitment that would become its second strongest vertical segment (after production PL2). In April 1915 it formed the British Tanker Company (R15), a wholly owned subsidiary to own and manage its own tanker fleet (PL12), in order to obtain secure and reasonable marine transportation.[33]

Disputes between the company and the Persian government (PL13) led to the "interpretative" agreement of 1920 (R20) which in effect revised the D'Arcy concession to reduce the base for royalty payments to the Persian government to the profits of the producing company in Persia. The D'Arcy concession had originally specified the Persian government's share to be 16% of net profits generated by all subsidiaries in exploiting Persian oil (including refining, marketing, and transportation).[34]

On occasion the company did not pay this amount. The Persian government was weak, inefficient, and corrupt (EC4). It did little except to organize bribery for its rulers and provided no protection in areas where its authority was nominal. According to one account the local press referred to them as the "robber princes." The company had already had to give £3,000/year and a 3% equity share in all companies working on their land to the Bakhtiari tribesmen for "protection" in 1905 (H2). In 1914 it unilaterally deducted these payments from the Persian government's share. In 1918 it suspended payments over a war damage claim against Bakhtiari tribesmen for blowing up its pipeline in 1915. It refused to arbitrate this dispute despite explicit provisions for arbitration in the concession (PL13).[35]

On these and other smaller matters in its dealings with the Persians the company took advantage of the strength of the British government (EC6) and the weakness of the Persian government (EC4), fostering a tradition of inflexible and arbitrary dealings (PL13) which later caused it great difficulty. It refused to allow the Persian govenment any participation in management or equity (PL14) and it tried to induce the Bakhtiaris to sell their shares to British interests. Bribery was the accepted method of getting things done. Both the company and the British government subsidized Persian officials when they wanted something (PL3).[36]

With the Persian treasury and rulers dependent on Britain, a treaty was signed in 1919 giving British advisors the right to negotiate for the Persians. Three Persian officials received £131,000 for signing the agreement. To settle the disputes between APOC and the Persian govenment, negotiations were held between a British Treasury official on behalf of the Persians and APOC to make the company live up to the payments promised and give up artificial bookkeeping methods unfavorable to the interests of Persia.[37]

The result of these "negotiations" was the "interpretative" agreement of 1920 (R20) which excluded from the calculation of royalty payments profits of the newly formed tanker subsidiary, subsidiaries trading non-Persian oil, large amounts of the profit of subsidiaries refining and marketing oil outside Persia, and all profits of less than majority owned subsidiaries. Freed from the restrictions of the D'Arcy concession APOC could now expand outside (PL15) Persia. Persia received £1 million in settlement of all grievances.[38]

By 1920 the major elements of Anglo-Persian's strategy were in place. It was an independent but quasi-govenmental company which was an arm of

Burmah Oil and the British Empire (PL1,5,8). It was organized to find and produce Persian oil, refine, transport, and sell it as fuel oil to the British Navy (PL2,4,7,12). It was a strategy quite different in geography, vertical integration, product, and administration from any other Major. From these commercial and national strategic purposes, the other policies follow.

10.2 Between the Wars

The next period in BP's history was marked by steady growth in its volume of business and some additions to its geographical areas of activity. There was one major shift in strategy (PL17): to integrate vertically from production through marketing, rather than simply produce and sell fuel oil to the Navy. The important additions were production in Iraq and Kuwait, refining in Britain and France, and joint marketing arrangements in Europe (R17-32).

Having found oil in Persia, the company resolved to seek additional concessions and find it elsewhere as well (PL16). This was partly due to its success in exploration (R8) and partly to diversify its sources of supply to carry out its mission for the British Navy (PL4). It was also to reduce its dependence upon Persian politics (PL15), which had become more difficult with the disputes cited above and less predictable as the Shah and, with him, British influence became weaker (EC4,5). It may also be attributed to the far-sighted and global views of Sir John Cadman, who became Deputy Chairman in 1924 (PV5).[39] The "interpretative" agreement of 1920 (R20) cleared the way of obstacles.

In a defensive move it first attempted to secure the rest of Persia but its efforts came to nought. An oil concession similar to D'Arcy's, covering three of the five northern provinces, had been given to a Russian with czarist support in 1916. APOC bought it for £100,000 in 1920 and began negotiating with the Persians for the other two provinces in 1921. Following a new treaty with the Soviet government (EC17) renouncing all previous ties including concessions, the Persians cancelled the concession on November 22, 1921 and on the same day offered Jersey a concession covering all five northern provinces. As Jersey could not build pipelines across the Soviet Union, it accepted APOC as an equal partner to gain pipeline rights across Persia and to improve its chances of entry into Iraq where APOC was considering sale of half of its interest. Then in 1922 the Persian parliament annulled Jersey's concession, allegedly because it had declined to pay *backsheesh* (bribes) to political circles in Teheran (EC4).[40]

APOC had long been interested in Iraq (EC5, PL4,5). D'Arcy had been negotiating with the Turks for a concession, as had others, since 1901. In 1914 it had obtained a 47-1/2% interest in the Turkish Petroleum Company (R14e), a company which had been put together by C.S. Gulbenkian in that year from competing bidders for concessions in Iraq. But it had no concession. Iraq, which had been part of the Ottoman Empire until after World War I, became a separate state in 1920, administered as a British mandate until 1932. In 1925 a concession was arranged which called for a royalty of 4s gold per ton and a minimum rental of £400,000/year in gold. Bribes were paid (PL3) and threats were made by the British government (EC6) before the concession agreement was signed. It was the first tonnage-based royalty in the Middle East (R25).[41]

Oil was discovered at Kirkuk on October 14, 1927 (EC27). It was APOC's third major discovery (R8,23c,27a) and it demonstrated its growing expertise in finding oil (R27c). A local refinery was built but an export pipeline to the Mediterranean was not completed until 1934.[42]

In 1928 APOC sold half its share of the Turkish Petroleum Company to a group of American companies led by Jersey when they agreed to limit themselves to a joint efforts within the boundaries of the old Ottoman Empire, an area marked by C.S. Gulbenkian, an Armenian with political connections in Turkey who acted as broker in the sale, with a red line during negotiations. It was a containment policy. As Gulbenkian noted, "... it was sounder and higher policy to admit the Americans into the Turkish Petroleum Company, instead of letting them loose to compete in Iraq for concessions when in reality the Company had a very weak grip there" (PL21).[43]

APOC began to explore for oil worldwide, that is, outside the British Empire as well as in it (PL16). It made a small discovery (R24b) in Argentina in 1924 and a noncommercial strike in Albania in 1926. Both were abandoned in 1930. It searched between the wars in Canada, Mexico, Australia, Papua (1938), and England (1936) without discovery. It made a discovery in Qatar just before the war (R39).[44] It was pushed into Kuwait by the British government, though its geologists did not believe there was oil on the Arabian side of the gulf until 1931. With Gulf it made a major discovery there in 1938 (R38a), giving it positions in three of the four huge Middle Eastern producing areas, but nowhere outside the Middle East.

Once it began to explore worldwide it is curious that Anglo-Persian did not establish itself in the other three major producing areas in the free world: the U.S., Mexico, and Venezuela (H4). The writer was unable to determine why APOC did not enter the U.S., nor why it was not successful in Mexico. But the government of Venezuela was not interested in a company with government ownership, so in Venezuela APOC's special relationship became a handicap (H1).[45]

One of the unfulfilled desires of both the company and the government, until discoveries in the North Sea, was to find oil at home. As early as 1936 the company explored in England, making a small discovery in 1943 (R43).[46] Its main home source was the acquisition of the once prosperous Scottish shale oil companies (R19b,32a) in 1919 and 1932, its motive being "to keep alive a home industry giving employment to many thousands of men" (SR3, PL10).

Scotsmen, one might add. One of them, William Fraser, who was associated with the four acquired in 1932, became APOC's chairman in 1941, succeeding Sir John Cadman.[47]

Two events triggered a renegotiation of APOC's Persian concession (R33) in 1933. One was the tonnage royalty agreement negotiated with Iraq in 1925 (EC25) which set the pattern for other concession agreements in the Middle East until 1948. It was far simpler to administer and did not provoke disagreements as did the net profits type of concession. It freed the company from obligations on profits earned downstream or in other places. It gave the government a more dependable income. The level (4s gold/ton) was attractive, being equivalent to about one-eighth of the estimated value of Middle East crude in the Persian Gulf, a level similar to the one-eighth royalty in general practice in the U.S.[48]

The other was a change of government in Persia (EC21). In 1921 a revolution overthrew the old regime. It was led by a soldier, Brigadier Reza Khan, first of the Pahlevi dynasty, who established a strong military government in Persia and was proclaimed Shah. In spite of their subsidy of the former government the British and APOC remained neutral and enjoyed cordial but distant relations with the new Shah until 1928 when profits fell with the worldwide crude glut (EC28a). Payments to the Persian government fell to £500,000 from £1.4 million the year before. They rebounded in 1930 but again fell from £3 million in 1930 to £300,000 in 1931. In 1932 the Shah annulled the D'Arcy concession unilaterally (EC32). The British Navy started patrolling the Persian Gulf (EC5,6, R14c).[49]

Just as talks in Persia were about to break down in April 1933, Cadman and, especially, Fraser were successful in renegotiating the concession. Cadman described his bargaining tactic as a "shock decision." After congratulating the Shah on the anniversary of his coronation, he announced that, as no satisfactory end to negotiations was in sight, he proposed to leave the next day.[50] A new concession (R33) was quickly arranged. The extreme measures taken provide evidence of APOC's difficulties in dealing with the Persian government (PL13).

Replacing the 16% of net profits was a 4s gold/ton royalty, a guaranteed minimum annual payment of £1,050,000, 20% of the dividends on common stock above a minimum, reduction of the concession area (done in 1938) to 100,000 square miles of the company's choice, an extension of the concession to 1993, and a £1 million payment to settle past claims as in 1920. In 1935, to please the Shah, the company's name was changed to Anglo-Iranian Oil Company (AIOC).[51]

There was one major shift in policy in the interwar period. It was to integrate forward all the way from production to marketing (PL17), rather than simply produce crude, sell it as fuel oil (PL4) to the Navy, and dispose of other products. In 1920 Lord Greenway, the former Burmah marketer (PV4), declared that integration forward into marketing was necessary to avoid forced sales of surplus products to other oil companies on disadvantageous terms. Only three years before he had stated, "Rate wars, or no rate wars, we shall always be certain of some profit, no matter at what prices we may have to sell our benzine and kerosene, and we can therefore regard with perfect equanimity any competition that may present itself in these products."[52] If APOC had marketed its products directly in Europe, he revealed, profits would have been twice as high (H3).[53]

Refining near markets was one step (PL15). During the war Admiral Slade, one of the two government directors, had urged APOC to build a refinery in Britain to supply the Navy. With the Admiralty's support the project was classed as a "war emergency" and begun only to be dropped in December 1917 when it was no longer essential to the war. This small refinery in south Wales at Llandarcy (Llan + D'Arcy) was completed in 1922 (R22). Refineries were built in Scotland at Pumpherston (1923) (Scottish shale) and Grangemouth (1924) (R23b, 24a). None of them processed more than 7 TBD before World War II. Small market refineries were built in Melbourne, Australia (1923, 3 TBD) and Dunkirk (1923) and Marseilles (1932) in France (R23-32). Small source refineries were built in Argentina (1924), Iraq (1927), and Persia (1935). Even taken together the output of these eight refineries was small. The refinery at Abadan (R13a) continued to be the jewel of the APOC empire (PL7).[54]

If its commitment to market refineries was relatively modest, its commitment to marketing was even more so (PL9). But it was establishing an Eastern Hemisphere marketing network (PL18). In 1917 APOC acquired from the U.K. Trustee for Enemy Property a small German-owned but British-registered marketing company (R17), the British Petroleum Company. APOC began marketing under the name "BP" in Europe in the 1920's and adopted the name for the parent company in 1954.[55] To expand quickly, at minimum expense, and without price cutting APOC formed joint marketing arrangements with existing marketers, the Whalers' Association in Norway and Denmark for example, buying in with crude rather than cash (PL19). It joined Shell in India and Egypt and in 1928 in the Mediterranean and east and south Africa (R28a). It is curious that Burmah Oil played no role with APOC. In 1932 APOC joined Shell as the junior partner in a 60:40 arrangement (R32c) in Britain under the Shell-Mex and BP names. Using its strength in fuel oil (PL4) in 1919 it began a chain of bunkering installations (R19a) to supply fuel oil to ships (PL20) like the whaling fleet in Scandinavia.[56]

In terms of the resulting vertical balance of operations, total refinery throughput in 1918, 1928, and 1938 nearly matched production. Only the scantiest data on product marketing volumes prior to World War II are available in published sources, but in 1938 AIOC produced 234 TBD (= thousand barrels per day) and refined 230 TBD and in 1939 sold 200 TBD, making it the third largest Major by volume.[57] These numbers indicate considerable success under Cadman's administration in the policy of forward integration (PL17).

More so than any other Major, APOC, supported by the British government (PL5,8, EC5), has pursued openly the minimization of competition (PL21) through cartelization, control of production, price, concession areas, and cooperative rather than competitive ventures in production, refining, and marketing. Control of all oil in Persia was a clear goal of the provisions of the D'Arcy concession granting the largest concession area prior to Saudi Arabia and limiting pipeline access to the Persian Gulf. It was also the motive behind the attempt to secure concessions covering the rest of Persia after World War I. It was a goal in Iraq in 1925, as Iraq Petroleum's chief executive remarked: "Competition for oil in Iraq was not economically sound."[58] When an independent British and Indian group, the British Oil Development Syndicate (BOD), acquired a 75-year concession in Iraq in 1932, Iraq Petroleum (formerly Turkish Petroleum) with government approval attained control in 1937 after BOD had been unsuccessful and established the Mosul Petroleum Company (R37) to operate it in 1941 following a strike in 1940 (R40).[59] To complete its control of Iraq, an Iraq Petroleum affiliate, the Basrah Petroleum Company (R38c) obtained a 75-year concession on the remaining portion of Iraq which bordered Kuwait without competing bids.[60] The Red Line agreement (R28b) of 1928 was a British inspired device to contain American competition. APOC participated in the Achnacarry agreement (EC28c) with Shell, Gulf, and Jersey, and Socony-Vacuum to divide markets and forestall price competition downstream as well as three other similar agreements between 1928 and 1934.[61] It was frank about its purpose; as its chairman stated in 1927, "It is in the interests of Persia, as it is of this company, that production should be steadily controlled—that is to say, regulated in conformity with the world's demands."[62]

In light of the wasteful results of competition under the rule of capture and the distress of the Great Depression (EC29,30) one cannot entirely condemn APOC's open aversion to competition (PL21) which was born in its identity (PL1,4) and its special relationship with the government (PL5,8, R14c). The American tradition of antitrust and adversary relationships between business and government was absent. In its view its close relations with the British government and its identity as an instrument of social policy of that government, rather than competition, ensured socially responsible behavior.

10.3 1940-1956: War, Nationalization, and Suez

William Fraser, who had negotiated the revised concession of 1933 with Persia, led Anglo-Iranian through World War II (EC39) and the Mossadegh nationalization in 1951 (EC51b), the most difficult period in the company's history.

The events of war (EC39) brought great changes in the company's environment but none in its strategy. Before the war a large expansion intended for completion in 1943 had begun. Drilling continued in new fields discovered since Masjid-i-Sulaiman in 1908: Naft Khaneh (1923), Haft Kel (1928), Gach Saran, Agha Jari, Pazanun (gas), Naft Safid (all discovered after Haft Kel), and Lali (1934) (R27c).[63] After Italy's entry into the war closed the Mediterranean, the "short haul" policy (EC40) adopted in mid-1940 to minimize the danger from submarine attack meant supply from Venezuela and the U.S. Persian production was cut from 234 TBD in 1938 to 130 TBD in 1941 and Iraqi production from 80 TBD (1938) to 30 TBD (1941). Reza Shah had attempted to remain neutral but had allowed Germans to infiltrate Persia. When Hitler attacked Russia on June 22, 1941 Britain came to her aid, desiring to use Persia to supply the Russians from the south. When the Shah refused to expel the Germans in August, Russia invaded from the north and Britain from India and the south. In three days the reign of Reza Shah came to an end (EC41a). He was exiled and died. The British and the Russians governed during the war.[64]

When the Japanese attacked Pearl Harbor later that year (EC41b), bringing the Americans into the war, sources of supply in Burma and the Dutch East Indies fell to the enemy. But Abadan, the largest refinery in the world, was ready and running at half steam. It became the principal source of fuel, especially aviation fuel, with which the war in the Pacific was fought.[65] During the war the British Tanker Company, AIOC's shipping arm, lost 44 of its 93 ships.[66] By 1945 Persia was producing 336 TBD, a volume that would double by 1951.[67]

With the rise of Middle Eastern production to prominence (EC43) as DeGolyer had predicted in 1943, AIOC's production began to exceed what it could refine and market. It became a large-scale seller of crude (PL22). AIOC expected the prewar crude glut (EC28a) to continue. Kuwait and Saudi Arabia had just begun commerical production (EC44, R46). A big increase was needed in Iraq to appease the government. Discoveries in Qatar and the Mosul concession before the war awaited development (R39,40). A strike was made on the Basrah concession in Iraq in 1948 (R48a). To secure markets for its crude AIOC signed long-term contracts (R47a,b) with Jersey and Socony-Vacuum in 1947. The Jersey contract was for 50 million barrels over 20 years at cost plus a shilling (about 20 cents in 1947) per barrel (7.33s/ton). This was about 1/3 the prevailing level of profits per barrel in the Middle East. The Persians complained about the price. Jersey and Socony diversified their sources of production cheaply. But since their new position in Saudi Arabia could have met all their needs even more cheaply, it was a buyer's market. The only leverage AIOC had was Jersey's and Socony's desire to be released from the Red Line agreement in order to enter Saudi Arabia. The company certainly missed an opportunity as Shell, its partner elsewhere, had met its requirements through a contract with Gulf.[68]

Refining was expanded geographically as quickly and cheaply as possible (PL17). At the end of the war AIOC had eight refineries in Iran (Abadan and Kermanesh), Iraq, Australia (Melbourne), France (Marseilles), and Britain (Llandarcy, Grangemouth, Pumpherston (Scottish shale). The refinery at Dunkirk (R23b) was rebuilt and opened in 1951. An Iraq Petroleum refinery at Haifa (Israel) (R48b) was in operation briefly by 1948. In 1948 AIOC purchased 49% of a refinery in Venice, one in Hamburg, and another in Hamburg in 1951 (R48c,d,51a). With Petrofina BP began constructing a refinery at Antwerp in 1948 and with Gulf in Kuwait in 1949 (R48e,49a).[69]

Marketing was similarly expanded to fill in the gaps in the Eastern Hemisphere (PL18) after 1945: Turkey, New Zealand, Algeria, Morocco, Greece, and Malta (R45a-f). Portugal and west Africa were added in 1954 (R54a,b); Canada, in 1957 (R57a); and Italy, in 1958 (R58a).[70]

But it was a blizzard of politics rather than oil which dominated the postwar decade at AIOC. After the war the British pushed the Russians out of Iran with difficulty and installed the 22-year-old son of Reza Shah as a puppet (EC45). In Churchill's words, "We have chased a dictator into exile, and installed a constitutional sovereign pledged to a whole catalogue of long-delayed, serious minded reforms and reparations."[71] But the new Shah felt differently: "We were an independent country, and then all of a sudden the Russians invaded our country and you British took my father into exile. Then we were hearing that the oil company was creating puppets—people just clicking their heels to the orders of the oil company—so it was becoming in our eyes a kind of monster—almost a kind of government within the Iranian government."[72]

Similar anti-British (EC46a) and anticompany (EC46b) sentiments were expressed in Iraq where it was widely believed that "the offices of Iraq Petroleum and the offices of the British consulate were one and the same."[73] These frictions over national pride have always been a source of difference between foreign investors and host governments, but the nationalism which followed the war exacerbated them. There was also resentment that the U.S. and U.K. governments received more in taxes from Middle Eastern oil than did Middle Eastern governments (EC46c).[74]

The initial dispute concerned, as usual, money. Limitations on dividend payments by British companies were imposed by the British government in 1947 to fight inflation (EC47). This limited payments to the Persian government based on dividends which had amounted to about one-quarter as much as the tonnage royalty. A supplemental agreement was reached in July 1949 (EC49) and submitted to the Iranian parliament for approval.[75]

External events were moving quickly. The granting of independence to India in 1948 (EC48a) had weakened the British military position east of Suez. On December 30, 1950 Aramco had signed a 50-50 profit sharing agreement with the Saudi Arabian government (EC50), replacing the tonnage royalty. Coupled with the tax credit on foreign income given by the U.S. government in 1951 (EC51d), which in effect indirectly subsidized the Saudi government, it was clear that the Americans were offering better deals than the British.[76] This did not strengthen British influence (EC6).

For the company, events were changing more rapidly than were perceptions of them. The war years were tranquil by comparison to the changes from 1945 to 1951 (EC45-51). AIOC, the leading producer in the Middle East, was slow to react, as the entrenched often are. The 50-50 idea had come from a Venezuelan law enacted in November 1948 (EC48b).[77] No doubt AIOC had hoped to lead a united front of resistance to its spread to the Middle East. Acceptance of 50-50 by Aramco's American partners pulled the rug from under AIOC leadership, British influence in the Middle East, and the supplemental agreement. It lingers bitterly in British memories.

Management was at fault (PL13). The narrow-mindedness of William Fraser (PV6), AIOC's charman, was criticized, where his predecessor, John Cadman, had been praised for his breadth of view (PV5). Fraser, wrote one critic in the British Foreign Office, "does not think politics concern him at all. He appears to have all the contempt of a Glasgow accountant (which he was) for anything which cannot be shown on a balance sheet."[78] Neither did the two government directors exercise their authority on strategic issues. These posts had become sinecures for retired civil servants, being occupied in 1951 by a field marshall and a former head of the post office.[79]

Internal events in Iran departed from rationality. With the new Shah in a weak position (EC45) and dependent on the British (EC6,45), a fanatic, Dr. Mossadegh, came to power as Prime Minister on April 28, 1951 (EC51a), riding a wave of anti-British (EC46a) and anticompany (EC46b) feeling. The company had offered to change the supplemental agreement along the lines of the 50-50 agreements but to Mossadegh nationalization was the only solution. The supplemental agreement was never considered. "What the Anglo-Iranian Company did was sheer looting, not business," cried Mossadegh, who had never set foot in the oil fields in his life.[80]

Facts refute the charge. In addition to its commercial activities, which employed 70,000 Persians at £20 million a year and produced £16 million for the Persian government in 1951, the company had built 30 schools, a technical college, three hospitals, 35 dispensaries and nearly all the public health services in southwest Persia, 1250 miles of roads, 40 bridges and from 1945 to 1950 alone £28 million in housing.[81] It was probably more than the government had done for the area and it showed a clear commitment to social as well as commercial responsibilities (PL6,10, SR1,2). The company was justifiably proud of its accomplishments and thought appreciation was in order. It thought of itself as more than a commercial enterprise which found and produced oil. It was an empire builder which brought civilization to the natives (PL1).

Neither were the financial results unreasonable. From Mikdashi's figures on profits and payments and dividends from annual reports, a calculation not adjusted for inflation shows that from 1909 to 1950 of the £492.2 million in profits before taxes and royalties, 22% went to the Iranian government, 35% to the British government in taxes, 20% to shareholders including the British government as dividends, and 22% was retained by the company.[82] In 1950 of £99.8 million, 16% went to Iran, 51% to Britain in taxes, 7% in dividends, and 26% was retained. By relative comparison the terms Iran received were not less favorable than other Middle Eastern governments received. They were considerably better than the one-eighth royalty to leaseholders customary in the United States. With the supplemental agreement or a 50-50 agreement they would be better still.

There appears to be no factual basis for the charge of looting by the company. Quite the contrary, it was clearly doing more for Iran than was required to conduct its business (SR1,2). In terms of services rendered for payments received, only the British government and perhaps the Iranians appeared to be engaged in looting.

The cry for nationalization could not be defused by these facts. It was, perhaps more than anything, a reaction against British colonialism (EC6). The company's special relationship (R14C) became a severe handicap (H1). The colonial mentality, the national role the company felt it had played in Persia, was a unique combination of profit and public service (PL1,6,10). But it did not contemplate stirrings of independence, of relinquishing the white man's burden. Some of the policies which expressed this philosophy (PL13,14) produced as much or more resentment as others (PL6,10) did praise.

The result was that on April 30, 1951, two days after Mossadegh became prime minister, the Majlis voted to nationalize the oil industry (EC51b). The National Iranian Oil Company (NIOC) was created to market the oil (EC51c).[83]

To the company it seemed a devastating blow. Overnight it lost three-quarters of its production and refining capacity, including the largest refinery in the world at Abadan.[84] Its unique political business identity (PL1,5,6,10,13,14,15, SR1,2,R4,H1) had been ingloriously rejected.

What troubled AIOC most was the sheer ingratitude of the Persians.[85] As Longrigg wrote, "the whole of this effort, the whole of a half-century of generous and enlightened treatment of its own workers and the public was, in the final destiny of the company in Persia, not only treated as of no account, but attacked in terms suggesting not mere neglect but the crudest exploitation."[86] By October 4 the British were gone.[87]

Fraser was confident that Mossadegh would fall. The British had been dissuaded from military intervention by the Americans. The other Majors pledged not to deal in Iranian oil and a State Department announcement to that effect for all American companies was issued in May 1951. By July 1 oil exports had ceased. AIOC expanded production in Kuwait, Iraq, and Qatar (R46,34B,49b). It built refineries in Kent (England) (1952), Aden (1954), and Australia (1955) (R52, 54c,55). The Americans were happy to expand in Saudi Arabia. Iran was cut off.[88]

Mossadegh fell in July 1953 in a CIA coup (EC53).[89] The Shah was restored to power but determined to assert his independence. The restoration of a British monopoly of the Iranian oil industry was unacceptable. The U.S. State Department persuaded the other Majors, who in view of a worldwide crude surplus were surprisingly reluctant, to join a consortium to operate the Iranian fields and refinery on a 50-50 profit sharing basis with NIOC. Agreement was reached in August 1954 (EC54). A secret "participants' agreement" was signed among consortium members to allocate production and restrict it to avoid a glut.[90]

AIOC received a 40% interest in the consortium (R54C) and compensation for its loss. Honor was preserved. In December 1954 with the change in its position irrevocable, the company changed its name to British Petroleum. Fraser retired.[91]

10.4 Aftermath

The Iranian nationalization in 1951 was the most traumatic political event in the history of the industry since the breakup of Standard Oil in 1911 and prior to the OPEC embargo of 1973 (EC11,73). In its aftermath relations between the company and the government became more distant. The company's response was a wave of exploration that produced a string of major discoveries unmatched by any other company.

In the pinch of 1951 it was apparent that the special relationship (R14c) with the British government did not work. Government ownership did not ensure protection or support. In spite of the millions of pounds drained away by the government in taxes, the British government could not or would not intervene militarily in the company's hour of need. The cost of the CIA coup is reported to have been $700,000. The two government appointed directors did not function. The attitude of the Foreign Office, rather than lamenting its responsibilities, seemed to be, in Sampson's words, that they had created a "Frankenstein monster, beyond control of its makers."[92]

Indeed it is widely acknowledged that much of the company's plight was of its own making. The desire to control, to be the sole company in Iran (PL21), to be an extension of the Empire (PL1,5), made it an inviting target. Its crimes were imperialism (PL1,5), arrogance (PL13,14), and the misfortune of circumstance, rather than avarice. The American Secretary of State, Dean Acheson, wrote later, "Their own folly had brought them to their present fix which Aramco had avoided by (in Burke's phrase) graciously granting what it no longer had the power to withhold." And again, "Never had so few lost so much so stupidly and so fast."[93]

It is remarkable that the careers of the executives responsible for this criticism were unaffected by such a colossal loss. Chief executives of three of the seven Majors, Teagle of Jersey, Deterding of Shell, and Rieber of Texaco, had been disgraced and retired in the 1930's by political miscalculations about Germany which were of little economic consequence to the performance of their executive responsibilities. Fraser retired but Neville Gass, who had negotiated the supplemental agreement, succeeded him in 1954. Eric Drake, general manager in Iran in 1951, became chairman from 1968 to 1975.[94]

Certainly they were men of talent with other accomplishments, but it is worth noting that the upheaval in Iran did not produce a noticeable shift in policy or personnel. Lacking evidence to the contrary, a plausible conclusion is that their careers offer indirect evidence of the continuation of their policies and inflexibility which subsequent events have not tested. With majority ownership and director representation the British government could have expressed its dissatisfaction on this important strategic matter but it did not; nor did other shareholders. Survival of the crisis, if anything, increased their independence from government constraint. Shortly thereafter, with the Suez Crisis in 1956 (EC56), British government influence (EC6) ended. Having been let down by the British government, the company took steps to outgrow it.

For the company the story had a happy ending. Their response was to double their efforts at what they did best, explore for oil (PL16), with results that were the envy of the industry. Sir Eric Drake said, "The 1951 nationalization of our Iranian properties seemed like a deadly blow at the time, but it caused us to engage in an unprecedented wave of new exploration. It turned out to be the freshest breeze that ever blew through these corridors. I guess a nasty jolt every now and then doesn't do any harm."[95] Iran did not want to be dependent on them; they did not want to be dependent on Iran either (PL15).

The search was worldwide (PL16). In the Middle East BP began exploring in the sheikdoms along the Trucial Coast in 1952 and made a major strike in Abu Dhabi in 1958 (R58b). Two more fields, Magwa and Ahmadi (R58c,d), were found in Kuwait the same year. In Africa they had begun exploring in Nigeria in 1949 (R49c), made the first discovery there in 1953 (R53a), and began export production in 1958 (R58e).[96] A small strike made in Egypt (Anglo-Egyptian Oil Fields—15-1/2% BP) (R54d) was confiscated during the Suez Crisis (EC56) and returned in 1959.[97] BP was among the first companies granted concessions in Libya in 1955 (R55b) and in 1961 discovered the Sarir field (R61a) in a joint venture with Nelson Bunker Hunt.[98] The acquisition of 42% of Triad Oil Company in 1953 (R53b) gave BP leases in Alberta on which it developed production by 1955 (R55c).[99] In 1954 and 1955 it acquired interests in two companies with production in Trinidad (R54b,55d).[100] In 1958 through BP North America (in Canada) when the current chairman, David Steel, headed it, it acquired Alaskan leases in Prudhoe Bay (R58f) which contained over half the reserves of the North Slope discovery in 1968 (R68).[101] It made a discovery in Colombia in 1961 (R61b). BP explored in most of Europe, purchasing a small field in Germany in 1958 (R58g).[102] It began exploring in the North Sea in 1962 (R62) and has made four discoveries with proved reserves of two billion of the thirteen billion barrels discovered to date, making it the largest holder there also.[103] Only in the Far East has it been less successful than other companies. Thirty years of searching in Papua (New Guinea) has yielded only traces of oil, though small discoveries in Indonesia and Australia were made in 1971 and 1972 (R71,72).[104] In addition to these successes it has also explored numerous other places.

BP's success must be largely attributed to skill (R27c). The number of major discoveries it has made exceeds what may be attributable to luck. Neither does evidence on activity support a timing advantage from a head start caused by the Iranian nationalization which might have allowed it to lock up promising areas while other Majors were preoccupied with trying to market a surplus of Middle Eastern crude. BP had more surplus crude than any of the others as a consequence of its position in the Middle East and limited integration (PL9,22). And others were exploring worldwide at least as vigorously according to my data on capital expenditures and numbers of countries in which activity occurred. Competitors note the unusual independence allowed the exploration group at BP to take risks and back commitments. One noted, "They seem to have a genius for finding large, simple structures,"[105] and he went on to note that they get in quickly and get out quickly when results are discouraging. Looking for only the largest structures is a luxury appropriate to a company with large resources elsewhere.

BP expanded downstream (PL17), predominantly in the Eastern Hemisphere (PL18), as its crude supplies grew. It added refineries in Northern Ireland, Germany (Ruhr), the Netherlands (Rotterdam), Italy (Turin), Switzerland, Cyprus, Ivory Coast, Nigeria, Rhodesia, South Africa, Lebanon, and Singapore (R60a,61). Nearly all these refineries were small, local market refineries and nearly all were jointly owned. The use of joint ventures reflected the effort to find outlets for its crude while limiting downstream commitments (PL19). Longer than any other Major, BP continued to rely on its huge source refineries in the Middle East. As late as 1960 at least half its products came from its refineries in Iran, Kuwait, and Aden (PL7). In the next decade this policy shifted as its nine large European refineries expanded to 2/3 of of its total and the Middle East declined to 1/5 (PL23).[106] In 1976 95% of its products were refined and 85% marketed in the Eastern Hemisphere (PL18).[107]

Marketing volumes have remained approximately even with refining levels and with similar geographic distribution. In 1960 56% of its products were marketed in Europe.[108] By 1976 the proportion rose to 71% (PL18).[109] Product distributions are not given in BP's annual reports but it is likely they would show continued reliance on fuel oil and wholesale marketing (PL4,11,19,20). This one can infer from the reliance on source refining (PL7) (lighter products are more difficult to transport and require market refineries), the products required by the European market (more heavy fuel oil and less gasoline than the U.S.) (EC60), and the limits on BP's commitment to refining and marketing measured by the relatively small scale and partial ownership of its refineries and marketing organizations. Also, after Europe, BP's second largest outlet, accounting for 23% of its product sales in 1960 and 10% in 1976, was the bunker trade (PL4,20).[110]

Another indication of BP's relative lack of commitment (PL9) to refining and marketing is that after Shell Mex-BP was dissolved at the end of 1975 BP was not number one in any of its product markets, even Britain, where its 16% was second to Shell's 24% of the market.[111] This contrasts sharply with its producing positions where it is the leader in the Middle East, the North Slope, and the North Sea.

The downstream policies discussed above add up to a continuation of the policy of the 1920's to use crude rather than cash to buy outlets for crude (PL19). This, in turn, reinforces the strategy of a company primarily interested in finding and producing crude (PL2,9,16). As refining and marketing lagged behind production after the end of World War II BP became the largest crude seller (PL22). Between 1971 to 1974 it was selling as much crude to companies like Exxon and Petrofina as it refined and marketed itself.[112] It admits taking downstream losses to move crude. The policy of moving the maximum volume of crude (PL24) was pursued partly because profits were highest at the production stage and often zero or negative in refining and marketing and partly to satisfy the demands of the Arab governments. As Robin Adam, a BP managing director, put it, "We were taking losses downstream in order to get out goods upstream because that's what the Arabs wanted."[113]

An interesting question about the behavior of the Majors during the 1950's and 1960's is to what extent this strategy of moving the maximum crude at minimum expense was profit maximizing or whether it was done at the sacrifice of profit to keep the producing country governments happy or both. If the path to maximum profit were to move the maximum volume of cheap Middle Eastern crude with the least investment downstream, then BP should have been the most profitable Major. But as shown in Table 11-5, it in fact has often been the least profitable since the 1950's. This fact may be explained by any of several possibilities: (a) the maximum crude volume theory is incorrect and downstream operations have not been unprofitable (as Adelman suggests);[114] (b) the theory is incomplete, having omitted possibly important factors like the profitability of the U.S. gasoline market compared to the European fuel oil market; (c) BP has been poorly managed; (d) BP's tax burden has been higher than others; or (e) it suffered relatively more from the crude glut and price declines in the 1960's.

Because of its relative weakness in refining, its abundant resources upstream, and its crude moving rather than value adding strategy, BP was the last of the Majors to make a significant entry in petrochemicals. Though it has had a joint venture with The Distillers Company to manufacture basic petrochemicals at its Grangemouth refinery since 1947, it did not make an entry comparable to other Majors until 1967 when it acquired The Distillers Company (R67).[115]

The same factors have limited other diversification, making it one of the least diversified Majors. Its only other significant venture outside oil came in 1970 with the foundation of a subsidiary (R70b) to manufacture high yeast proteins for animal feed in 1970.[116]

These downstream and diversification patterns are the logical conclusions one might expect from a consistent set of policies beginning with selling fuel oil from Persia to the Royal Navy. BP unobtrusively did what it does best, though less profitably than others. Its latest, and probably most brilliant, strategic move, entry of the U.S. market through acquisition of Sohio in 1970 for North Slope crude, is a significant departure from the past.

Prior to its major discovery at Prudhoe Bay on the North Slope of Alaska, BP had few operations in the Western Hemisphere (PL18, H5) and almost none in the United States. It had obtained leases and some production in Canada by acquiring Triad in 1953 and had built a refinery at Montreal in 1956. It had some production in Colombia. But it had not even explored in the United States (H5).

Looking for outlets after the discovery in Alaska it bought (R69) 9700 Sinclair stations from Atlantic Richfield for a $400 million note from ARCO in 1969. Then in 1970 it bought what will become (R70a) majority ownership of Sohio, a crude short Midwest refiner and marketer (Standard Oil of Ohio), with the first 600 TBD of its North Slope production, acquiring a U.S. presence and a strong downstream organization without spending a drop of cash. With a debt free balance sheet Sohio can borrow to finance development of North Slope production and its own purchase.[117]

Even more amazing is that the Sohio purchase puts it well on the way to becoming the sixth U.S. Major. After its Eastern Hemisphere reserves have been lost through nationalization (EC76), it will be a vertically balanced U.S. and European company with the majority of its assets in the U.S. by the 1980's.

Its geographical, vertical, and product policies as well as its identity (PL1-11,13-15,18,19,21,24) will all change. It will have broken its last link with the Empire. In the process it has rescued itself from decline. Perhaps it will be renamed American Petroleum if the British government ever divests its shares.

This concludes the narrative of the strategy of British Petroleum. A classification and summary of evidence for each policy is presented in Exhibit BP-1.

Exhibit BP-1
BP Policies

PL 1	Identity as a British (and then colonial) company. Regimental character (a) G,A; (b) 1909-present.
PL 2	To produce Persian oil (a) G,V; (b) 1909-71.
PL 3	Subsidy of host government officials (a) A; (b) 1901-25?
PL 4	To refine oil for the British Navy (a) V,P: (b) 1914-present.
PL 5	Cooperation and close relations with the British government. Extension of the Empire (a) A; (b) 1904-present.
PL 6	Provide medical care in Persia to anyone in the area (b) 1904-1951.
PL 7	Source rather than market refining (a) G,V; (b) 1913-60.
PL 8	Government ownership but managerial independence over commercial affairs (a) A; (b) 1914-present.
PL 9	Not to develop a large marketing organization to sell products. Minimal integration into retail marketing (a) V; (b) 1914-present.
PL 10	To make public service expenditures (a) P,A; (b) 1914 or earlier-present.
PL 11	Wholesale marketing in large-volume contracts (a) S,V,P; (b) 1914-present.
PL 12	Own and manage its own tanker fleet (a) V; (b) 1915-present.
PL 13	Tradition of inflexible and arbitrary dealings with Persian government (a) A; (b) 1915-51.
PL 14	Deny Persian government any participation in management or equity (a) A; (b) 1901-51.
PL 15	Diversification of production away from dependence on Iran (a) G,V; (b) 1920, 1955-present.
PL 16	Explore for oil worldwide (a) G,V; (b) 1921-present.
PL 17	Integration forward from production to marketing as quickly and cheaply as possible (a) V; (b) 1920-present.
PL 18	Eastern Hemisphere refining and marketing; little or none in Western Hemisphere (a) G,V; (b) 1920-70.
PL 19	Joint ventures in refining and marketing. Buying in with crude rather than cash (a) V,A; (b) 1920's-present.
PL 20	Supply fuel oil to ships (a) P; (b) 1914-present.
PL 21	Control production and price. Regulate through cooperation and cartelization. Minimization of competition (a) H,A; (b) 1920-present.
PL 22	Be a large-scale crude seller (a) S,V,P; (b) 1947-present.
PL 23	Shift from source (Middle East) to market (Europe) refining (a) G; (b) 1960-present.
PL 24	Maximize crude volume moved (a) S,V,P; (b) 1933-present.

Notes:
(a) type of strategic choice

S — Scale	PL11,22,24
G — Geography	PL1,2,7,15,16,18,23
H — Horizontal concentration	PL21
V — Vertical integration	PL2,4,7,9,11-12,15-19,22,24
P — Product diversification	PL4,10,11,20,22,24
A — Administration	PL1,3,5,6,8,10,13-14,19,21

(b) duration

Exhibit BP-2
BP Environmental Conditions

EC1	Discovery of Spindletop in Texas, January 10, 1901.
EC2	Favorable oil prospects in the Middle East: 1892.
EC3	Russian influence in Persia: before 1917, 1941-46.
EC4	Weakness and corruption of Persian government: before 1921, 1941-45.
EC5	Interest of the British government in a British controlled source of oil: 1904-World War II.
EC6	Power and influence of the British Empire: 19th century-Suez Crisis (1956).
EC7	Remoteness of Persia from markets.
EC9	High quality, unlimited quantity, and low cost of Persian oil.
EC11	Breakup of Standard Oil: 1911.
EC14a	Churchill is suspicious of Shell and favors APOC: 1911-14.
EC14b	Parliament invests 2 million in APOC for 50+% ownership: 1914.
EC15	World War I: 1914-1918.
EC17	Russian revolution: 1917-29.
EC21	Revolution in Persia establishes Pahlevi dynasty: 1920-present.
EC28a	Worldwide crude glut begins: 1928-39.
EC28b	Red Line agreement: Americans enter Middle East through purchase of 23.75% of IPC and agree to act as a unit within Red Line area.
EC28c	Achnacarry agreement: market sharing in Eastern Hemisphere.
EC29	Great Depression: 1929-40.
EC30	Rule of Capture: prior to prorationing in 1930's.
EC39	World War II: 1939-45.
EC40	British government adopts "short haul" policy to minimize submarine risk: Mid-1940.
EC41a	End of Reza Shah's reign: August 28, 1941.
EC41b	Pearl Harbor: Americans enter World War II: December 7, 1941.
EC43	DeGolyer predicts dominance of Middle Eastern production: 1943.
EC44	Commercial productionin Saudi Arabia.
EC46a	Anti-British sentiment: post World War II.
EC46b	Anti-BP sentiment: post World War II.
EC46c	Resentment of Middle Eastern governments that U.K. and U.S. governments receive more in tax on Middle Eastern oil than they do.
EC47	Limitation on dividends by British government: 1947.
EC48a	India becomes independent, changing military position of British east of Suez: 1948.
EC48b	Fifty-fifty law enacted in Venezuela: November 1948.
EC50	Fifty-fifty adopted by Saudi Arabia: December 30, 1950.
EC51a	Mossadegh comes to power in Iran: April 28, 1951.
EC51b	Persian parliament votes to nationalize the oil industry: April 30, 1951.

EC51c	NIOC created to market Iranian oil: 1951.
EC51d	Tax accord of 1951 making foreign income tax creditable against U.S. income tax.
EC53	Mossadegh falls to a CIA coup and the Shah is restored to power: July 1953.
EC54	Iranian consortium formed: August 1954.
EC56	Suez Crisis: 1956.
EC60	Product demand pattern of European market compared to U.S.: 25% gasoline, 50% middle distillates, 25% HFO versus 50% gasoline, 25% middle distillates, 25% HFO in U.S.
EC73	OPEC embargo: October 1973.
EC76	Nationalization by OPEC countries completed: 1973-76.

Exhibit BP-3
BP Resources

R1	The D'Arcy concession: May 28, 1901-April 30, 1951.
R4	Interest and support of the British government: 1904-present.
R5	Capital from Burmah Oil: 1905, 1909 (1 million).
R8	Discovery at Masjid-i-Sulaiman (Mosque of Solomon): 1908-1951.
R9	£1 million capital from the public: 1909.
R11	Pipeline to coast in Persia: a11-1951.
R13a	Refinery at Abadan: 1913-1951.
R13b	Commercial production in Persia begins: 1913-51.
R13c	High quality, large quantity, and low cost of Persian oil.
R14a	£2 million capital from the British government: 1914.
R14b	Oil supply contract for the Royal Navy: 1914-present.
R14c	Special relationship with the British government: government majority ownership.
R14d	Large scale in production, refining, and bulk marketing: 1914 present.
R14e	47-1/2% interest in Turkish Petroleum (Iraq Petroleum after 1929): 1914-28.
R15	Tanker fleet.
R17	Acquisition of British Petroleum: 1917-present.
R19a	Bunkering installations: 1919-present.
R19v	Acquisition of Scottish shale oil companies: 1919.
R20	"Interpretative Agreement": 1920-32.
R22	Refinery at Llandarcy, England: 1922-present.
R23a	Refinery at Melbourne, Australia: 1923-.
R23b	Refinery at Dunkirk, France: 1923-40 (destroyed in war); 1951-.
R23c	Discovery at Naft Khaneh in Persia: 1923-51.
R23d	Refinery at Pumpherston, Scotland (shale): 1923-.
R24a	Refinery at Grangemouth: 1924-present.
R24b	Discovery in Argentina: 1924-30.
R24c	Refinery in Argentina: 1924-30.
R25	Iraq concession and tonnage royalty (4s gold/ton) agreement: 1925-52.
R27a	Discovery in Iraq: October 14, 1927.

R27b	Local refinery begun in Iraq: 1927-28.
R27c	Recognition of expertise in finding oil: 1920's.
R28a	Joint marketing ventures in Europe and the Eastern Hemisphere: 1920's.
R28b	Discovery of Haft Kel field in Persia (1928).
R30	Discoveries in Persia of Gach Saran, Agha Jari, Pazanun (gas), Naft Safid, Lali (1934) fields: 1928-30's.
R32a	Acquisition of four Scottish shale oil companies: 1932.
R32b	Refinery at Marseilles, France: 1932-.
R32c	Formation of Shell-Mex and BP joint (60:40) venture to market in U.K.
R33	Revised concession in Persia: April 29, 1933.
R34a	Kuwait concession obtained with Gulf Oil: 1934.
R34b	Export pipeline from Iraq to Mediterranean (Haifa and Tripoli) (12" diameter); export production from Iraq begins.
R35	Kermanshah refinery in northern Persia: 1933-51.
R37	Mosul concession obtained by IPC.
R38a	Discovery of Burgan field in Kuwait: 1938.
R38b	Alkylation process for making aviation fuel developed at BP's Sunbury labs, patented: 1938.
R38c	Basrah concession awarded to IPC: July 29, 1928.
R39b	Haifa refinery: 1939-48 (closed by 1948 war: sold in 1959).
R40	Discovery on Mosul concession:—1940.
R43	Discovery in England: 1943.
R45a-f	Marketing in Turkey, New Zealand, Algeria, Morocco, and Greece.
R46	Production in Kuwait: 1946-.
R47a	Long-term contract with Jersey (cost + 1s/bbl.): 1947.
R47b	Long-term contract with Socony: 1947.
R48a	Discovery on Basrah concession: 1948.
R48b	49% of Venice refinery: 1948.
R48c	Eurotank refinery in Hamburg acquired: 1948.
R49a	Concession in Nigeria: 1949.
R51a	Schindler refinery in Hamburg acquired: 1951.
R51b	Refinery in Kuwait: 1951.
R51c	Refinery at Antwerp with Petrofina: 1951.
R52	Refinery in Kent, England: 1952.
R53a	Discovery in Nigeria: 1953.
R53b	Acquisition of Triad Oil Company with leases in Alberta, Canada: 1953.
R54a	Discovery in Egypt by Anglo-Egyptian Oil Company (15-1/2% BP).
R54b	Acquisition of Trinidad Northern Areas (33%); production 1954.
R54c	Refinery in Aden: 1954.
R54d	Marketing in Portugal: 1954.
R54e	Marketing in west Africa: 1954.
R54f	40% interest in Iranian consortium: 1954.
R55a	Refinery in Freemantle (western) Australia: 1955.
R55b	Concession in Libya: 1955.
R55c	Production in Alberta, Canada: 1955.

R55d	Acquisition of 77% interest in Trinidad Petroleum Development Company: 1955.
R57	Marketing in Canada: 1957.
R58a	Marketing in Italy: 1958.
R58b	Discovery in Abu Dhabi: 1958.
R58c	Discovery of Magwa field in Kuwait: 1958.
R58d	Discovery of Ahmadi field in Kuwait: 1958.
R58e	Export production from Nigeria: 1958.
R58f	Obtained Prudhoe Bay leases in Alaska: 1958.
R58g	Small field in Germany purchased: 1958.
R60a-1	Refineries in Northern Ireland (1960), Germany (Ruhr-1960), Turkey (1960), Montreal (1960), Netherlands (Rotterdam), Switzerland, Cyprus, Ivory Coast, Nigeria, Rhodesia, South Africa, and Italy (Turin).
R61a	Discovery of Sarir field in Libya in 50-50 joint venture with Nelson Bunker Hunt: 1961.
R61b	Discovery in Colombia: 1961.
R62	Exploration begun in North Sea: 1962.
R67	Acquisition of The Distiller's Company: 1967.
R68	North Slope discovery at Prudhoe Bay, Alaska: 1968.
R69	Acquisition of 9700 Sinclair stations from ARCO: 1969.
R70a	Acquisition of Sohio for North Slope crude: 1970.
R70b	Animal feed subsidiary organized: 1970.
R71	Discovery in Indonesia: 1971.
R72	Small discovery in Australia: 1972.

Exhibit BP-4
BP Handicaps

H 1	Special relationship with British government: majority government ownership with commercial autonomy: 1914-present.
H 2	Payments to Bakhtiari tribesmen: 1907-29.
H 3	APOC's lack of a marketing organization: 1920-.
H 4	No U.S. or Western Hemisphere operations.

Exhibit BP-5
BP Personal Values

PV 1	D'Arcy: speculator, fortune seeker with great perseverance.
PV 2	G.B. Reynolds: leadership, scientific ability, devotion to work, determination to find oil.
PV 3	Scots' characteristics: engineering skill, accounting skill, shrewdness, mendacity, adventurousness.
PV 4	Greenway: marketing background with Burmah in India.
PV 5	Cadman: Professor of petroleum, Britain's leading oil expert, civil servant, negotiator, diplomat, unusual breadth of view.
PV 6	Fraser: narrow-minded, financially oriented Scotsman.

Exhibit BP-6
BP Social Responsibilities

SR 1	Housing, education, roads, and communications for Persia.
SR 2	Provide medical care, clinics, hospitals for non-employees and employees: 1901.
SR 3	Purchase of Scottish shale oil companies to maintain employment: 1919, 1932.

11

Analysis of Policy Formation and Effects of Strategy on the Firm

11.1 Summary

This chapter presents a discussion and analysis based on the data in Chapters 4-10 of the origin and evolution of the strategies of the Majors and the effects of strategy on them. The results of this analysis may contribute some insights to the process of strategy formation in business policy and the role of strategy as a variable influencing structure, conduct, and performance in these seven firms in the petroleum industry.

In the preface this study was addressed to three purposes. The first, to identify and document a record of the policies which formed the strategies of each of the Majors, has been accomplished in the preceding company narrative chapters. The second purpose, to contribute to the theory of strategy formation by analyzing how these policies resulted from conscious decisions and natural events in the oil industry over long periods of time, is addressed in sections 11.2-11.4 on the analysis of policies, as well as in Chapters 4-10. In these sections policies are used as data points to analyze the origins of the Majors' strategies. The third purpose, to analyze the importance of strategy as a determinant of firm and industry structure, conduct, and performance, is addressed in sections 11.5-11.9 on the analysis of strategies. In these sections the seven strategies of the Majors, rather than their policy components, are compared and contrasted and their effects on performance are considered. A summary of the main results follows:

(1) Internal factors were more frequently causes of policy formation than were external factors overall. This result shows the dominant role of factors internal to the firm, which were unique to each firm, as determinants of policy and strategy.

(2) Environmental conditions, resources, and personal values were each involved in 45 to 55% of all policy origins studied. Social responsibility causes appeared in 10%. These results showed that all four business policy variables were important in policy formation, with three being involved in roughly half of all policy formation.

(3) The Majors' policies were caused by combinations of the four business policy variables more often than by single variables. This supports the view in economic and business policy literature that policy formation is a complex organizational process, rather than a simple constrained maximization.

(4) Only a small percentage of policies were formed in response to single traumatic events in the environment. This refutes the view that strategy was formed as a response to immediate short run events and reinforces the conclusion that policy making was a complex long run process not easily disturbed by events of the moment.

(5) Most of the policies of the Majors in effect today were formed early in the history of the organization, either close to the date of origin of the company or within its first independent decade, and have remained in effect thereafter. This implies that early events in a firm's history were of greater significance to its strategy than more recent events of similar magnitude. This result shows the great influence of internal factors and inertia upon strategy and the resistance of this equilibrium to change.

(6) A related result is the longevity of the policies studies. They were rarely terminated and often outlived the conditions which caused them to be adopted. This longevity makes them useful predictors of each firm's future conduct and performance.

(7) The strategies of the Majors differed between companies sufficiently to be identifiable sources of strength or weakness. This implies that there are effects on behavior and performance from these differences.

(8) There were broad similarities among the Majors in size, worldwide position, and vertical integration which were descriptively useful in distinguishing them from non-Majors. But the similarities were not so great as to permit accurate prediction of their behavior as a strategic group on more than a small number of issues. Their actual behavior on most issues was more accurately predictable from differences in their strategies than from the similarities. Strategic differences were good predictors of the profitability of each Major relative to the others in the environments throughout the long period covered by this study.

(9) As to why strategies differed, permanent differences among the seven Majors arose and have persisted because there was no single optimum strategy toward which all competitors were forced, the internal dynamics of each firm which determined its strategy was unique, and internal constraints made it difficult and costly to change an established strategy.

(10) The principal mechanisms which account for the longevity of policies and the stability of strategy, despite large recurring changes in the environment and in the optimum strategy, were: assets with long productive lives were difficult to acquire or replace quickly; there were few instances where an efficient competitor which had established a low cost position in a market was dislodged; experience and organizational learning added to the comparative

advantage a firm enjoyed in its areas of strength while the opposite was true of its weaknesses with the result that differences in strategy among firms did not narrow over time; and more capital expenditures were allocated to reinforce success than to remedy weaknesses.

(11) Although strategies once established were highly persistent over time within each company, there were changes in strategy when sufficiently large changes in the environment occurred and persisted over a sufficiently long period. The following strategic changes were much more likely to occur than others. It was possible to lose an area of strength. It was possible to enter new geographic, industry, or product markets successfully when entrenched competitors did not have to be dislodged. But it was very difficult to ameliorate or eliminate strategic weaknesses. And to develop or switch major strategic strengths from weaknesses was almost impossible.

11.2 Causes of Policy Origin

This section analyzes the mix of environmental, resource, personal value, and social responsibility factors that were responsible for each of the policies of the Majors. The purpose of the analysis is to make for the first time some empirical estimates of the relative importance of the four business policy variables. The analysis shows the large influence of internal factors and joint causes on policy origin. This supports the view, expressed elsewhere in this chapter, that the unique strategies of the Majors are partly due to the internal complexity of the large firm and the criterion of strategy that policies be consistent with each other.

By constructing a distribution of the frequency in which each of the four business policy variables appears both alone and in combination with others as a cause of origin of a policy one can answer some interesting questions about the business policy model as applied to the strategies of the Majors. (1) Were policies formed primarily in response to forces external to the firm, internal to the firm, or a combination of both? (2) Which of the four variables was most/least important? Was the social responsibility variable a significant cause at all of policy? (3) Were policies more often the result of a single variable cause or of combinations of variables? (4) Did single traumatic events in the environment cause policy to be formed?

To answer these questions I compiled from the company narratives a list of causes which seemed to be most clearly responsible for the origin of each policy. These causes are summarized in the first exhibit at the end of each company chapter. Each cause is classified as one of the four types of business policy variables: environmental conditions, resources, personal values, or social responsibilities. Then I calculated the frequency with which each of the business policy variables appeared as a cause of policy origin both alone and in combination with the other three types of variables (Exhibit 11-1). This method assumes that all policies are of equal importance in the sample and is appropriate for the study of policy origins in general since the unequal impact of policies on strategy is not the concern here.

Classified as environmental conditions are forces of a political, economic, or technological nature. To the usual matters of social conscience included as social responsibility I have added the social ethics of the day, including attitudes about the British Empire and laissez faire capitalism, and I have classified it as an external rather than an internal variable. The resources category includes both physical and organizational assets and handicaps. Those influences counted as personal values were not just individual beliefs but also judgments which could easily have been different if made by someone else. Resources and personal values are considered internal factors.

The frequencies which result are not interesting for their precision, which is subject to errors of measurement inherent in a judgmental process and to any biases peculiar to this sample of companies. But what they indicate qualitatively in answer to the four questions posed is likely to be a significant and reproducible result.

The first question was, "Were policies formed primarily in response to forces external to the firm, internal to the firm, or a combination of both?" An answer is tabulated in Table 11-1: 87 of the 180 or 48% of the policies formed involved combinations of both internal and external variables; 73 of 180 or 40% involved internal factors only; and 20 of 180 or 11% involved only external factors. The conclusion is that a new policy resulted from a combination of internal and external variables about half the time in this sample. The other half of the time new policies resulted more often from internal (40%) than external factors (11%), when only internal or external factors were involved. In total, internal factors were involved in 88% (48% + 40%) of these 180 policy origins while external factors were involved only 59% (48% + 11%) of the time. Since 58% of these policies were formed in the first decade of the organization's existence (see Exhibit 11-1, line g), this result shows the heavy constraint, even in the early years, of internal factors on policy choice.

The second question was, "Which of the four variables was most/least important? Was the social responsibility variable a significant cause of policy at all?" Results are summarized in Table 11-2 below. Environmental conditions were involved in 55% of all origins (46% in combination with others, 9% alone). Resources appeared in 53% (36% in combination, 17% alone). Personal values contributed to 45% of all policies (31% in combination, 14% alone). Social responsibility appeared 10% of the time (9% in combination, 1% alone).

The third question was, "Were policies more often the result of a single variable cause or of a combination of variables?" The results, also shown in Table 11-2, show that 59% of all policies were caused by combinations of variables and 41% by single variables. The policies were the result more often of variables in combination with others. This supports the view of policy formation in the business policy model, Cyert and March, and elsewhere as a complex organizational problem solving process.

The fourth question was, "Did single traumatic events in the environment cause policy to be formed?" Table 11-3, below, shows that single events, mostly economic events, were involved in only 14% of all origins. This refutes the view that policies were formed in response to short run events and reinforces the conclusion that policy making was a complex long run process not easily disturbed by the course of events at the moment.

There is no standard or theoretical expectation to be used in interpreting these results. This is a first attempt to calculate joint and separate frequency distributions of the business policy variable categories, even for a small sample of seven companies, as a way of testing the validity of that model. Since all four categories were present and there were no causes which did not readily fit into this model, it seems to be sound.

What is most interesting about these results is the heavy influence of the internal factors, resources and handicaps and personal values. As shown in Table 11-1 they were involved in 88% of policy origins. If all firms were homogeneous internally, one would expect to find much closer to 100% external origins for policy. The internal complexity of the modern firm, as reflected in the heavy internal influence on policy, clearly differentiates it from the entrepreneurial firm which characterized business before the Industrial Revolution. Also the small impact of specific events indicates the difficulty of responding to events by altering policy. This reflects a complex internal organization which cannot respond effectively in the short run and pursues a long run equilibrium of stable policies.

11.3 Early Formation of Strategy

Perhaps the most important conclusion about policies from this study is that most of the policies in effect today became evident early in the history of the organization. Evidence for this conclusion comes from the frequency distributions of policies presented in Exhibit 11-2. As line (f) shows, 69% of all policies in effect in 1976 for the Majors were either present at origin from predecessors or were formed in the first decade of a firm's existence. Only 31% on average of all policies in effect in 1976 were formed in the remaining 65 years. Of those policies present at origin which were inherited from predecessors the same is true. By tracing all policies to their dates of origin and comparing the dates of policy origin with dates of origin of the organization one can see from line (g) that 31% of all policies were formed in the first year and 58% were formed in the first decade of the organization's existence. It is remarkable that so many of the policies which guide each of the Majors were recognizable so long ago.

The early formation of policies in the life of each Major made early events in its history of greater importance to its strategy than they were to economic history in general. Often events of prime importance to a company's strategy were of little significance historically. For some examples, consider that the instability of the crude oil producing industry after the Civil War and the nineteenth-century system of merchandise distribution through the "general store" confined Standard Oil to refining, transportation, and wholesale marketing. The births of Gulf and Texaco were in no small part due to an obscure antitrust decree which temporarily forced Standard out of Texas, coupled with the very important event of the discovery of oil there. Both Gulf's strategy of finding crude first and then integrating and Texaco's nationwide gasoline marketing strategy were responses to the dominance of the Standard Oil companies. Socal's strength upstream was due to the discovery of oil in California and the ability of a gifted producing executive stranded in California by the 1911 divestiture of Socal from Standard. A technical innovation, the bulk tankship, brought Marcus Samuel and Shell into the oil business.

These early events have also been much more important than more recent events in shaping the policies of the organization. The basic vertical, geographical, scale, horizontal, product and administrative strengths and weaknesses of each company are nearly all the same as in its early years. The early events which gave it success or failure in given areas have become cornerstones of its strategy: Exxon's strength in refining, transportation, scale, worldwide exposure, wholesale marketing, massed financial reserves, purchased producing properties and committee management, as well as its weaknesses in exploration, retail marketing, light and specialty products, is an example of the importance of early events which can be repeated for each company.

11.4 Longevity of Policies

A conclusion related to the early formation of policies is that policies, once formed, were rarely terminated and often outlived the conditions which caused them to be adopted. Without question they were modified over time according to necessity, but in the lifespan measured in this study they have remained recognizably the same as at their origin. Evidence is shown in Exhibit 11-2 line (c). Approximately 79% of all policies studied for the Majors are still in effect. This percentage for each company ranges from 71% for Gulf and BP to 93% for Shell. Those Majors which were more stable or better managed had fewer changes than those which experienced more upheavals.

Policies were rarely terminated except under duress. It would be equally accurate to say that they were rarely terminated even under duress. Only 38 of the 178 policies documented in this study have been terminated in a time span which includes nearly all of this century and in some cases longer. Duress implies that policies were terminated only under external pressure. This is true although catastrophes were far from the sole cause for termination. The most important cause for termination was a perception that the environment had changed, making a policy obsolete. For example, Socal ceased to be a price leader as it lost market share.

The causes for termination, like those for origin, seemed to be long run forces. Perhaps the conclusion should be simply that policies were rarely terminated. For whatever reason it is the conclusion of this study that the difficulty of implementing a strategic change is surely one of the most imposing obstacles to economic efficiency in an organization, even after the need for change is recognized and the direction of change is clear. Often the need for and direction of change was recognized years before changes were implemented. The larger and older the corporate mass, the greater was its inertia.

11.5 Strategies Differed between Companies

The company narratives document that in each major industry function (upstream, downstream, transportation, non-petroleum, and administration) and for each of the six types of strategic decision categories studied the strategies of the Majors were sufficiently different to be identifiable sources of strength or weakness between companies.

As examples of differences in the major industry functions, BP, Gulf, and Socal have repeatedly been the most successful explorers worldwide; Shell under Deterding once had a lead in exploration but lost it; Texaco has done well in the U.S.; and Mobil and Exxon have not done well at wildcatting. Exxon has been the shrewdest purchaser, producer, and evaluator of reserves. It has also been the strongest in pipeline transport and, after Shell and BP, in marine transport. Exxon and Shell excelled in refining while Shell, Mobil, and Texaco (in the U.S.) have excelled in retail marketing. Exxon has done well at wholesale marketing. BP, Socal, and Gulf have been the largest crude wellers while Exxon, Shell, and Mobil have been the largest crude and products buyers.

As examples of differences in the six types of strategic decisions studied, Exxon and Shell have benefitted particularly from the scale of their operations. The worldwide geographic presence of Exxon and Shell has been a strength as has the nationwide marketing of Texaco in the U.S. Lack of downstream position abroad has handicapped Socal and Gulf. Lack of a U.S. position limited BP until recently and lack of nationwide U.S. marketing limited Exxon. Large market shares by Exxon, Socal (U.S. West), Mobil (New York and New England), and Shell in local markets were once strengths which became indefensible. Texaco's mile-wide, inch-deep U.S. marketing strategy was long a pillar of its strength but has lately fallen into disrepair. Control of low cost crude sources both in the U.S. and abroad has long been a crucial strength of all seven to different degrees. The vertical balance of Texaco and Socal in the U.S. and lack thereof for Mobil (crude short worldwide), Exxon (crude short), Shell (crude short since the 1930's), BP (crude long), Gulf (crude long abroad, crude short in the U.S.), and Socal (crude long abroad) has been the stimulus for billions of dollars of investment. In product lines, Mobil has led in lubricants and diversification; Exxon, in fuel oil; Texaco and Shell, in gasoline; Socal and Shell, in asphalt; Gulf, BP and Socal, in crude sales; Gulf, in uranium; Shell, in lubricants and bulk chemicals; Socal, in insecticides; Mobil, in fertilizer; and Exxon and Gulf, in coal. Important administrative differences are Gulf's family ownership, Exxon's committee management, Shell's national decentralization and dual nationality, BP's ownership by government, Socal's conservatism, Mobil's reorganization and planning, and Texaco's centralization. Exxon's financial strength and Shell's success in dealing with governments are especially noteworthy.

The last part of section 11.6 will show that these strategic differences have had significant impact on the relative profitability of the Majors.

11.6 How Much Did Strategies Differ?

This topic examines the Majors' similarities, the question of whether they can be usefully included in a strategic group, and the use of strategic differences as a predictor of relative profitability. The conclusions are that similarities are useful in description but strategic differences make better predictors of behavior and performance.

Similarities

The company narratives describe each of the Majors as having a definite personality, a strategy distinctly different from the other six, and as being about as similar and different as the same number of human beings for many of the same reasons. By no means were there only differences. An appropriate description is that they demonstrated a pattern of overall similarity with significant individual differences.

At the broadest level all were integrated oil companies and were the largest competitors in the industry. Traditional economic models which assumed homogeneity among competitors in an industry did not have a role for individual differences to play. Constrained by the profit motive and facing the same environment, they are presumed to share very similar interests on every issue, interests that are often mutually exclusive. The principal way seen to avoid the competitive way of life was to collude to raise prices to the public. On this base antitrust theory was built.

This view from traditional economic models of competitors being substantially similar, at least to the extent that behavior was determined by competitive simiarities rather than differences, was only partially true of the Majors. All strategies were not rewarded. Market forces did force some conformities in strategy in the traditional manner. The Majors shared five strategic similarities not possessed to any similar degree by any other oil companies. These similarities or conformities were:

(1) All possessed a worldwide organization which limited vulnerability in any geographic area.

(2) All had large positions in the lowest cost crude sources in the Middle East and Venezuela.

(3) All had a high degree of vertical integration from crude oil production through retail marketing if one defines "high degree" as substantial operations at each functional level both in absolute size and relative to the size of its other parts.

(4) All had a large position in the United States gasoline market, the largest and most profitable single market in the industry. One may correctly argue that BP was a Major long before it entered the U.S. market due to its dominant position in crude oil production in the Middle East. However, it is also true that its U.S. entry remedied its greatest strategic weakness.

(5) All the Majors have been significantly larger in size than the next largest competitors, although this is less true today than one or two decades ago.

These were the most important similarities of the Majors. On other strategic issues their differences appeared to be wider. The relevant question is how much similarity in behavior and performance one can predict from these structural similarities. One way to approach this is to consider whether they can usefully be included in a single strategic group.

Strategic Groups

The concept of strategic groups is currently popular in the literature (see Hunt [1972], Newman [1973], Porter [1980, Ch. 13]). A strategic group is a group of companies with similar strategies or structures in an industry. Strategic groups form a kind of substructure within the industry of firms which compete with each other but which may not compete directly with firms or other groups in all markets: full line versus specialty manufacturers, for example. There may be mobility barriers to entry or exit which separate groups. A useful question about the value of the strategic group concept is whether the descriptive similarities in strategy or structure which define the group have any predictive power to explain behavior or performance of the group.

If the strategic group concept were applied to the petroleum industry, the Majors, with other vertically integrated competitors, might be included in one group. Independent producers, refiners, marketers, and government owned companies might form other groups. The concept of a strategic group of companies which are worldwide, diversified, and vertically integrated is useful in describing the competitive structure of the oil industry by a measure other than the conventional measures of market structure like distribution of sellers (concentration), height of barriers to entry, and degrees of product and industry diversification.

However, the degree of variation in strategy among the Majors in this study shows that some care must be taken not to overstate the degree of strategic similarity within a group. Descriptive similarities in structure (i.e., worldwide, diversified, vertically integrated) may be inadequate descriptions of strategy and misleading guides to behavior and performance.

The actual behavior and performance of a vertically integrated company, for example, was likely to be much more sensitive to its individual degree of vertical integration overall (crude long/crude short) and in each of its operating subsidiaries than to its membership in the strategic group. Some vertically integrated Majors (BP, Socal, Gulf) have historically been crude long, while others (Exxon, Mobil, Shell) were crude short. Their actual behavior on most issues was more accurately predictable from these differences in their individual positions than from the similarity of vertical integration. The Majors are successful companies of enormous financial size, whose size relative to the other largest U.S. companies has been rising. But their financial performance over a period of years has not been necessarily superior to that of other strategic groups in the petroleum industry. Otherwise they would not have suffered steady erosion in their market positions in production, refining, and marketing in the period since World War II (Table 11-4).

The Majors had significant differences in strategy in each of the six types of strategic decisions studied, as the company chapters and the previous section of this chapter show. If one were to try to predict behavior or performance for a strategic group on any issue, the prediction would very likely be wrong because it would overstate the degree of strategic similarity within the group. As an example, consider the simplest of all issues, a price increase from price decontrol of U.S. crude. Three Majors would benefit, three would be hurt, and one would be relatively unaffected.

The finding of this study is that the strategies of the Majors were sufficiently unique to make it difficult to predict accurately their behavior or performance as a group, even excluding other vertically integrated oil companies. Their resources, values, social responsibilities and resulting strategies were so diverse that every issue or event affected all or some of them differently and each responded in its own fashion. The company narratives document such differences in response on all of the following major issues and events: thermal cracking (1913), post World War I feared crude shortage (1918), application of science to petroleum exploration (1910-1930), rise of the gasoline market (1920s), the Great Depression and crude glut (1930s), discovery of oil in the Middle East (1920s-1950s), closing of the Suez Canal

(1956), petrochemicals (1960s), crude glut (1960s), Libya and Teheran agreements (1969-1971), move toward diversification (late 1960s). There are few, if any, issues which have evoked a uniform response in direction, much less in degree. Perhaps the only one at present would be regret at the OPEC nationalizations since 1973.

The measures which define strategic groups are usually descriptive structural measures which reflect some of the components of a strategy. They are very useful in structural description and in broad descriptions of industry trends where all the members of the group are similarly affected. The number of issues where the similarities outweigh the differences appears to be limited for the Majors over a long time span. Consequently, the ability to predict behavior and performance for the Majors as a group is also somewhat limited. Strategies vary enough among the Majors to place significant limits on the usefulness of the strategic group as a predictor of behavior and performance.

Strategies as Predictors of Behavior and Performance

If strategies varied too much on most issues to be able to predict behavior and performance as a strategic group, can they be used to predict individual behavior and performance?

Since strategies were deduced from behavior they are by definition good predictors of individual behavior. And since the industrial organization model supports a strong link between behavior and performance, one should likewise be able to predict economic performance from strategy.

The effect of strategy on performance can be stated as a hypothesis which can be tested: because strategies and resources varied among firms and were relatively fixed for each firm (making it unable to adjust rapidly to new conditions), while market conditions changed frequently, the relative profitability of firms can be ranked by how closely they approached the optimum strategy for any given economic environment. Those whose match was best will rank highest in profitability. For example, this hypothesis predicts that since 1973 firms which had efficient refining and marketing organizations and ownership of or access to large amounts of crude would be the most profitable. A decade earlier, by contrast, efficient refining and marketing was much less important, purchased crude was a disadvantage, and ownership of low cost crude and outlets to dispose of it was the optimum strategy.

If one takes the optimum strategy since 1973 to be ownership of large amounts of non-U.S. crude or, as the second best strategy, access to large amounts of purchased crude and an efficient refining and marketing organization, one might predict the relative profitability of the Majors as follows. None has had the optimum strategy since losing their ownership of OPEC oil. So none would be as profitable as the OPEC producers. Of the Majors, Shell (predicted rank #1), which is strong in refining and marketing and has long been dependent on purchased crude, would appear to be best matched to the environment since 1973. Exxon (#2), strong in refining, Socal (#3) with new large-scale refineries, and Mobil (#4), a capable competitor and aggressive marketer, would have the next best matches. Texaco (#5), a strong marketer with small-scale refineries, and Gulf (#6), more dependent on the Middle East, with weaknesses in refining and marketing, and plagued by management and legal problems, are less well situated. BP (#7), weak downstream with the greatest dependence on Middle Eastern crude, would be least well matched.

These predicted rankings of relative profitability of the Majors were correlated with rankings of actual return on assets to test the hypothesis. A significantly positive correlation will be required to accept the hypothesis. Table 11-5 shows the results.

The actual rankings of return on assets were computed by ranking the average annual return on assets for the period. That is, the Majors were ranked by return on assets from one to seven for each year. These rankings were aggregated to a ranking from one to seven for the period by ranking the averages of the annual ranks. This has the effect of giving each year an equal weight in the rankings. Rankings of return on equity were compared with rankings of return on assets and were found to be very similar. The use of either measure would make very little difference to the results.

The time periods used are periods during which the environment was relatively stable and, therefore, during which the optimum strategy would be the same. The end points mark major shifts in the environment. The characteristics and major events of each period which affect the optimum strategy are shown in Table 11-6.

An optimum strategy was calculated for each period from the characteristics of the environment in Table 11-6 and the strategies of each company were ranked by the author on the basis of how well they matched the optimum. These rankings are the predicted profitability for each Major relative to the others. A narrative of those rankings for the first six periods is presented below so the reader can judge the basis for the rankings. Knowledgeable people with similar information should arrive at a similar set of rankings. A narrative for the seventh period was presented four paragraphs above. The order in which the companies are mentioned is the order of ranking.

From 1911 to 1918, during the prewar and World War I years, companies with secure or expanding crude supplies had the optimum strategy. The Burton-Humphreys refining patents were held by Standard of Indiana, which licensed them for a stiff fee. All seven Majors were at a similar disadvantage and the patents were not a factor in their rankings. Socal, isolated in the West in the premier producing state at this time (California), seemed the best match. Gulf, with discoveries in Mexico and the mid-continent (Texas, Oklahoma, Louisiana), was close. Shell was expanding worldwide under Deterding's leadership, but its major discovery in Venezuela did not come until 1922. Mobil and Exxon were short on crude (but Mobil needed less crude) and Texaco underwent a management upheaval deemphasizing crude production. BP was just getting started.

In the 1920s (1918-1927) the largest rewards went to companies which were expanding production, especially foreign production, and which were building their position in the U.S. gasoline market. Shell did both; Texaco did so in the U.S. Socal, with an abundance of heavy crude not well suited for making gasoline, slipped a little as it was unsuccessful in expanding outside the West or abroad. BP came on stream with Iranian crude. Gulf expanded foreign and domestic production but had limited outlets. Exxon, with limited retail outlets, expanded production through exploration (unsuccessfully) and purchase (successfully). Mobil was a leading gasoline marketer with limited crude in a period of high crude prices.

In the 1930s (1927-1938) debt and market outlets were the deciding factors. BP was less hard hit because it was not exposed in the U.S. (where the depression was worse than in Britain); it had little debt and a long-term supply contract with the British Navy. Western isolation and little debt paid off for Socal. Shell's position deteriorated from its worldwide exposure and management troubles. Mobil had opted for marketing strength by merger with Vacuum in 1932. Exxon reorganized and bought crude producing properties at depressed prices. Texaco and Gulf, with the highest debt-equity ratios, suffered. Gulf, in particular, paid the price for overexpanding in the late 1920s.

During the World War II years (1938-1950), the effects of the war were paramount. BP, the Navy supplier with Iran secure, benefitted. Socal was not negatively affected. Exxon and Texaco, the two American Majors who greatly expanded reserves during the 1930s, moved up in rank. Gulf came next. Mobil's global marketing network was disrupted. Shell, exposed globally and in the Far East, was shattered.

The postwar decade (1950-1960) was dominated by expanding production from the Middle East as Europe recovered. Exxon, with worldwide markets, along with Socal and Texaco, all three equal partners in Saudi Arabia, prospered. So did Gulf, except that it sold much of its Kuwait and Venezuelan production under long-term contract. BP stumbled in Iran when Mossadegh took over. Mobil and Shell were crude short.

During the 1960s (1960-1969), vertical balance was required as crude markets became glutted. Ranking the companies by degrees of vertical balance from the data in Table 1-3.7, Texaco was best matched to this environment with an even balance of operation; then Socal, Exxon, Mobil, Shell, Gulf, and BP.

The positive correlations in Table 11-5 confirm the hypothesis that relative profitability can be predicted from the goodness of fit between a firm's strategy and its present environment. All except the last correlation are significant at the

95% level. The problem in the last case is probably the period (1970-1976) rather than the method. Beginning this period in 1974 might correct the problem.

With a stable strategy and changing environment, relative profitability for each of the firms has shifted a great deal in this century. Periods of environmental stability seemed rarely to last more than a decade or so. One can conclude that for these seven firms there was no strategy which would maximize profits, at least relative to those achieved by competitors, over any "long run," which is longer than these periods of relative stability in the environment. There was no single strategy which could take the greatest advantage of all opportunities. A firm's profits were maximum only when its strategy was the one most favored by events.

It seems clear that the performance of the Majors described in this study could not ever achieve a profit maximum in the long run that is the sum of maximum profits in each short run period. On the other hand, there have been multiple "second best" results of more or less similar performance; that is, similar long run profitability achieved by firms following different strategies, some prospering relatively more when crude is short and some when crude is in surplus, for example. This is one explanation for the existence of permanent differences in strategy observed in this study.

11.7 Why Did Strategies Differ?

Permanent differences in strategy among these seven companies arose and have persisted for at least three reasons: (1) there was no single optimum strategy toward which all competitors were forced; (2) the internal dynamics of each firm which determined its strategy was unique; and (3) internal constraints made it difficult and costly to change an established strategy.

As Table 11-6 shows, the scope of opportunities in the environment was not static but constantly shifting with events, sometimes gradually and sometimes precipitously. The petroleum industry has had to cope with (a) shifts from crude glut to scarcity and back again, (b) advances in technology for finding and processing oil, (c) increases in minimum efficient scale in refining, transportation, and service station volume, (d) shifts in product demand from the rise of the automobile and electricity which require gasoline and fuel oil rather than kerosene, (e) technological advances producing new products like petrochemicals and better products like higher octane gasoline and better lubricants, and (f) general economic and political conditions external to the industry including financial and economic cycles, wars, laws, public opinion and nationalizations of property.

The result of shifting environmental opportunities was that optimum strategy also shifted. Firms with different strategies were affected differently and responded at different rates, giving them different strategies at any moment. Payoff on new opportunities was affected by a firm's existing resources. Firms differed in their financial and managerial ability to change. Mobil, beginning after 1911 with no crude, found it very difficult to build an exploration capability. Socal, Texaco, and Gulf found it difficult to enter Eastern Hemisphere markets with Middle Eastern crude. The result of shifting environmental opportunities was that there was no single optimum strategy toward which all competitors were forced as was shown in Section 11.6 above.

Since differences from an optimum strategy were not entirely eliminated, strategies were free to reflect the compelxity of the companies and their environments. Complexity was the result of the large number of choices to be made and their interrelations. No two firms ever faced exactly the same opportunity sets. For example, if one firm chose to enter market A, a second entrant may have found it optimal to choose an unoccupied market rather than also enter market A. Standard Oil often had this effect on potential competitors. In Texas in 1901 Standard's absence encouraged the founders of Gulf and Texaco to take the risks of entry. Of all the opportunity sets faced by the Majors in this study, the ones faced by Gulf and Texaco's founders in 1901 were the most nearly identical. Yet their strategies evolved quite differently. The result of the complexities in the environment and the internal dynamics of each firm was a strategy that was unique.

The policies and strategies which resulted from the internal dynamics of each Major are evidence of the uniqueness of their situations. As shown in Section 11.6 there were only a few descriptive structural similarities in strategy among the Majors. The similarities which existed did not permit accurate prediction of behavior or performance of the Majors as a strategic group. The range of policies of each Major was different from every other. There was no standard set of policies possessed by all seven. Very few individual policies on similar strategic topics would result in similar behavior or performance. Each policy almost invariably reflects a relationship to a Major's other policies which makes it quite different from any comparable policy at another Major. The unique opportunity sets and internal dynamics which resulted in the formation of each policy in this study are detailed in the company narrative chapters.

The following comparison of the entrepreneurial firm of economic theory and the large number of interrelated choices the Majors had to make is intended to show how unique strategies result from complex environments and internal dynamics at each firm.

If these complexities were not present it is likely that modern business enterprises like the Majors would have resembled the entrepreneurial firm more closely. If the Majors had sold their products in a single auction market, they would not have needed a geopraphical strategy. If there had been constant returns to scale, they could not have attained a cost advantage in some functions by growing larger. If disproportionate rewards did not flow to holders of large market shares through experience curves and abuses of market power, market share would have been unimportant. If the Majors could have gained the same security, risk reduction, and flexibility, and if they could have provided the same after-sale services and financing through an auction market, there would have been no need for vertical integration. Each of the Majors could have competed at only one vertical level without disadvantage. If consumer demand had not been so varied and changeable and if markets, products, and technologies had not changed so rapidly, the Majors would not have needed to diversify their product lines or portfolios of businesses. If each Major had been guided by a single entrepreneur (like Gus Long of Texaco) with brains, capital managerial ability, an infinite span of control, and heirs, a large bureaucratic technostructure of managers and coordinators would not have been needed. Outside finance would not have been required, given sufficient profitability. There would have been no separation of ownership and control. Complicated administrative policies would not have been necessary. If all these variables did not change and were not subject to uncertainty, optimal solutions could have been more easily identified. If all Majors could have adjusted instantaneously to new conditions without cost, they would have responded to opportunities more similarly. If all these complexities had not existed or had had uniform distributions of consequences for all, there would have been much less opportunity or need for strategies to differ or even exist. Simple profit maximization would have served as a goal. But these complexities existed and the unique strategies of the Majors were rational responses to them.

A third reason strategies differed was the difficulty of overcoming internal constraints. These seven firms became more and more constrained by internal factors over time rather than less and less. The internal consistency of the organization which allowed it to function smoothly placed a high cost on change which disturbed interdependent links. The larger the organization became, the larger was the effort required to change it. For example, Exxon spent $100 million to change its name in 1972. As a result of these internal constraints the Majors were unable or found it unprofitable to adjust to the rapid shifts in the external environment. These seven firms became fixed in their strategies and characters by their early experiences. Since they started at different times when optimal strategies were different, and since their initial circumstances and resource endowments were different, their strategies were different from the beginning. Since internal constraints made it difficult to change strategy, they remained different forever after.

Evidence for the persistence of strategic differences due to internal constraints comes from previous results in this chapter. The results in Section 11.2 show the heavy influence of internal factors on strategic choice. The results in Sections 11.3 and 11.4 show that the rate of policy formation tapered off rapidly after the firm's first decade even though the environment, shown in Table 11-6, continued to change frequently. Some strategic differences arose as the result of adaptation to the environment in each firm's early years. But external factors do not explain why differences in strategy have persisted. The results above are evidence that differences in strategy among the Majors have persisted due to the difficulty of overcoming internal constraints which changed little once the internal equilibrium of a successful strategy was formed.

11.8 How Strategies Remain Different

Section 11.7 identified two reasons why differences between the Majors' strategies arose (lack of a single optimum strategy towards which all competitors were forced, and unique environment and policy interrelationships at each firm) and one reason why they have persisted (inability to overcome internal constraints). This section examines the mechanisms by which they remained not only different but also relatively unchanged over time.

This study finds several mechanisms by which strategy perpetuated itself and successfully resisted change.

(1) The first is that while the environment caused shifts in the profit maximizing strategy from year to year or decade to decade, the productive life of assets in the petroleum industry was often thirty to forty years or more for oil fields, pipelines, and refineries. Replacing or changing these assets or strategy based on them was a long run process done only at great cost. Therefore, strategy did not shift rapidly because it could not acquire or replace these crucial assets quickly.

For example, after posted price reductions signaled the onset of a crude oil glut about 1960, Majors like Gulf, BP, and Socal with more crude than outlets found it very difficult to respond by expanding market outlets against entrenched competitors like Exxon, Mobil, and Shell. Just over a decade later in 1973-1974 all the Majors were stuggling to buy enough crude to keep their downstream organizations in operation. Outlets and crude supplies were difficult to acquire or replace quickly.

The fact that it has not been possible to acquire or replace assets as rapidly as optimum strategy changed has placed a constraint on strategic change which has forced departures from existing strategy to be incremental.

(2) A second reason strategy was slow to change was that once a competitor had established a low cost position in Middle Eastern crude or a geographical market like Europe, for example, there was little another competitor could do to dislodge him so long as he remained efficient and not too greedy. Inherent long-lived cost differences between competitors created natural barriers to entry and to strategic change. Opportunities to make major changes in market share occurred rarely, perhaps once in a generation with

changes like the rise of the automobile and gosoline market or discoveries in the Middle East. At other times, existing market arrangements constrained the ability to change strategy radically.

(3) A third mechanism for retaining strategy was organizational learning. This appeared to occur in two ways.

One way organizational learning occurred was through people. The people in these organizations passed along expertise and attitudes to their successors. The organizations which took the opportunities to accumulate this experience preserved their leads over competitors. This organizational learning embodied in people with special expertise has been particularly evident in technical areas at various times.

But it appears to have been the less important of the two mechanisms. There are no identifiable areas of long-term comparative strength among the Majors which resulted from this type of learning curve. Proprietary technical expertise has not been as responsible for the successes of the Majors as have major discoveries, large-scale activities, and market positions. Experience gained at the Majors has often been an important asset of executives at non-Majors. But experience gained at another Major has been rare among executives at the companies in this study.

If all organizational learning were embodied in people, it would have been possible to acquire such expertise by hiring successful people from competitors. But this has not happened among the Majors, where moves from one company to another have been unusual.

This implies that the enduring strengths of the Majors relative to each other may be due to a second organizational learning mechanism which does not reside in people but is a separate corporate asset of strategic attitudes toward risks in various activities. An organizational attitude toward the risk in some activity which was learned from experience appeared to become a part of the company's strategic equilibrium of policies and thereby an enduring characteristic of that organization which permanently differentiated it from others. The power of the company's strategic equilibrium both preserved such attitudes and prevented the adoption of others. This mechanism may help explain the endurance of both strengths and weaknesses in strategy which have lasted at least since the first two decades of this century. Some examples are BP's strength in new field exploration and weaknesses downstream, Shell's strengths in chemicals, tankers, and refining, and Exxon's strength in refining and weaknesses in exploration and retail marketing. Once an attitude toward risk in some activity was formed it was sustained not only by its own record but also by its contribution to the identity and strategic equilibrium of the organization.

Is it possible that continued success in an activity was due to the chance occurrence of a sequence of talented people and not to organizational factors?

Though it is possible for luck to favor a company in exploration for oil over a period of time, for example, the generally pervasive pattern of continued success for activities of all seven companies and the inability to remedy weakness by hiring expertise make it improbable that strategies remain unchanged and different due to chance.

(4) The fourth mechanism which perpetuated stategic differences was probably the most important. It is that capital expenditures were allocated to reinforce success. The results of these differences in expenditure pattern are apparent in the cumulative positions shown in Table 1-2.2-1-2.5 at the end of Chapter 1. Keeping these investments profitable and productive placed a major constraint on new directions.

These four explanations of how strategies remain not only different but relatively unchanged are offered to show that similarities between actions of present management and those taken by the same organization some years ago are not likely to be coincidental, despite every good effort of present management to be objective in its consideration of courses of action. Continued similarity in assets and expertise dictated quite objectively similar decisions.

11.9 Changes in Strategy

A large contributor to the persistence of strategic differences has been the high rate of failure of efforts to change or to match the the results of the leading competitor in activity. Why have conscious efforts to change so often failed? Was it because the Majors were bound by the past? The answer seems to be both yes and no. There appear to have been four areas where strategic change was or was not possible.

(1) One change in strategy that was made was the loss of an area of competence or strategic strength. As examples one can cite the loss of Shell's position as the leading producer in the 1920s and Texaco's loss of refining leadership between the 1920s and 1940s.

(2) Another change was that companies entered new areas of opportunity in geography, industries, or products where entry did not conflict with existing vested interests. There are numerous examples of this in the company narratives, including the discoveries in Mexico, Venezuela, and the Middle East, offshore drilling technology which allowed Socal to establish itself in the Gulf of Mexico and thereby east of the Rocky Mountains, and Texaco's entry into the U.S. gasoline market. It may not have been possible to take volume away from an entrenched competitor but a new entrant was often able to grab a piece of the growth in a market or new product.

(3) It is also apparent from this study that it was very hard for an organization to eliminate or ameliorate its weaknesses. It appears to take a combination of five factors to make a successful change: environmental

opportunity, money, willpower, organizational acceptance, and intelligent execution. If any one was missing the effort faltered.

When Gulf and Socal tried to create downstream market positions in the Eastern Hemisphere in the 1960s against entrenched competitors or when Exxon and Gulf tried to create nationwide U.S. retail marketing organizations in the 1960s the efforts were less than successful because the environmental opportunity for entry was lacking. Mobil took only 10% of Aramco, which would have alleviated its shortage of crude, because it was not willing to spend the money for more. The Majors generally have recognized their strategic weaknesses but rarely have they had the willpower to carry through the changes required to remedy them. One of the biggest obstacles to strategic change has always been the lack of acceptance by the organization of changes in their identities and practices. Some examples of this factor and the last one, intelligent execution, are given in area (4) below.

(4) Finally, the hardest strategic change of all to make was to develop or switch major strategic strengths. The same five factors as in (3) were required to take advantage of any opportunity. To change the vertical balance of such large organizations requires the discovery of a new Saudi Arabia or a major new market like the automobile or electricity generation, events that are rare. Mobil lacked the willpower and thought it lacked the money to take a larger share of Saudi Arabia. Socal lacked the money to develop Saudi Arabia and the markets for its products alone. Exxon has had difficulty thinking of itself as a retail marketer; Mobil, as an explorer; and Gulf, as a worldwide company. Gulf sincerely desired to build an Eastern Hemisphere downstream organization and a new kind of nuclear reactor but did neither one well.

In this study there are only two cases where any kind of effort at all was successful in turning weakness into strength: exploration at Socal after 1911 and management at Mobil after 1959. The BP-Sohio merger is a third possibility, which certainly remedied a weakness but may or may not eventually turn it into a strength. On the other side of the ledger where efforts to remedy weakness have not been successful are two billion-dollar fiascos (Gulf's downstream move into the Eastern Hemisphere in the 1960s and the Gulf-Shell nuclear reactor in the 1970s), several efforts where large investments of money have not yielded comparable results (the exploration programs of Exxon, Shell, and Mobil), numerous unsuccessful efforts on a smaller scale, and several instances where hiring people with outside expertise may have made a positive contribution but did not really turn a weak activity into a competitive strength.

11.10 Recommendations for the Management of Strategy

As a result of this study of the seven largest oil companies what can one recommend to general executives of similar organizations to improve the management of strategy?

(1) Identify present strategy and desired strategy and recognize that the two may be different. These differences provide one source of strategic goals.

(2) Identify internal constraints and overcome or adjust for them. Otherwise, implementation of change is likely to be ineffective.

(3) Recognize that the environment will change. Periods of environmental stability in the oil industry in this century have lasted only about a decade each. The last upheaval was the nationalization of production by OPEC in 1973-1974, almost eight years ago. That environment probably ended its life cycle in 1982.

(4) Recognize the requirements for successfully implementing changes in policy. Successful changes in strategic weakness have been rare because one or more of the critical elements for successful change was lacking.

(5) Establish a strong planning function. These recommendations based on historical results imply that a strong corporate planning function is required to accomplish the recommendations above. Plans for change implemented no more carefully than some of the examples above will only waste scarce resources.

Exhibit 11-1

Analysis of Policy Origin and Termination Causes

Numbers of Policies

Company		Ex	Gf	Mo	Sc	Tx	Sh	BP	T#	Ex	Gf	Mo	Sc	Tx	Sh	BP	Ave	%T
										Percentage of Total for Each Company								
CAUSES OF ORIGIN (180 policies originated)																		
Ext T	J	19	16	10	5	11	14	12	87	65	47	40	29	37	67	50	48	48
	A	2	4	2	3	2	3	4	20	7	12	8	18	7	14	17	12	11
EC	J	17	16	10	5	11	11	12	82	59	47	40	29	37	52	50	45	46
	A	2	4	2	3	2	3	1	17	7	12	8	18	7	14	4	11	9
SR	J	7	0	0	0	0	4	5	16	24	0	0	0	0	19	21	9	9
	A	0	0	0	0	0	0	2	2	0	0	0	0	0	0	8	1	1
Int T	J	19	16	10	5	11	14	12	87	65	47	40	29	37	67	50	48	48
	A	8	14	13	9	17	4	8	73	28	41	52	53	57	19	33	40	41
R	J	9	18	7	6	10	6	9	65	31	53	28	35	33	29	38	35	36
	A	1	6	7	6	3	2	5	30	3	18	28	35	10	10	21	18	17
PV	J	16	11	3	5	7	10	4	56	55	32	17	29	23	48	17	32	31
	A	4	3	5	0	10	1	2	25	14	9	20	9	33	5	8	13	14
Specific Event Causes																		
T		5	8	5	0	1	3	1	23	17	24	20	0	3	14	4	12	13
Pol		1	1	1	0	0	0	1	4	3	3	4	0	0	0	4	2	2
Tech		2	1	0	0	0	0	0	3	7	3	0	0	0	0	0	1	2
Ec		2	8	4	1	1	3	0	18	7	24	16	0	3	14	0	9	10
Total #		29	34	25	17	30	24	21	180									
CAUSES OF TERMINATION (39 policies terminated)																		
Ext T	J	0	6	2	0	1	0	1	10	0	67	33	0	14	0	17	19	26
	A	3	2	1	1	1	2	5	15	43	22	17	50	14	100	83	47	38
EC	J	0	6	2	0	1	0	1	10	0	67	33	0	14	0	0	16	26
	A	3	2	1	1	1	2	5	15	0	22	17	0	14	0	0	8	38
SR	J	0	0	0	0	0	0	0	0	0	0	0	0	0	0	0	0	0
	A	0	0	0	0	0	0	0	0	0	0	0	0	0	0	0	0	0

								T								Ave	%T
Int T	J	0	6	2	0	1	0	1	8	0	0	33	0	14	0	17	9	21
	A	2	0	3	1	5	0	1	12	0	0	50	50	71	0	17	27	31
R	J	0	6	2	1	1	0	0	10	0	67	33	50	14	0	0	23	26
	A	0	0	2	0	4	0	0	6	0	0	33	0	57	0	0	13	15
PV	J	0	0	3	1	2	0	1	7	0	0	50	50	29	0	0	18	18
	A	2	0	0	0	0	0	1	3	29	0	0	0	0	0	0	4	5
Unknown		2	0	0	0	0	0	0	2	29	0	0	0	0	0	0	4	5
Specific Event Causes																		
T		3	2	1	0	0	0	1	7	43	22	17	0	0	0	17	14	18
Pol		2	2	0	0	0	0	1	5	29	22	0	0	0	0	17	10	13
Tech		0	0	0	0	0	0	0	1	14	0	0	0	0	0	0	2	3
Ec		0	1	1	0	0	0	0	1	0	0	17	0	0	0	0	2	3
Total #		7	9	6	2	7	2	6	39									

Key:

Ext—External to firm
Int—Internal to firm
J—Joint
A—Alone
T—Total
%T—% of Total Policies

EC—Environmental conditions
SR—Social Responsibilities
R—Resources
PV—Personal Values
Ave—Average of seven Majors

Key to Causes

Ext	J	Total policies with *both* internal and external causes.
	A	Total policies with *only* external causes, either EC or SR.
EC	J	Policies with EC causes *and* other causes (R, PV, SR).
	A	Policies with *only* EC causes.
SR	J	Policies with SR causes *and* other causes (EC, R, PV).
	A	Policies with SR causes *only*.
Int	T	Total policies with *both* internal and external causes.
	A	Total policies with *only* internal causes, either R or PV.
R	J	Policies with R causes *and* other causes (EC, SR, PV).
	A	Policies with R causes *only*.
PV	J	Policies with PV causes *and* other causes (R, EC, SR).
	A	Policies with *only* PV causes.

Exhibit 11-2

Distribution of Dates of Origin of Policies Still in Effect

	Exxon	Gulf	Mobil	Socal	Texaco	Shell	BP	Total
(a) Total Number of Policies Studied								
	27	34	25	17	30	21	24	178
(b) Number Still in Effect in 1976								
	22	24	19.5	15	23	19.5	17	140
(c) Percentage Still in Effect:								
(b) ÷ (a)	81%	71%	78%	88%	77%	93%	71%	79%

(d) Dates of Formation of Policies Still in Effect in 1976 Relative to Dates of Origin of Company (Number of Policies Formed in Each Period)

	Exxon	Gulf	Mobil	Socal	Texaco	Shell	BP	Total
Inherited	15	0	10.5	5	0	14.5	0	45
Years 0-1	0	9	1	3	6	2	3	24
Years 2-10	4	4	2	1	7	1	7	26
After Year 10	3	11	6	6	10	2	7	45

(e) Inherited Policies (from d): Year of Policy Origin
Relative to Year of Company Origin
in the Companies from Which the Policies were Inherited

	Exxon	Gulf	Mobil	Socal	Texaco	Shell	BP	Total
Years 0-1	2	0	6	3	0	9	0	
Years 2-10	6	0	0	0	0	5.5	0	11.5
After Year 10	7	0	4.5	2	0	0	0	1
Total in (d) above	15	0	10.5	5	0	14.5	0	45

(f) When Policies Still in Effect in 1976 were Formed: Cumulative (d) ÷ (b)

	Exxon	Gulf	Mobil	Socal	Texaco	Shell	BP	Total
Inherited	.68	.00	.54	.33	.00	.74	.00	.32
Thru Year 1	.68	.38	.59	.53	.26	.84	.18	.50
Thru Year 10	.86	.54	.69	.60	.57	.90	.59	.69
After Year 10	.14	.46	.31	.40	.43	.00	.41	.31

(g) When Policies Still in Effect in 1976 Were Formed Relative to Company Origin: ((d) + (e)) ÷ (a)

	Exxon	Gulf	Mobil	Socal	Texaco	Shell	BP	Total
Thru Year 1	.09		.36	.40		.56		.31
Thru Year 10	.55		.46	.47		.90		.58
After Year 10	.45		.54	.53		.10		.43

References below are policy numbers for: (a) Policies terminated; (b) Policies inherited; (c) Policies formed in years 0-1 (both inherited and not); (d) Policies formed years 2-10; (e) Policies formed after year 10. These are the policies summarized in (a) through (g) above. The policy numbers come from the exhibits at the end of each company chapter.

Exxon: (a) 4, 9, 10, 13, 17 (20,23 possible but excluded); (b) 1-20; (c) 7-10; (d) 1, 3, 6, 15, 19, 20-23, 25; (e) 2, 4, 5, 11-14, 16, 18, 24, 26, 27.

Gulf: (a) 8, 12, 19, 23, 24, 27, 29, 30, 33, 34; (b) None; (c) 1-8, 17, 28; (d) 9-13; (e) 14-16, 18-27, 29-34.

Mobil: (a) 4(1/2), 5, 7, 8, 12, 13; (b) 1-7, 14-19; (c) 1, 3, 7, 8, 11-13, 15-18; (d) 9, 10; (e) 2, 4-6, 14, 19-25.

Socal: (a) 5, 8; (b) 1-5, 8, 9; (c) 1, 3, 6-10; (d) 17; (e) 2, 4, 5, 11-16.

Texaco: (a) 1, 6, 9, 16-18, 27; (b) None; (c) 1-4, 6, 14, 15, 23; (d) 5, 7-13, 16; (e) 17-22, 24-30.

Shell: (a) 13(1/2), 21; (b) 1-14, 18; (c) 1-9, 15, 16; (d) 10-14, 17, 18; (e) 19-21.

BP: (a) 3, 6, 7, 13-15, 18; (b) None; (c) 1-3, 5, 6, 14; (d) 4, 7-13, 20; (e) 15-19, 21-24.

Table 11-1
External and Internal Causes of Policy Origin
(Numbers of Policies)

Type of Cause	Ex	Gf	Mo	Sc	Tx	Sh	BP	Total	% Total
External & Internal	19	16	10	5	11	14	12	87	48%
External Alone	2	4	2	3	2	3	3	20	11%
Internal Alone	8	14	13	9	10	4	8	73	40%
Total # Policies	29	34	25	17	30	24	21	180	100%

Source: Exhibit 11-2 and Chapters 4-10. Abbreviations (Ex,Gf,Mo,Sc,Tx,BP,Sh) are of the names of the Majors.

Table 11-2

Frequency of Each Variable as a Cause of Policy Origin
Alone and in Combination with Others

(Numbers of Policies)

Policy Variables	Acting	Ex	Gf	Mo	Sc	Tx	Sh	BP	Total	% Total
Environmental Conditions	Alone	2	4	2	3	2	3	1	17	9%
	With Others	17	16	10	5	11	11	12	82	46%
Social Responsibility	Alone	0	0	0	0	0	0	2	2	1%
	With Others	7	0	0	0	0	4	5	16	9%
Resources	Alone	1	6	7	6	3	2	5	30	17%
	With Others	9	18	7	6	10	6	9	65	36%
Personal Values	Alone	4	3	5	0	10	1	2	25	14%
	With Others	16	11	3	5	7	10	4	56	31%

Source: Exhibit 11-2 and Chapters 4-10.

Table 11-3

The Impact of Specific Events on Policy Origin
(Numbers of Policies Affected)

Type of Event	Ex	Gf	Mo	Sc	Tx	Sh	BP	Total	% Total
Politics	1	1	1	0	0	0	1	4	2%
Technology	2	1	0	0	0	0	0	3	1.7%
Economics	2	8	4	1	1	3	0	19	10.6%
Total	5	10	5	1	1	3	1	26	14%

Source: Exhibit 11-2 and Exhibits 1(d) in chapters 4-10.

Table 11-4
Combined Volumes and Worldwide Market Shares of the Majors
(000 b/d)

	Net Production			Refinery Runs			Product Sales			Vertical Integration
	Majors	World	%	Majors	World	%	Majors	World	%	
1912	100	966[1]	10	215[9]	NA[8]	22	179[4]	NA[8]	19	47:100:83[4]
1918	276	1379	20	453[2]	NA[8]	33	317[5]	NA[8]	23	61:100[2]:70[5]
1927	1053	3456	30	1392	NA[8]	40	1117[6]	NA[8]	32	76:100:80[6]
1938	1996	5446	37	2626	5368	49	2731	NA[3]	51	73:96:100
1950	4996	10407	48	5504	10537	52	5118[7]	NA[3]	49	91:100:93[7]
1960	9810	21348	46	10274	21308	48	10267	NA[3]	48	95:100:100
1970	21787	46130	47	20352	45625	45	20655	NA[3]	45	100:93:95
1976	10657	59555	18	19194	56700	34	21304	58790	36	50:90:100

Notes: NA—not available
1. Excluding Mobil whose volume is NA
2. Excluding Gulf and Mobil whose volumes are NA
3. Refinery runs used as an approximation
4. Excluding Gulf, Mobil, and Texaco whose volumes are NA
5. Excluding Mobil, Texaco, and BP whose volumes are NA
6. Excluding Mobil, Socal, and BP whose volumes are NA
7. Excluding BP whose volume is NA
8. Net production used as an approximation
9. Excluding Gulf whose volume is NA

Vertical Integration Index = Max (P:R:M)   P:R:M

Data are from Table 1-2 line 1 and D.E.F.

Table 11-5.
Expected and Actual Relative Profitability Rankings
(Numbers rank profitability from 1 to 7)

		Exxon	Gulf	Mobil	Socal	Texaco	Shell	BP	Correlation
1911-1918	P	3	2	4	1	6	3	7	.61
	A	5	3	4	1.5	6	1.5	NA	
1919-1927	P	6	5	7	3	2	1	4	.84
	A	5	6	7	2.5	4	1	2.5	
1928-1937	P	5	7	4	2	6	3	1	.82
	A	5	7	6	3	4	2	1	
1938-1949	P	3	5	6	2	4	7	1	.82
	A	5	4	6	3	2	7	1	
1950-1959	P	1	4	6	2	3	7	5	.66
	A	4	4	6.5	1	2	6.5	4	
1960-1969	P	3	6	4	2	1	5	7	.61
	A	4	3	6	2	1	7	5	
1970-1976	P	2	6	4	3	5	1	7	.46
	A	1	5	5	3	2	5	7	

P = predicted relative profitability
A = Actual relative profitability
Correlation is between P and A for each period
1 = most profitable
7 = least profitable

Table 11-6
Environmental Changes

Period	Events and Expected Characteristics
1885-1901	Large crude oil discoveries in Russia (1885) Invention of the marine bulk tanker (1888) Demand for kerosene in the Far East Competition from Standard Oil becomes worldwide Zenith of British Empire enforced by British Navy Golden age of Dutch science
1901-1911	Discovery of oil in Texas (1901) Wars between Standard and Shell Discovery of oil in Oklahoma (1905)
1911-1918	Discovery of oil in Mexico (1910) Standard Oil dissolved (1911) Burton-Humphreys patents for thermal cracking (1912) World War I (1914-1918)
1918-1927	Crude oil shortage feared after the war Rise of the automobile and the gasoline market Discoveries in Venezuela
1927-1938	Crude shortage turns to glut Great Depression
1938-1950	World War II Major discoveries in the Middle East
1950-1960	Pax Americana Europe and Japan recover from World War II U.S. becomes a net importer of crude oil Center of gravity of petroleum production shifts from Western Hemisphere to Eastern Hemisphere
1960-1970	Glut from Middle Eastern production Golden age of the automobile and cheap energy Rise of petrochemicals
1970-1980s	Producing country governments take control of production and pricing (Teheran, 1971; OPEC, 1973) U.S. production and reserves peak and begin to decline (1970)
1980s-2000	?Transition to new energy sources ?Major new discoveries ?Economic stagnation ?Soviet Union becomes dominant world power

Appendix A

Review of Theoretical and Empirical Literature

This appendix outlines the development of theoretical and empirical literature on the firm in both business policy and industrial organization. This study is an incremental contribution to these literatures with the goal of adding to the understanding of corporate strategy and behavior.

As is appropriate to an interdisciplinary study, the literature is drawn from both economics and business administration. The theories which describe how the firm should behave come from literature in microeconomics, industrial organization and business policy, while the empirical literature and methodology of this study come from the literatures of industrial organization, business policy, and business history. It will be observed that, although the literatures of economics and business administration developed somewhat independently, a synthesis is emerging.

Theoretical Literature

There are several models of the process by which a firm, such as one of the Majors, makes decisions which determine its behavior and performance. These models were developed at different times with different purposes and assumptions. In a descriptive role they are only different and increasingly complementary approaches to the complex empirical question of how and why firms actually behave as they do.

Similarly, several different models of the structure and characteristics of the markets in which firms operate have been developed, and each of these has been used to analyze the consequences of the firms' actions in each of these respective market contexts. The different models developed at different times on different assumptions have in part reflected the increasing power and sophistication of economic analysis and theory building over the years. But it will also be apparent that to a considerable extent the different assumptions used at different times reflected changing conditions in the external business world and shifts in the issues and concerns posed for analysis. This review will turn first to different theories of market structure.

A century ago there were only two recognized models of firm behavior, the models of perfect competition and pure monopoly. These models represented possible extremes in industry structure and logically derived behavior. Both models incorporated streams of ideas about competitive behavior (or its absence) which could be traced at least as far back as Adam Smith's *Wealth of Nations* in 1776. Succeeding generations of economists gradually formalized the analysis into increasingly sophisticated models derived from more refined and demanding sets of assumptions. [1]

The development of general equilibrium theory, and later the refinement of "welfare economics," solidified the role of the competitive model as the desirable structural norm for firms and industries, while the monopoly model was shown to lead necessarily to inefficiencies and distortions in the allocation of resources. These simple polar models reflected the concerns of the public in the late nineteenth century with the formation of monopolies and the response in the form of regulatory and antitrust legislation and subsequent litigation.

Several events in the first two decades of this century caused economists to become dissatisfied with both models and focused economists' attention on forms of competition other than perfect competition and pure monopoly.[2] Weaknesses in both models were recognized. Perfect competition was seen to be an idealized limiting case as was the idea of a "pure" monopoly unaffected by the actions of other firms. Scale economies were recognized as a factor in the large combinations at the end of the nineteenth century and appeared to offer the theoretical possibility of potentially infinite expansion in the size of the firm.[3] This cast doubt on the likelihood of ever reaching the kind of perfectly competitive general equilibrium economists had envisioned where each factor earned a similar risk adjusted rate of return. The rise of national markets, instead of approaching the ideal of perfect competition as expected when the size of the market and number of competing firms expanded,[4] produced large vertically integrated oligopolies of firms selling multiple lines of brand name products.[5] These firms, fortified by economies of scale and barriers to entry, formed into more or less stable groups of interdependent enterprises, none of which could be characterized as having a monopoly in the traditional sense. On a different dimension, the functions of ownership and control became separated and a new type of economic person, the manager or administrator, evolved to run these giant organizations.[6] The seven oil companies in this study were all formed and grew large during this period and are prime examples of several of these trends.

The impacts of these events were far reaching. Business schools were formed to train managers in the new discipline of administration. More directly relevant at the moment, these disturbing events also raised new questions for theoretically oriented economists.

New models of imperfect competition and the organization of industry resulted. Robinson and Chamberlin simultaneously brought forth in the early 1930's a new model christened "monopolistic competition."[7] This hybrid of the perfect competition and pure monopoly models addressed the new realities of specialized firms, differentiated products, brand names or trademarks, and a market structure where each differentiated competitor possessed at least some monopoly power to raise prices in relation to its marginal costs or production and distribution.[8] But this power was limited by the elasticity of substitution between each market and the others. Many other models of imperfect competition were developed by making different assumptions on the number of relevant competitors (as in models of duopoly and oligopoly, as distinct from the "large number" case of monopolistic competition), on the source and relative height of barriers to entry, and on whether each firm's decision at a given time assumed that others' prices or quantities were "given." But in the structuring and analysis of all these models, the mutual interdependence between the firms was emphasized. A main theme common to all this work was recognition of the dependence of one firm's costs and demand curves upon those of other firms.[9]

The issue of market structure above (oligopoly, vertical integration, scale economies, differentiated products, separation of ownership and control, barriers to entry, and other non-competitive elements) and the worldwide depression of the 1930's brought about a reassessment of the type of competition that was desirable and the market or antitrust mechanisms which might produce it. As Mason noted, "This is the core of the difficulty of devising standards of public action in the antitrust field. None of the markets encountered meet the tests of pure competition; at the same time they fall short of a degree of monopoly justifying public utility regulation. What is a suitable test of effective competition?"[10] Having rejected perfect competition as a standard, much of the profession turned to an alternative, termed "workable competition" by J.M. Clark, which would adjust for these technological and institutional issues of market structure which prevented perfect competition.

The field of industrial organization was born as economists began to work out the probable empirical consequences of the existence of these market structure elements upon economic performance. Bain's model of industrial organization linked performance to market structure, omitting behavior.[11] Market structure had great advantages over behavior as a policy variable. It could be more readily measured and controlled. Led by Bain in the 1950's, industrial organization economists employed a new methodology by applying statistical techniques to cross-industry data to avoid the biases in a single industry structure. The new empirical tools of analysis included four and eight firm concentration ratios, height of entry barriers, minimum optimum scale of plants, price/cost ratios, and degrees of product diversification. Later scholars extended these empirical techniques to behavioral variables as well and the elements of the industrial organization model became structure, conduct, and performance.[12]

In the area of behavior, the sheer size and multi-level bureaucracy of the modern large-scale firm led scholars to examine whether it departed from the models of competition in the pursuit of goals other than profit maximization. Cyert and March (1963) were among the first economists to model the firm as a large complex organization whose behavior was not pure profit maximization by an owner-entrepreneur but a joint preference function formed by structures of internal control.[13] Sales, growth, and managerial utility maximization have also been suggested as alternatives to profit maximization.[14] Cohen and Cyert have suggested that managerial slack may lead to behavior where monetary profits are not maximized but are "satisfied" in favor of non-monetary (and non-taxable) rewards, a variant of Hicks' "quiet life" as the best of all monopoly profits.[15] Galbraith substituted survival, independence, and technological virtuosity for profit maximization as determinants of firm behavior.[16] Bower (1970) demonstrated that information flows and incentive structures within an organization can affect its capital investment behavior, producing a different set of choices than those predicted using a profit maximizing expected rate of return model.[17] This stream of research into corporate behavior led to recognition that the firm's structure was an important determinant of its behavior and performance. Other research on the importance of the firm's structure is discussed below with empirical research.

At roughly the same time that economists were actually developing these more complex analyses of the determinants of the behavior of firms scholars of business administration were separately formalizing a rather different model of the determinants of firm behavior from case studies and other empirical data. It was begun in the policy group at the Harvard Business School in the 1950's by Edmund P. Learned and C.R. Christensen, but the first comprehensive written statement of the model by Andrews came only in 1971.[18] The business policy model was a tool for determining the best strategy and the component policies (behavior) for an enterprise that are required to achieve its long-term goals and objectives.

The methodology (individual case studies rather than cross-industry statistical studies) and the terminology (policies instead of behavior) of the business policy model differed from its counterparts in economics. But the phenomena of corporate behavior being described were not dissimilar. The process of choosing a strategy by matching its strengths and weaknesses against its risks and opportunities is analogous to the constrained optimization process in economics, though less precise, less structured, and more dynamic. And the business policy model developed quite separately from the events taking place in economics.

This empirical research in industrial organization documented some statistically significant effects of market structure elements on conditions which could facilitate oligopolistic coordination or contribute to mutual dependence recognized among competitors. But until relatively recently most models of economic theory and analysis, including industrial organization, were done in a static, cross-sectional framework that analyzed what would be the optimal position under a given set of conditions. This methodology emphasized a single predicted optimum behavior in a given circumstance or set of structural conditions, or for any firms a significant probability that a given market structure would have a predicted impact on performance.

More recently, further analysis of the behavior of individual firms within oligopolistic industries has introduced more dynamic considerations which suggested that the same behavior or course of action might not be appropriate for all competitors. If a firm was number three in an industry, its behavior was likely to be different than if it was number one or two. And the result of its efforts to increase market share could be to change the structure and performance of the industry as has been the case, for example, in the 1970's with foreign competition in the auto, steel, textile, shoe, and electronics industries.

The introduction of dynamic considerations into the study of firm and industry structure, conduct, and performance has been an important step because it has brought economists studying competition and practicing managers seeking solutions to study the same issues. Research on competition, feeding from fact to theory and back again repeatedly, has advanced microeconomic theory over the last century through the milestones outlined above: from perfect competition and pure monopoly to imperfect and monopolistic competition to "workable competition" to industrial organization to the present issues. This accumulated research has made it possible to study dynamic issues of competition such as how specific firms and industries will evolve, how to recognize and interpret the actions which signal competitors' intentions, and how to determine the best course of action for an individual firm.

The study of these dynamic issues of firm behavior and competitive interaction required new tools. Economists have begun searching both inside and outside the profession for applicable models as work by Porter on the analysis of firms and industries and Spence on market signaling shows. [19] On the issue of determining the best course of action for the individual firm, the business policy model has an important contribution to make and a synthesis of research on business policy with that of industrial organization has begun. As Caves noted, "Industrial organization economists have only begun to incorporate strategic choice into their analyses of market structure, conduct, and performance." They have both "something to learn" and "something to offer in this line of research." [20] The same may be said of business policy scholars.

The study of the dynamic issues of competition is still in an early stage and the taxonomy of corporate strategies and structures, as Caves also noted, is incomplete. Eventually a practical model of dynamic competition may result which can be used by both economists and managers to predict and analyze firm and industry structure, conduct, and performance.

One of the common threads in the literature above is that technological and institutional conditions of market structure and conduct systematically affect performance. This has been affirmed by both economists and scholars of business administration for many years now for both industries and individual firms. However, the process by which structure and conduct affect performance is not always clear and the outcome of a given realistic situation is often not predictable.

This study is addressed to just this issue. It traces the known determinants of the major strategic decisions made by the seven firms which are alleged to form the dominant group in the petroleum industry to their origins and from their origins over time to the present. The result is a narrative for each company of all its major decisions and an analysis of their structural and behavioral determinants. One interesting question addressed is how strong a predictive model of the future performance of an individual firm can be built from a knowledge of the determinants of its past decisions. To economic models of firm behavior this study hopes to contribute some insight about the origin and stability of strategy as a variable determining firm behavior over time.

This study is also addressed to another issue. Both Chandler and Andrews defined strategy in terms of its long-term outcomes. However, most empirical literature consists of either case studies or cross-section studies covering only relatively brief segments of time. These studies leave unanswered the question of whether and to what extent the long-term goals defined in a strategy actually persist in affecting behavior over the long term. The narratives of these seven companies from their origins to the present provide an answer to this question.

Empirical Literature

The pioneering study in the empirical literature on strategy by Chandler (1962) documented that changes in strategy caused changes in the firm structure. Subsequent studies concentrated on the relation between types of strategic decisions and types of organizational structures.

Chandler identified the two most important structures as functional and multidivisional. Williamson (1970) formalized this analysis, describing the functional structure as the unitary or "U" form and the multidivisional as the "MD" form. He concluded that organizational form, as a method of controlling the firm, was an important determinant of efficient firm size—that is, how large a firm could be without losing efficiency.

The strategic decision which most affected organizational structure was the number and variety of the firm's activities. Thus early classifications of strategies emphasized the firm's degree of diversification. In a cross-section study Wrigley (1970) classified the strategies of the 1967 Fortune 500 by degree of product diversification into single product, dominant product, related product, and unrelated product strategies. In another cross-section study, Rumelt (1974) subdivided Wrigley's dominant and related categories further. Both studies documented Chandler's strategy-structure relationship for large numbers of firms. Newman (1973) explored the effect of strategic groups on the structure-performance relationship of the industrial organization model in thirty-four chemical process industries, concluding that the degree of strategic diversification was a significant variable in determining how much variation in profit margin could be explained by industry structure. Porter (1976) found that strategic variation was limited by a high degree of product differentiation in convenience consumer goods.

Other studies have addressed other strategic decisions in addition to degree of product or industry diversification: geographic expansion, vertical integration, large scale, and horizontal combination. Wilkins (1970, 1974) examined the economic and political forces which caused firms to enter production abroad and become multinational enterprises (geographic expansion). Hunt (1972) examined strategic groups in the major home appliance industry which differed significantly in their degrees of vertical integration. Chandler (1977) in *The Visible Hand* traced the changes in markets and technology which made options for large scale, horizontal combination (within an industry), and vertical integration attractive.

This study will examine these strategic decisions from prior studies (product and industry diversification, geographic expansion, horizontal combination, vertical integration, and choice of scale) plus an administrative category which includes organization (structure) and financial decisions. No previous study has examined all of these strategic decisions simultaneously for a sample of individual firms.

Appendix B

Strategic Changes during the Growth of the Large-Scale Firm

This appendix describes strategic changes in six areas which occurred during the growth of the large-scale firm. These changes occurred in administrative structure (organization - O), scale (S), geographic expansion (G), horizontal combination within industries (H), vertical integration (V), and product and industry diversification (D). The letter in parentheses following each area of change is used as a reference in succeeding paragraphs to trace each area of change through the narrative. This study uses these areas of change as yardsticks to measure the strategies of the Majors as they changed through time and for purposes of comparison between companies. The development of these six strategic changes for U.S. firms is described below.

Before 1850 the economic activities of even the largest firms with few exceptions could be directed by two or three men, partners or perhaps family members (O).[1] Structure was simple. Scale was small. The firm rarely participated in more than a single business or vertical segment (D, V). Some trading, shipping, and banking firms had export or sales agents or associates in other locations but few engaged in production operations in more than one location (G). Because economic concentration is meaningful only relative to its market and because markets were largely local, it is difficult to assess the degree of economic concentration or its consequences. Monopoly power was not unknown. But concentration at the national level was generally low (H).

The rise of large-scale business in the United States began after the Civil War (1865) with construction of the trunkline and transcontinental railroads. Cheap transportation provided by the railroads greatly increased the size of markets in which a firm could successfully compete. Operations spread geographically (G). Expansion of volume accompanied the growth of markets, especially the less self-sufficient and more interdependent urban markets. It led to specialization and scale economies (S). Specialization lowered unit costs, standardized products, improved quality and increased variety. Quality and features were communicated to the large number of customers through brand names and advertising of differentiated products (D). Manufacturing firms vertically integrated forward into marketing and backward into raw material acquisition to capture these savings (V). Manufacturing economies of scale required high marketing volumes. Technologically advanced products required new methods of distribution and improved communication with consumers. The threat of overcapacity among other factors led to horizontal combinations, ineffectively in the trusts and trade associations of the 1880's and more effectively in the consolidations of the first great merger wave from 1887 to 1904 (H).

Changing demography, purchasing power, and technological progress produced rapid change in product quality, features, and variety (O). To keep its products from becoming obsolete, to keep the resources of a large organization steadily employed, and to maintain the volume required by heavy fixed costs of manufacturing plants and distribution facilities, the firm invested in research and development, developed new products, and diversified into multiple product lines. New products came from byproducts, applications of existing processes to produce new products, and from technological advance.

The single entrepreneurial decision maker of classical economic theory gave way to a chain of command of line managers and staff specialists (O) organized into a pyramid of functional and geographical departments connected to a central office.[2] The line and staff organization resulted in new occupations, the professional administrator or manager, and a host of specialists with detailed knowledge of day to day activities.

The results of these changes are that the American economy today has sectors composed of industries dominated by huge firms where assets, sales, and profits are measured in millions and billions of dollars (S). Manufacturing is the largest sector in the American economy, generating 28% of the gross national product in 1967. The vertical chain of manufacturing, transportation, and distribution generated over 50%.[3] In 1954 Kaysen and Turner found that industries where the eight largest firms accounted for over 50% of industry sales made up 23% of all manufacturing sales.[4] In 1964 corporations with assets of $100 million or more controlled over 50% of all non-financial corporation assets and 66% of the assets of all financial corporations.[5]

The geographical spread (G) of large-scale business has been well documented in individual cases, although it has not been studied to any extent in the aggregate. There is little aggregate data on expansion within the United States. The phenomenon of multinational enterprise, the establishment of manufacturing operations as opposed to export sales or raw material purchases in a foreign country, has received much attention. Wilkins (1970 and 1974) has written about U.S. enterprise abroad. Graham (1974) studied European investment in the U.S. There have also been many studies of foreign enterprise in particular areas.

The growth of horizontal concentration (H) at the national level, from some low but unmeasured amount before the combinations and mergers at the end of the nineteenth century to a significant fraction of assets, sales, and earnings by the 1930's, is generally established from the 1932 study of Berle and Means. Since then the trend to higher aggregate asset concentration seems to have slowed[6] if not stopped.[7] These studies do not adjust for the concurrent growth trend in the geographical coverage of competitors from regional to national and beyond, so one can deduce little from these data about concentration in particular markets.[8] Nor do they take into account the effects of increased competition from larger numbers of substitute products due to diversification.

Vertical integration (V) has presented difficult measurement problems so that, although its importance has been demonstrated in individual cases, aggregate measurement of its extent or trend is scarce. Gort (1962) found that firms in the petroleum industry were by far the most vertically integrated of all industries, using the ratio of employment in functions upstream or downstream from the primary business to total employment as his measure.[9] Adelman and Laffer, using the imperfect value added/sales ratio, found no discernible trend toward increased or decreased vertical integration over the periods 1849-1939 and 1929-1965 respectively.[10]

The operations of firms today are diversified across a number of product lines and industries (D). Of the 1000 largest industrial corporations in 1962, only 4.9% participated in only a single 5 digit SIC code product line; 62% participated in six or more; and diversification since 1950 has been increasing.[11] Of the 90 largest industrial firms in 1960 only four (all oil companies) participated in only one four-digit SIC industry group; only 31 (17 of them oil companies) in five or fewer; and 59, in six or more.[12]

The administrative structures (O) documented by Chandler and other business historians, which Scott, Wrigley, and Rumelt have analyzed, can be roughly described as the single owner-manager-entrepreneur, the central office with functional or geographical departments composed of line and staff positions, and the multidivision structure with each division having departments of its own. An even newer form, the matrix, has met with some success in project management.[13]

All six of these changes in the firm are well established in individual cases. At least four (scale, geography, product diversification, and structure) have clearly affected the characteristics of the aggregate population of firms. It is interesting, as Scherer notes, that none of the well-known nineteenth-century economists, save Karl Marx, foresaw the rise of the large-scale enterprise.[14]

As an economic organization the firm is no longer confined to a single industry, product, geographical area, small size, or single decision maker as in the perfectly competitive model. As Coase observed, "the distinguishing mark of the firm is the supersession of the price mechanism . . . there is a cost to using the price mechanism . . . a firm will tend to expand until the costs of organizing an extra transaction within the firm become equal to the costs of carrying out the same transaction by means of an exchange on the open market. . . ."[15]

Notes

Chapter 1

1. Andrews, 36.

Chapter 2

1. Harold F. Williamson and Arnold R. Daum, *The American Petroleum Industry, volume 1, The Age of Illumination 1859-99*, Northwestern University Press, Evanston, Ill., 1959, 57. In 1860 there were 400 manufactured gas plants compared to a handful of refineries. The value of gas output was $17 million compared to $5 million for coal oil.

2. Ibid., 32.

3. Ibid., 33, 39-40.

4. Williamson and Daum, 38.

5. Ralph W. and Muriel E. Hidy, *Pioneering in Big Business 1881-1911*, volume 1 of the *History of Standard Oil Company (New Jersey)*, New York: Harper and Brothers, 1955, 5.

6. Williamson and Daum, 59.

7. *U.S. Consular Reports*, "Petroleum and Kerosene Oil in Foreign Countries," no. 37 (January 1884), 406.

8. Williamson and Daum, 56; Hidy and Hidy (abbreviated H&H hereafter) 5. In 1858 at the latest Young licensed his patents, for a royalty of 2¢ per gallon.

9. H&H, 5.

10. Williams and Daum, chapter 4.

11. Ibid., chapter 1.

12. Williamson and Daum, part 4. H&H, chapter 1.

13. H&H, 9.

14. H&H, 32.

15. H&H, 7.

16. H&H, 32.

17. Gerald T. White, *Formative Years in the Far West*, New York: Appleton-Century-Crofts, 1962, 495.

18. Ibid.

19. H&H, 8.

20. Williamson and Daum, 159.

21. Ibid., 123-26.

22. Williamson and Daum, 124.

Chapter 3

1. Wilkins, *Emergence*, 84-85.

2. McLean and Haigh, 74, 384.

3. Sampson, 50-51.

4. H&H, 13.

5. See Exhibits S-1 and S-2. Approximately 2/3 by volume and value of Standard's refined products were kerosene. Even as late as 1910, nearly 2/3 of Standard's kerosene and a sizable proportion of its other products by volume were exported.

6. *Business Week*, 28 February 1977, p. 10.

7. McLean and Haigh, 60; H&H, 14.

8. H&H, 34.

9. H&H, 22.

10. H&H, 34.

11. H&H, 120.

12. H&H, 23.

13. H&H. 46.

14. See Exhibit S-1.

15. H&H, 37-38.

16. Letter to J.A. Bostwick, 3-6 July 1885 in H&H, 607.

17. H&H, 611-12.

18. H&H, 35.

19. H&H, 37.

20. See Chandler, *Strategy and Structure*.

21. H&H, 44-45.

22. Ibid.

23. H&H, 46 ff.

24. H&H, 57.

25. H&H, 65.

26. H&H, 58.

27. Ibid.

28. H&H, 65.

29. H&H, 80.

30. H&H, 79.

31. H&H, 86.

32. H&H, 81.

33. H&H, 83-84.

34. H&H, 83-84.

35. H&H, 88.

36. H&H, 85.

37. H&H, 177.

38. H&H 88.

39. H&H, 177-78.

40. H&H, 177.

41. H&H, 157.

42. H&H, 158.

43. H&H, 160-61.

44. H&H, 165.

45. H&H, 176,187.

46. H&H, 167.

47. H&H, 84.

48. H&H, 46, 222.

49. H&H, 129.

50. H&H, 124.

51. H&H, 132.

52. H&H, 144.

53. H&H, 146.

54. Ibid.

55. H&H, 148.

56. H&H, 147-52.

57. H&H, 143.

58. H&H, 126.

59. Wilkins, *Emergence*, 84-85.

60. H&H, 196-97.

61. Letter to Archbold, 20 August 1888, H&H, 194n.

62. Exhibit S-1.

63. H&H, 129.

64. H&H, 124.

65. H&H, 129.

66. W.L. Mellon, *Judge Mellon's Sons*, 272-73.

67. Ibid.

68. H&H, 230.

69. H&H, 314.

70. H&H, 114.

71. McLean and Haigh, 77; Thompson, 18.

72. H&H, 256, 497.

73. H&H, 518.

74. H&H, 75, 613.

75. H&H, 64, 126.

76. H&H, 219.

77. H&H, 221-23.

78. H&H, 219-27.

79. H&H, 312.

80. H&H, 713; White, 385, 646 n.71.

Chapter 4

1. Gibb and Knowlton, 6-7.

2. Ibid.

3. See Exhibit S-3 (Ch. 3) for what happened to these companies.

4. Gibb and Knowlton, 47,49,55.

5. Ibid.

6. Ibid., 59-67.

7. Ibid., 49,67-70.

8. Gibb and Knowlton, 89-90.

9. Ibid., 598,676-81.

10. Ibid., Ch. 8.

11. Statement of Fritz von Friedlanderfuld, czar of German coal, in the *New York Times*, 15 December 1912.

12. Gibb and Knowlton, 201 ff.

13.   Ibid., 678, 460, 166-68.

14.   Ibid., 472, 154-55, 164-65.

15.   Gibb and Knowlton, 80,84,687-89.

16.   Ibid.

17.   For example from 1915 to 1918 Carter's profits were 30¢/bbl. (mainly production); Louisiana Standard's were 6¢ (mainly purchases). Producing and purchasing profits averaged 7% of Jersey's net income from 1912 to 1918, compared to 30% in 1911, showing the difficulty of profiting from purchased crude. Calculations by author from Gibb and Knowlton, 32, 445, 458, 676.

18.   Gibb and Knowlton, 106,410.

19.   Ibid., 107-8.

20.   Ibid., Ch. 13.

21.   Gibb and Knowlton, 384.

22.   Ibid., 390-91.

23.   Ibid., 323-27, 331-36.

24.   Ibid., 283.

25.   Ibid., 305-6,309-13.

26.   Gibb and Knowlton, 676-77,393.

27.   See Ch. 3,6 of this study; H&H, 696.

28.   Gibb and Knowlton, 428-29.

29.   Ibid., 428-29, 411-12.

30.   Ibid., 414-17, 436-37.

31.   Ibid., 676-79.

32.   Gibb and Knowlton, 501, 561-65.

33.   Ibid.

34.   Ibid., 536, 561, data from 678-79.

35.   Ibid., 520-21.

36.   Gibb and Knowlton, 535,553.

37.   Ibid., 541-45.

38.   Ibid., 495-96.

39.   Ibid., 489-91.

40.   Gibb and Knowlton, 398.

41.   Ibid., 480-81.

42.   Ibid., 468.

43.   Ibid., title.

44.   Chandler, Ch. 4 "Standard Oil Company (N.J.)—*Ad Hoc* Reorganization."

45. Larson, Knowlton, and Popple (abbreviated LK&P hereafter), ch. 3.

46. See Socal chapter for details.

47. LK&P, ch. 5.

48. LK&P, ch. 4.

49. LK&P, 148,200-201,296,324.

50. LK&P, 148; or see Table 1-2 (1938).

51. Wilkins, *Maturing,* 322.

52. Gibb and Knowlton, ch. 17.

53. LK&P, 160-63.

54. Sampson, 78-79; LK&P, 405-8.

55. Wall and Gibb, ch. 16.

56. LK&P, 433 ff.

57. Wall and Gibb, ch. 16.

58. Interview, Harold W. Fisher, 17 January 1978, Contact Director for chemicals, 1959-68.

59. Harold W. Fisher, 17 January 1978; LK&P 159, 414 ff.

60. McLean and Haigh, 104; LK&P 264.

61. See Socal chapter.

62. LK&P 266 ff.

63. Ibid., 280 ff.

64. See Mobil chapter.

65. McLean and Haigh, 104.

66. Computed from data in Annual Reports and BP *Statistical Review.*

67. Interview, 1 May 1977, with Bennett H. Wall and Gerald Carpenter, authors of forthcoming volume 4 of the history of Standard Oil of New Jersey.

68. Ibid.

69. Ibid; also Harold W. Fisher, 17 January 1978.

70. Harold W. Fisher, 17 January 1978.

71. Wilkins, 316; Annual Report 1959 (1,20).

72. Statistical Supplement to 1977 Annual Report(hereafter abbreviated as AR), p.22; Table 1-1.

73. AR1959 (12-16).

74. AR1977 (17).

75. Table 1-1 in Ch. 1.

76. AR1958 (18).

77. See Kaufman for a description of events from the government point of view.

78. AR1965 (3-5).

79. Interview, Harold W. Fisher, 17 January 1978.

80. Interview, Dr. Donald L. Guertin, Sr., Planning Advisor at Exxon, 27 October 1976.

81. LK&P, 771.

82. Ibid., 765-70.

83. AR1958 (24).

84. Harold W. Fisher, n 79 above.

85. AR1963 (2); Harold W. Fisher, n 79 above.

86. LK&P, 153-54.

87. AR1959 (23).

88. Annual reports; *Forbes* 15 August 1977, pp.37-41, 12 June 1978, pp. 115-18.

Chapter 5

1. Policies are listed in Exhibit G-1.

2. McLean and Haigh, 72ff.

3. The only exceptions are 1908 and 1918 when net production and refinery runs were nearly equal.

4. Interview, E.D. Brockett, Chairman 1965-71, 11 August 1975. AR 1937 (5,6); Wilkins, *Maturing*, 14,15,214,217-18.

5. 1976 estimate by company official.

6. Andrew Mellon was later Secretary of the Treasury from 1921 to 1933 under Presidents Harding, Coolidge, and Hoover and is fairly regarded as possibly the most distinguished holder of that office in U.S. history.

7. Thompson, 9.

8. McLean and Haigh, 75.

9. Thompson, 12.

10. Ibid., 13. Andrew Carnegie, the steel man, had maintained an employee stock sharing plan for some time. When J.P. Morgan and Company bought the Carnegie interests to form United States Steel the employees received cash for their stock. One bookkeeper received $1.9 million.

11. Thompson, 18-19.

12. Ibid., 14-15.

13. Thompson, 19; McLean and Haigh, 76.

14. Thompson, 15.

15. Ibid., 17.

16. See Mobil chapter; H&H, 610,696,796n4.

17. Thompson, 18.

18. Ibid., 20-21.

19.   Thompson, 19-22.

20.   Ibid., 61-63.

21.   Ibid.

22.   Ibid.

23.   Ibid. 19-22.

24.   Ibid.

25.   Ibid., 25.

26.   Thompson, 22-23.

27.   Ibid.

28.   Ibid., 26-27.

29.   Ibid.

30.   Ibid.

31.   Ibid., 33-35.

32.   AR 1911; McLean and Haigh, 697.

33.   McLean and Haigh, 98; Exh. S-1; API *Petroleum Facts and Figures*, 1950, p. 148.

34.   McLean and Haigh, 697.

35.   Gulf was the second Major to use geologists beginning in 1911. Shell was first in 1898. Others began after World War I: Socal about 1920 and Exxon in 1924 (Gibb and Knowlton, 429).

36.   W.L. Mellon, 272-73.

37.   Ibid., 208.

38.   McLean and Haigh, 72-73.

39.   Thompson, 96.

40.   Ibid., 71.

41.   Wilkins, 118.

42.   Schoenleber, 271; Gibb and Knowlton, 387-88.

43.   Wilkins, 116.

44.   Ibid., 72n.

45.   Wilkins, 118-20; Thompson, 76-77.

46.   Stegner, 7; confirmed by Warren B. Davis, Director of Economics, Gulf Oil.

47.   Sampson, 67 ff.

48.   Wilkins, 118-20.

49.   Thompson, 74.

50.   McLean and Haigh, 96,98,378.

51.   Thompson, 40.

52. Ibid.; McLean and Haigh, 104. Standard of Indiana was #1 (10.6% of U.S. market); Socony, #2 (9.6%) and Sinclair, #3 (7.8%). See Table M-5 in Mobil chapter.

53. ARs.

54. AR 1926, 1927; Thompson, 66, 70.

55. Estimate from AR 1930 (2).

56. Thompson, 80.

57. McLean and Haigh, 97-99, 379.

58. Thompson, 49; ARs 1929-34.

59. Author's calculation, McLean and Haigh, 86; Thompson, 53; ARs.

60. ARs.

61. Thompson, 52-54.

62. McLean and Haigh, 377-80; AR 1934.

63. Ibid.

64. ARs 1937, 1948 (12); McLean and Haigh, 380. Calculation by author.

65. Mikadashi, 81.

66. Anglo-Persian, renamed British Petroleum in 1955.

67. Mellon's influence on Gulf's behalf is documented by Chisolm, 134-35; Sampson, 92; Mikdashi, 81.

68. Sampson, 93; Anderson, 151.

69. Mikdashi, 83, Appendix III.

70. Anderson, 152.

71. Thompson, 77.

72. AR 1961 (5); Anderson, 152.

73. AR 1942 (8), 1945 (10); Thompson, 77.

74. AR 1947 (12); FTC, *The International Petroleum Cartel*, 150.

75. AR 1950; *BP Statistical Review* 1959, p.21.

76. Interview, J.H. McDonald, VP Refining and Marketing, and D.T. Parks, U.S. Refining Coordinator, 19 August 1976.

77. Interview, E.D. Brockett, former Chairman, 11 August 1975.

78. ARs.

79. John J. McCloy, *The Great Oil Spill*, New York, 1976; interviews with company officials.

80. AR 1951 (8).

81. McCloy; interviews.

82. AR 1958 (22).

83. Interview, Warren B. Davis, 28 April 1976.

84.   AR 1956 (25).

85.   Calculation by author.

86.   AR 1954 (20).

87.   AR 1956 (15).

88.   AR 1956 (13).

89.   Ibid.

90.   Ibid.

91.   ARs.

92.   AR 1957 (6).

93.   Calculation by author from ARs.

94.   AR 1958.

95.   AR 1957.

96.   AR 1957 (5).

97.   ARs.

98.   AR 1960 (19).

99.   *Oil and Gas Journal*, 27 November 1961, p. 57.

100. *Business Week*, 25 November 1961, p. 84.

101. *Oil and Gas Journal* pp. 55-57; Warren B. Davis, Director of Economics, Gulf Oil, interview 28 April 1976.

102. AR 1960 (11).

103. AR 1963 (18).

104. AR 1965 (1).

105. AR 1966 (20).

106. Annual report data.

107. Calculation by author.

108. Estimates by author from annual report data.

109. Kaufman, 47.

110. Adelman, chapter 6.

111. Sampson, 60ff; Mikdashi, 33. As events attest this bitterness did not go away. It was still evident in every syllable spoken by Faud Rouhani, the first Secretary-General of OPEC, in a speech at Harvard, spring, 1975.

112. Calculation by author from ARs.

113. Calculation by author from ARs.

114. Annual Reports; interview, Warren B. Davis, 28 April 1976.

115. Interview, Sam L. Sugarman, EVP Gulf Oil Transp. and Trading Company, 19 August 1976.

116. AR 1969 (13).

117. Interview, J.H. McDonald and D.T. Parks, 19 August 1976.

118. At one time Gulf lost money at AFRA rates. Warren B. Davis, fall, 1975.

119. U.S.: see Gulftane incident above; foreign: see *Forbes*, 17 April 1978, p. 49.

120. ARs.

121. Ibid.

122. AR 1964 (6).

123. *Forbes*, 15 October 1977, p. 97.

124. AR 1976 (9).

125. J.A. Strand, 18 August 1976.

126. Ibid., ARs.

127. AR 1967 (21), 1974 (12).

128. Calculation by author.

129. Interview, J.H. McDonald and D.T. Parks, 19 August 1976.

130. Interview, James E. Lee, President, Gulf Oil Corporation, 17 August 1976.

131. Interview W.C. Rohrer and Frank Pyle, Pres. and VP, Gulf Oil Chemicals Company, 18 August 1976. Petrochemicals achieved recognition as a separate business about 1970.

132. AR 1942 (9).

133. AR 1954 (27), 1955 (29).

134. AR 1950 (22), 1959 (5).

135. Interview, Rohrer and Pyle, 18 August 1976, (see note 131 above).

136. ARs.

137. Rohrer and Pyle, 18 August 1976.

138. AR 1968 (9), AR 1963 (16), *Forbes*, 15 October 1977, p.100.

139. AR 1968 (21).

140. Ibid, AR 1967 (18).

141. AR 1968 (21).

142. Ibid.

143. AR 1968 (22).

144. AR 1970 (7).

145. AR 1971 (11-2).

146. AR 1971 (13).

147. AR 1973 (9), 1974 (17); *Business Week*, 26 July 1976, p. 46.

148. AR 1971 (15-6).

149. Ibid.

150.   AR 1969 (8).

151.   AR 1972 (12).

152.   ARs; *Forbes*, 15 October 1977.

153.   *Business Week*, 31 January 1977, 7 November 1977.

154.   ARs.

155.   AR 1973 (15).

156.   McCloy; *Forbes*, 15 October 1977, p. 95.

157.   *Forbes*, 15 October 1977, p. 96.

158.   *Forbes*, 15 October 1977.

159.   *Business Week,* 31 January 1977, p. 78.

160.   *Business Week*, 31 January 1977, p. 80; *Forbes*, 15 October 1977, p. 97, 29 May 1978, p. 35.

161.   W.L. Mellon, 272-73.

Chapter 6

1.   Some information about the early years of Socony and Vacuum comes from two unpublished histories of Mobil prepared by and obtained from the company, hereafter designated as Mobil I and II. Mobil I (9) (cable address). SOCONY was registered as a trademark on 26 February 1908. Mobil II (5).

2.   Mobil II (1,6,7); Table 1-1 (Chapter 1).

3.   Mobil I (9-10), II (3); Henriques, 138.

4.   See Chapter 3.

5.   Mobil II (4).

6.   Mobil II (9).

7.   Mobil II (7).

8.   Mobil II (9).

9.   As its New York and New England territory was known in the industry. Mobil II (8).

10.   *Mobil World*, July-August 1976, 8.

11.   Mobil II (8).

12.   McLean and Haigh, 87.

13.   FTC, *The International Petroleum Cartel*, 38.

14.   Mobil I (17) says 45%; Mobil II (10) says 70%.

15.   Mobil I (18). For a number of years after this time a barrel of drinking water was worth more than a barrel of oil. Mobil II (11-12).

16.   H&H, 610,696,796n4. Secrecy had been one of Standard's most criticized policies. See Chapter 2 above.

17.   Mobil II (12); McLean and Haigh, 103.

18. Mobil I (2), II (12); White, 407,475, 646-47n3: E.K. Harington, "General Petroleum Corporation" in *California World and Petroleum Industry*, 5 July 1937 and three subsequent issues.

19. Mobil I (21), II (15).

20. Mobil I (22); Wilkins, 119.

21. Mobil II (16).

22. Calculation by author.

23. Mobil I (1,3), II (19-20); *Mobil World*, July-August 1976, p.7. The brand name Gargoyle was not added until 1904. See below.

24. Ibid.

25. Ibid.

26. Mobil II (21-24).

27. Ibid.

28. Ibid.

29. Mobil I (3), II (23,27,30).

30. Mobil II (25-30).

31. Ibid.

32. Ibid.

33. Mobil I (31).

34. LK&P, 316.

35. Wilkins, *Maturing*, 15-16.

36. LK&P, 311.

37. Wilkins, 119.

38. Calculation by author.

39. LK&P, 305, 316; Mobil II, 37; AR 1933 (4).

40. Mobil II, 37,41.

41. Ibid.

42. Ibid., 36, 41.

43. Interview, Larry H. Atherton, Mgr. Marine Planning, 13 September 1976.

44. AR 1946 (8).

45. AR 1947 (13).

46. Interview, Henry K. Holland, EVP Expl. & Prod., 14 September 1976.

47. Wilkins, 116-18; *Intl. Petroleum Cartel*, ch. 4.

48. LK&P, 51-58.

49. Lenzen, 36-9. Formal meetings with Jersey began on 29 July 1946, but discussions between Socal, Texaco, and other companies had been held as hearly as 1944.

50. LK&P, 51-58, 736.

51. Lenzen, 40. Socal and Texaco were initially unaware that Jersey was considering bringing Socony-Vacuum into the deal.

52. Socony-Vacuum's symbol.

53. Sampson, 102.

54. Interview, Albert L. Nickerson, Pres. 1955-60, Chmn. 1961-68, 6 January 1977.

55. Geo. V. Holton to Brewster Jennings, 28 October 1946 in *Multinational Hearings*, VIII.

56. 30+ from Stanvac, 10- from Iraq.

57. AR1946 (8).

58. *Multinational Hearings*, VIII, p. 85. Letter from R.C. Stoner to H.D. Collier and R.G. Follis, 10 June 1946.

59. BP *Statistical Review*, 1959, p. 21.

60. Ibid., 19.

61. Interview, Howard Bird, Jr. VP Intl. Marketing, 17 September 1976.

62. Calculations by author.

63. Interview, Albert L. Nickerson, 6 January 1977.

64. Interview, Bonner Templeton, VP U.S. Supply and Distribution, 17 September 1976.

65. Ibid.

66. Ibid., Nickerson.

67. Albert L. Nickerson, 6 January 1977.

68. Howard Bird, Jr., 17 September 1976.

69. Ibid., Nickerson.

70. Ibid.

71. ARs.

72. Ibid.

73. AR 1955 (7).

74. AR 1955 (16-19).

75. ARs.

76. Nickerson.

77. Ibid.

78. Rawleigh Warner, Jr. quoted by Irwin Ross in "Public Relations Isn't Kid-Glove Stuff at Mobil," *Fortune*, September 1976, p.8.

79. Nickerson.

80. Ibid.

81. Ibid.

82.   Stanvac was dissolved in 1962 as the result of a consent decree by Jersey in a civil antitrust case to which Mobil was not a party. Mobil I, 39. See Exxon chapter.

83.   Interview, Bonner Templeton, 17 September 1976.

84.   Nickerson, 6 January 1977.

85.   Ibid.

86.   Ibid.

87.   Ibid.

88.   Steven B. Packer, Chief Economist, MOBIL, interview, 15 September 1976.

89.   This story came from several sources and is something of a legend in the oral history of the company.

90.   With 40-50% of markets under the "As Is" agreement. Howard Bird, Jr., 17 September 1976.

91.   See Irvine H. Anderson, *The Standard-Vacuum Oil Company and United States East Asian Policy, 1933-1941*, Princeton Univ. Press, 1975.

92.   Chapter 7.

93.   Interviews, Howard Bird, Jr. 17 September 1976, and Albert L. Nickerson, 6 January 1977.

94.   Howard Bird, Jr., 17 September 1976. In the 1950's lubricants were shipped to Mexico in tank cars and packaged in drums and cans there.

95.   Table M-7B.

96.   Calculated from Table M-7A.

97.   Ibid.

98.   Bonner Templeton, 17 September 1976.

99.   AR 1945.

100. ARs.

101. AR 1947 (3).

102. Computed from annual report data.

103. Interview Howard Bird, Jr., VP Intl. Marketing, 17 September 1976.

104. Interview, W.H. Marshall, Manager METS Planning, 16 September 1976.

105. Ibid.

106. Interview, Howard Bird, Jr., ARs 1959 (18), 1960 (13).

107. Computed from AR data.

108. AR 1960 (14), for example.

109. Ibid., Howard Bird, Jr.

110. Interview, Bonner Templeton, 17 September 1976.

111. Ibid.

112. Howard Bird, Jr., 17 September 1976.

113. Ibid.

114. AR 1961 (13).

115. Interview, Larry H. Atherton, Manager Marine Planning, 13 September 1976.

116. Ibid.

117. Interview, Joe E. Pennick, VP U.S. Refining, 16 September 1976.

118. Table 1-1.

119. Interview, Joe E. Pennick, VP U.S. Refining, 16 September 1976.

120. Interview, Bonner Templeton, 17 September 1976.

121. Interview, Albert L. Nickerson, 6 January 1977.

122. Henry K. Holland, EVP N. Am. Expl. & Prod., 14 September 1977.

123. In 1976 it did not even rank in the top ten in U.S. wildcats. *Business Week,* 13 June 1977.

124. AR 1975 (4).

125. AR 1960 (19).

126. Interview, Arthur Biggs, Pres. Mobil Chemical, 14 September 1976.

127. *Business Week*, 13 June 1977, p.82.

128. Interview, Steven B. Packer, Chief Economist, Corporate Planning, Mobil, 15 September 1977.

Chapter 7

1. White, 93.

2. Ibid, 94.

3. White, 193.

4. Ibid.

5. White, 197; API, 145-48 (1950).

6. White, 211.

7. See Chapter 3, above, The Legacy of Standard Oil.

8. White, 195-96.

9. White, 195-96, 213-16.

10. Ibid., 333.

11. Ibid., 247-50.

12. Ibid., 284-87.

13. Ibid., 334-47.

14. Ibid., 576-80.

15. Std. Oil Co. (Iowa) properties renamed on 4 October 1906. White, 572.

16. Figures and computations based on White, 489,576-80.

17. White, 406.

18. Calculation by author from data in White, Ch. 16.

19. White, 576-80.

20. Ibid., 408-10.

21. Ibid., 568.

22. ARs; White, 558-59.

23. Ibid.

24. Wilkins (*Maturing*, 120) states that by the end of 1928 Socal had spent $50 million on foreign exploration. This may be a typographical error in date (1938) or amount ($5 million). From research in company records the amount should be nearer $5-10 million.

25. Interview, 30 June 1976.

26. White, 559.

27. AR 1921 (5).

28. AR 1922 (2); Gibb and Knowlton, 436.

29. AR 1922 (2).

30. Interview, 29 June 1976.

31. Ibid.

32. White, 464, 471-72; AR 1925 (10-13), 1927 (6), 1929 (13), 1949 (8).

33. Stegner, 5; Wilkins,, 118-20. Only Ibn Saud was an exception, his complete sovereignty over Saudi Arabia having been recognized in the Treaty of Jiddah in 1927, after having chased his rival off the Arabian peninsula into exile on Cyprus. All the others were bound by treaties signed with the British at the end of the nineteenth century. The Suez Canal was completed in 1869 and the British Navy cleared the sea lanes to the Far East of pirates from the sheikhdoms on the Arabian peninsula bordering the Red Sea and the Persian (Arabian) Gulf. A typical treaty bound the sheikh, his heirs and successors never to enter into any agreement or correspondence with any foreign power other than Britain; never to permit, without British consent, the residence of any foreign agent other than British; never to sell, cede, mortgage, or otherwise grant rights to any of his territory without British approval; to refrain from piracy in return for an annual bounty and British protection from foreign aggression.

34. Wilkins, 119.

35. This interpretation is supported by the FTC study, *The International Petroleum Cartel*; Standard Oil Co. (New Jersey), "Background Memorandum," 1954, p.7.

36. Wilkins, 119 (see Table M-6, Mobil chapter).

37. Wilkins, 119-20.

38. BP (with large reserves to develop), Shell (with heavy capital commitments in the U.S.), and Jersey (preoccupied with Venezuela) all rejected the concessions peddled by Holmes. Sampson, 88, LK&P; Stegner, 8; Wilkins, 120. Exact terms differ a little from account to account.

39. Stegner, 8.

40. Wilkins, 117.

41. Stegner, 8.

42.   Ibid., 10.

43.   Stegner, 11-12.

44.   Sampson, 89.

45.   Ibid., 90. Note: Britains's Middle Eastern areas were administered through India, where the currency was the rupee.

46.   Wilkins, 215. This was the same royalty as BP paid Iran under the revised concession of 29 April 1933. Longhurst, 75-80.

47.   Interview with R.G. Follis, 30 June 1976.

48.   See Mobil chapter.

49.   LK&P, 46.

50.   Bahrain was discovered 1 June 1932.

51.   Ibid. From press release, W.C. Teagle to Northrop Clarey.

52.   See Exxon chapter.

53.   LK&P, 47.

54.   Ibid., 268 from letters: F.H. Bedford, Jr. to W.C. Teagle, 20 January 1933, and Heinrich Riedeman to W.C. Teagle, 15 September 1933.

55.   LK&P, 47.

56.   Interview with Prof. Bennet H. Wall, biographer of W.C. Teagle, 1 May 1977.

57.   17 October 1933. LK&P, 823.

58.   LK&P, 52-57.

59.   AR1936 (40-45).

60.   Ibid.

61.   McLean and Haigh, 387.

62.   Lenzen, 45. According to the *BP Statistical Review* (1959, p.21) consumption in 1938 in Australasia and other Eastern Hemisphere was 620 TBD.

63.   AR1936 (5).

64.   Ibid., 4.

65.   Ibid., 6.

66.   AR1936 (7).

67.   Defined in the concession agreement to be 2000 tons/day (14.7 TBD). Stegner, 20-21.

68.   Stegner, 90-91.

69.   Ibid.

70.   Ibid.

71.   *Multinational Hearings*, part 7, 567.

72.   *Petroleum Facts and Figures*, 8th ed., 1947, 83.

73.   U.S. Secretary of the Interior and Petroleum Administrator for War.

74. AR 1943 (14).

75. Sampson, 97.

76. AR 1943 (14).

77. Sampson, 98; Stegner, 176-77.

78. Sampson, 95. Casoc (California Standard Oil Company) Socal's Arabian subsidiary formed 8 November 1933 was renamed Aramco Arabian America Oil Co. on 31 January 1944.

79. Stegner, 20.

80. *Multinational Hearings*, part 4, 87.

81. Feis, p. 129.

82. *Multinational Hearings*, part 8, 85. Memorandum 10 June 1946 from R.C. Stoner to H.D. Collier and R.G. Follis.

83. 20 billion barrels. *Petroleum Facts and Figures*, 8th ed., 1950, 182.

84. They had been expelled from Iran after the war with difficulty. Sampson, 116.

85. *Multinational Hearings*, part 8, p. 88.

86. Ibid., 86.

87. *Multinational Hearings*, part 8, pp. 89-94.

88. See n 101 below.

89. Ibid., 96.

90. Ibid.

91. Ibid., 94.

92. Sampson, 102.

93. Lenzen, 40.

94. Sampson, 102.

95. Lenzen, p. 40, says October. Sampson says Harry Klein, new Texaco president, was discussing the possibility in September.

96. Lenzen, 40.

97. R.G. Follis, interview with author, 30 June 1976.

98. Lenzen, 41.

99. Letter from Brewster Jennings, president of Socony Vacuum, to Harold Sheets, chairman of Socony-Vacuum, 24 July 1946. *Multinational Hearings*, part 8, 95.

100. George V. Holton to Brewster Jennings, 28 October 1946. *Multinational Hearings*, part 8, 116-17. Note that initials in hearings are incorrect. Holton's first name is George.

101. Sampson, 102. *Multinational Hearings*, part 8, 97. Mr. Follis indicated that this figure originated from Socal's chairman, Harry D. Collier.

102. Lenzen, 37.

103. Sampson, 104.

104.   Lenzen, 43.

105.   See *Multinational Hearings*, part 8, 87-88, for terms and magnitudes of some previous sales.

106.   Wilkins, 319.

107.   Interview with author, 30 June 1976.

108.   Sampson, 103.

109.   Interview, Kenneth H. Crandall, 29 June 1976.

110.   Ibid.

111.   AR 1946 (7).

112.   AR 1948 (8).

113.   Interview, George T. Ballou, VP Public Affairs, 24 June 1976.

114.   AR 1958 (13), 1960 (12).

115.   Interview, Robert K. Gordon, Asst. to the VP Public Affairs, July 1976.

116.   Interview, George T. Ballou, 24 June 1976.

117.   *Business Week*, 10 June 1961, p. 27.

118.   AR 1961 (13).

119.   Interview, George T. Ballou, VP Public Affairs, 24 June 1976.

120.   AR 1961(12). Nathan M. Adams, "A Tankful of Gas," *Reader's Digest*, May 1976 (condensation).

121.   Calculation by author.

122.   Ibid.

123.   Interview, George T. Ballou, Robert K. Gordon, 7 July 1976.

124.   Intrerview, Donald L. Bower, VP Marketing, July 1976.

125.   White, 499ff.

126.   Interview, George T. Ballou, VP Public Affairs, 7 July 1976.

127.   I am indebted to Mr. Thomas Buffalow, Socal Asst. Vice President for Engineering for this narrative. Interview 9 July 1976.

128.   AR 1951 (16); interview with George T. Ballou, 7 July 1976.

129.   AR 1948 (9), 1949 (15-6), 1950 (16).

130.   Computed from 1975 Annual Report, Statistical Supplement, 6-7.

131.   AR 1967 (2).

132.   AR 1967 (15).

133.   Interview, J.D. Bonney, VP Corporate Planning, 23 June 1976.

134.   AR 1935 (12).

135.   AR 1964 (17). In India and Iowa.

136.   AR 1936 (13).

137.   Toluene, AR1941 (2); butadiene, AR1941 (2). The butadiene plant, purchased from the government in 1955, employs a company-developed process in the manufacture of synthetic rubber. AR1955 (13).

138.   Interview, George T. Ballou, 24 June 1976.

139.   Ibid. Also AR1950 (90).

140.   Ibid.

141.   AR1954 (12).

142.   AR1952 (16); AR1955 (13).

143.   AR1961 (17).

144.   AR1946 (7); AR1968 (23); AR1971 (21); AR1974 (22).

145.   AR1975 (35).

146.   *Business Week*, 16 June 1975, p. 25. *Forbes*, 1 October 1975, p. 51.

147.   Data from annual reports and supplements.

148.   *Forbes*, 1 February 1977, p. 24; *Business Week*, 7 December 1974, p. 49.

Chapter 8

1.   James, 11. J.S. Cullinan and Company became part of Magnolia Petroleum and later Socony-Vacuum. See Chapter 5.

2.   James, 7,11; Sampson, 40.

3.   James, 4; King, 212; Sampson, 41.

4.   James, 6,7.

5.   Ibid.

6.   Ibid. 21-23,32,45.

7.   Ibid., 24-25.

8.   Ibid., 24,29.

9.   Ibid., 25,27.

10.   James, 25-34.

11.   Ibid.

12.   Ibid.

13.   James, 108 (Calculation by author); *Forbes,* 17 April 1978, p. 47. Texaco was #3 in 1935. See Table 5 in Mobil chapter.

14.   James, 36.

15.   Ibid., 116.

16.   Ibid., 36.

17.   Ibid., 45.

18.   Ibid., 41.

19. McLean and Haigh, 385; James 43-44, 47-48.

20. James 47-48.

21. McLean and Haigh 384, 683.

22. Ibid., 385.

23. Ibid.

24. McLean and Haigh, 385-86; James 54-55; AR1928 (8).

25. McLean and Haigh, 683.

26. AR 1928 (9).

27. McLean and Haigh, 683; James, 62.

28. McLean and Haigh, 698; James 68; ARs.

29. James, 63.

30. McLean and Haigh, 345.

31. James, 103.

32. James, 55-56, 80.

33. Ibid., 104.

34. Ibid., 65-66.

35. Ibid., 86.

36. McLean and Haigh, 387, 685; ARs 1928-35.

37. ARs.

38. AR1944 (11); James, 104.

39. James, 67-68.

40. James, 66-67; ARs.

41. Sampson, 81; James, 61, 67.

42. Sampson, 81-83; James, 70, 104.

43. James, 71-75; ARs 1940-45.

44. Year of discovery.

45. James, 60-62; AR1953.

46. See Socal chapter; James, 81.

47. Wilkins, 319.

48. James, 82-85.

49. ARs 1949,1950,1951,1952.

50. AR1953.

51. McLean and Haigh, 683; AR1952,1953; comparison with Socal figures.

52. *Forbes*, 17 April 1978, p. 47.

53. Calculation by author. Some figures are given in Table 11-5.

54. *Forbes*, 49; Sampson, 197.

55. Ibid.

56. *Forbes*, 51.

57. Comments by competitors to author.

58. Calculation by author.

59. Sampson, 196.

60. *Forbes*, 50.

61. Ibid., 52.

62. Ibid., 48.

63. Ibid.

64. See Gulftane incident in Gulf chapter; *Forbes*, 49.

65. Ibid.

66. ARs 1956,1959,1966,1967; *The Texaco Star*, January 1977; *Forbes*, 49.

67. Socal interviews; ARs, *Forbes,* 50.

68. *Forbes*, 49.

69. Ibid.

70. Ibid.; ARs.

71. Ibid.

72. Table 1-1; AR1947 (11), 1954, 1976 (23).

73. Author's calculation.

74. *Forbes*, 50-51, 54; ARs.

Chapter 9

1. Policies (PL#), Resources (R#), Handicaps (H#), Personal Values (PV#), Social Responsibilities (SR#), and Environmental Conditions (ECu#) are keyed to exhibits at the end of the text.

2. Anthony Sampson, *The Anatomy of Britain*, quoted in Sir David Barran, "How Shell Works," monograph, 13.

3. Interview, Gerrit A. Wagner, Chairman 27 September 1976.

4. Sampson, 11-12.

5. Interview, Dr. H.L. Beckers, Manager of Organizational Planning, 23 September 1976.

6. Tables 1-1 and 1-2, Chapter 1.

7. See BP chapter.

8. *Shell Magazine*, May 1934, 194.

9. *A Short History of the Royal Dutch/Shell Companies*, 3.

10. Ibid., 1.

11. Some oil companies name their tankers, like battleships, after places, famous men, directors or their wives, and other landmarks. Shell's tankers were named after varieties of seashells. Interview, E.E. Bowyer, Exec. Dir. of Marine Transportation, 21 September 1976.

12. Henriques, 138.

13. *A Short History,* 1-2.

14. Deterding, April 1934, 148.

15. *A Short History,* 2.

16. Ibid.

17. Ibid.

18. Henriques, 306.

19. Ibid., 229ff.

20. Ibid., 400-401.

21. Ibid., 40.

22. Ibid., 290ff, 267-68.

23. Ibid., 353-59.

24. Ibid., 369 (size); 267-68 (alliance); 293-96 (Boer war).

25. Henriques, 317,355.

26. Ibid., 400-401.

27. Ibid., 361.

28. Ibid, 390.

29. *A Short History,* 3.

30. Henriques, 455.

31. Deterding, *Shell Magazine,* May 1934, 193.

32. Henriques, 298.

33. This point was made repeatedly in interviews by Shell executives with the author.

34. Deterding, *Shell Magazine*, April 1934, 150-51: May 1934, 195.

35. Henriques, 389, 505.

36. Deterding, *Shell Magazine*, May 1934, 195.

37. Deterding, 77.

38. See Chapter 2.

39. Henriques, 476-90; the phrases "eastern wing" and "western wing" are Gerretson's.

40. Henriques, 499-505; 414-15, 427.

41. Ibid.

42. Henriques, 507.

43. Ibid., 501.

44. Deterding, *Shell Magazine*, April 1934, 151.

45. Henriques, 42.

46. *A Short History*, 4.

47. Henriques, 520.

48. See statement of W.L. Mellon, in Chapter 3.

49. When he lost control of Shell above.

50. Henriques, 521ff.

51. Deterding, 85ff. Beaton (p.57) records the incident from a reference by Ida Tarbell as taking place in 1904 to Philadelphia.

52. Henriques, 521 ff.

53. Henriques, 505.

54. Beaton, 59-60; Deterding, 83.

55. Churchill became First Lord of the Admiralty in October 1911.

56. See BP chapter for more details of these years. BP did not begin shipments until 1914.

57. Henriques, 551, quoted from Churchill's, *The World Crisis*.

58. Ibid. 591-92; Beaton, 99.

59. Sampson, 11-12,198-99.

60. *A Short History*, 5.

61. Gerretson, 273-81; Gibb and Knowlton, 389.

62. Gibb and Knowlton, 472.

63. *A Short History*, 6; Beaton, 3-5,755; LK&P, 55.

64. LK&P, 307-11.

65. Deterding, 76.

66. Beaton, 401-13,517; *A Short History*, 6; "The Lubricants Business," 1.

67. Beaton, 512-21; *A Short History*, 6; Prohibition stimulated the demand for a substitute for ethyl alcohol, Beaton, 536.

68. Interview, Gerrit A. Wagner, Chairman of the Committee of Managing Directors, Royal Dutch/Shell Group, 27 September 1976; also in "I remember ..." J.B. August Kessler (Managing Director 1923-61, Chairman, 1947) *Shell World*, 161-63 (son and namesake of his father).

69. Beaton, 512-21; "Chemicals Digest," Shell Briefing Service, November 1976, 2.

70. Beaton, 356-73, 427-69, 783 (net income).

71. Beaton, 315; Sampson, 80-81, 198.

72. *A Short History*, 7; Beaton, 560-70.

73. Gerretson, III 15,62, IV 286-92.

74. Gerretson, III 15,62, IV 286-92; Interview, Gerrit A. Wagner, Chairman, 17 September 1976.

75. Interview, Bruce T. Patterson, Treasurer, 5 October 1976; and Sir David Barran, 6 October 1976, Chairman (retired).

76. Interview, Sir David Barran, 6 October 1976. Second generations of the Loudons, Kesslers, and Samuels have served Shell as executives or directors. The father of Sir David Barran, contrary to Sampson's reference (p.198), was an outside director representing minority interests in Venezuelan Oil Concessions, Limited but was neither an oil man nor a Shell employee.

77. Interview, Dr. H.L. Beckers, 23 September 1976, Manager Organizational Planning.

78. Interview, C.C.P. Pocock, Vice-Chairman, 22 September 1976.

79. Interview L.C. van Wachem, Coordinator-Exploration, 28 September 1976.

80. See Table 1-2, 1938, 1950, 1960,1970

81. Interview, L.C. van Wachem; see n. 79 above.

82. ARs; *A Short History,* 7-9.

83. Interview, Gerrit A. Wagner, Chairman, 27 September 1976.

84. ARs.

85. Interview, E.E. Bowyer, 21 September 1976.

86. Interview, C.C.P. Pocock, 22 September 1976.

87. Gerrit A. Wagner, Chairman, 27 September 1976.

88. C.C.P. Pocock, Vice-Chairman, 22 September 1976.

89. Robert Burr, Exec. Dir. Mktg., 1 October 1976.

90. Ibid.

91. J.H. McDonald, Treasurer, 24 September 1976; Geoffrey Chandler, Group Public Affairs Coordinator, 24 September 1976.

92. Interview, Robert Burr, n. 89 above.

93. R.M. Hart, Oil Supply Coordinator, 22 September 1976.

94. Interview, Robert Burr, n. 89 above.

95. E.E. Bowyer, Exec. Dir. Marine Transp., 21 September 1976.

96. C.C.P. Pocock.

97. R.M. Hart.

98. C.C.P. Pocock; Wm. C. Thomson, Chemicals Coordinator, 6 October 1976.

99. Gerrit A. Wagner.

100. R.M. Hart.

101. Alan F. Peters, Manager of Public Affairs Extraordinary, 24 September 1976; J.H. McDonald.

102. J.H. Choufer, Manufacturing Coordinator—Oil (Chem. Engr. from Delft), 28 September 1976; Monroe E. Spaght, Managing Director (PhD chemist—Germany), Pres. Shell Oil, 27 October 1976.

103.   J.H. McDonald; J.H. Choufer.

104.   R.M. Hart.

105.   L.C. van Wachem.

106.   Robert Burr.

107.   Geoffrey Chandler.

108.   J.K.H. Choufer; C.C.P. Pocock.

109.   Dr. H.L. Beckers (a physicist from Delft), Organization and Planning Manager, Manager, 23 September 1976.

110.   Geoffrey Chandler.

111.   C.C.P. Pocock.

112.   H.L. Beckers.

113.   Ibid.

114.   Robert Burr.

115.   H.L. Beckers.

116.   Monroe E. Spaght.

117.   Geoffrey Chandler; D.F. Stephenson, Public Affairs Director, 6 October 1976.

118.   Robert Burr.

119.   H.L. Beckers.

120.   D.F. Stephenson.

121.   Ibid.

122.   H.L. Beckers; Alan F. Peters.

123.   Ibid.

124.   J.H. McDonald.

125.   Ibid.

126.   Sir David Barran.

127.   Ibid.

128.   Ibid.

129.   J.H. McDonald.

130.   Geoffrey Chandler.

131.   Gerrit A. Wagner, Chairman, Interview, 27 September 1976.

Chapter 10

1.   Longhurst, 17; Mikdashi, 9.

2.   Longhurst, 82-83,

3.   Sampson, 53; Longhurst, 25.

4. Longhurst, 19.

5. Ibid.; Mikdashi 10,11,20.

6. French (Rothschilds), German (Deutsche Bank), Swedish (Nobel), Dutch (Shell), or American (Standard).

7. Anderson, 31-32; Longhurst, 23-24.

8. Anderson, 35.

9. Mikdashi, 11,15; Longhurst, 28.

10. Anderson, 33.

11. Longhurst, 35-36. The first was gold in Australia.

12. Ibid., 68-69.

13. Ibid., 54.

14. Ibid., 42-43.

15. Sampson, 55; Longhurst, 36; Anderson, 58.

16. Mikdashi, 11.

17. Sampson, 51.

18. Rowland and Cadman, 99; ARs.

19. Mikdashi, 15 (56%), Anderson, 20 (51%), Longhurst, 50 (50%+).

20. Longhurst, 50-2; Mikdashi, 15.

21. Ibid.

22. Sampson, 55.

23. Mikdashi, Ch.1, n. 23.

24. Longhurst, 53.

25. Wilkins, 220.

26. Ibid., 48.

27. Ibid., 44.

28. Mikdashi, 31-32, 43.

29. Ibid., 32-33.

30. Ibid. 32, 47.

31. Ibid.

32. Longhurst, 60.

33. Longhurst, 57.

34. Mikdashi, 13-14, 17-18.

35. Ibid., 20, 36-39, 294.

36. Mikdashi, 16-17, 35-36.

37. Ibid., 16-19,

38. Ibid.

39. AR 1924.

40. Mikdashi, 29; Gibb and Knowlton, 312, 316.

41. Mikdashi, 69-71.

42. Longhurst, Chapter 9.

43. Mikdashi, 71; Hewins, 133.

44. ARs; Longhurst, 65.

45. Interview, Dr. R.W. Ferrier, BP Group Historian, 14 September 1977.

46. ARs 1936, 1943.

47. Anderson, 279; AR 1931 (5-6), AR 1941.

48. Mikdashi, 61-63.

49. Mikdashi, 45-46, 76.

50. Ibid., 75-76.

51. Mikdashi, 76; Anderson 47-48; Longhurst, 75-80.

52. Mikdashi, 32-33.

53. Ibid.

54. AR 1922, 1923 (8), 1931 (6), 1935 (5).

55. Longhurst 178; Wilkins, 238n.

56. Mikdashi, 33; Wilkins 238n; Longhurst, 178-81.

57. AR 1948 (11), Longhurst, 181.

58. Mikdashi, 72.

59. Ibid., 104.

60. Ibid., 72.

61. Sampson, 72; Wilkins, 233-38.

62. Mikdashi, 34; Chairman's statement at annual meeting, 2 November 1927.

63. Rowland and Cadman, 150-51.

64. Longhurst, 97-104.

65. Ibid., 107.

66. Ibid., 127.

67. Ibid., 136.

68. Longhurst, 211; Mikdashi, 96; International Petroleum Cartel, 137-62; Lenzen, 40. See Socal chapter. Mikdashi, 114.

69. Longhurst, 172.

70. Ibid., 181.

71. Sampson. 116-17.

72.   Ibid.

73.   Wilkins, 220.

74.   See Chapter 9 for figures of U.K. and Persia.

75.   Mikdashi, 112, 154-55.

76.   Ibid., 149-51.

77.   Sampson, 109.

78.   Ibid., 120.

79.   Ibid., 104.

80.   Longhurst, 139, 145.

81.   Ibid.

82.   Data from Mikdashi, 45-46, 109-10.

83.   Mikdashi, 155; Wilkins, 367.

84.   Calculation from ARs and Longhurst, 153.

85.   Interview, Dr. R.W. Ferrier, BP Group Historian, 14 September 1977.

86.   Longrigg, 157.

87.   Longhurst, 143.

88.   Mikdashi, 157; Longhurst, 142-43, 173; Sampson, 121.

89.   Sampson, 126-27.

90.   Anderson, 24; Longhurst, 156-59; Sampson 128-32.

91.   Ibid.; AR 1954.

92.   Sampson, 120, 127.

93.   Sampson, 121, 113; Acheson, 650.

94.   ARs.

95.   *Barron's*, 24 March 1975, p.9.

96.   ARs; Longhurst, 259.

97.   ARs.

98.   Anderson, 226-27.

99.   Longhurst, 264.

100.   ARs.

101.   *Business Week*, 6 October 1975, p.36.

102.   ARS.

103.   *Barron's*, 24 March 1975, p.9.

104.   Longhurst, Chapter 24; ARs.

105.   Interview with author, September 1976.

106. ARs.

107. Table 1-1, Chapter 1.

108. AR 1960.

109. Table 1-1, Chapter 1.

110. AR 1960; Table 1-1, Ch.1.

111. *Business Week*, 8 December 1975.

112. ARS.

113. *Forbes*, 15 March 1977, p.86.

114. Adelman, 166.

115. AR 1947 (10); Anderson, 20.

116. AR 1970 (20).

117. *Forbes*, 15 March 1977, pp 83-86; Ars.

Appendix A

1. Stigler (1957), "Perfect Competition, Historically Contemplated."

2. Samuelson (1967), "The Monopolistic Competition Revolution."

3. The debate over this issue is known in the economic literature of the 1920's as the "cost controversy." See Stigler and Boulding.

4. Samuelson, 41.

5. Chandler, *The Visible Hand*.

6. Berle and Means (1932).

7. J. Robinson, *Economics of Imperfect Competition*, London: Macmillan, 1933; E.H. Chamberlin, *The Theory of Monopolistic Competition*, Cambridge, MA: Harvard University Press, 1933.

8. Samuelson.

9. Cournot, von Stackelberg, Fellner.

10. Mason (1949), 1266-67.

11. Bain (1959).

12. Caves (1972), Scherer (1970), and others.

13. Cyert and March (1963).

14. Cohen and Cyert (1963), Marris (1963), Williamson (1966).

15. Cohen and Cyert (1963).

16. Galbraith, *The New Industrial State*.

17. *Managing the Resource Allocation Process*.

18. *The Concept of Corporate Strategy*.

19.   Porter (1980); Spence (1977).

20.   Caves (1980), pp. 88-89.

Appendix B

1.   See Chandler (1962) Ch. 2 for an outline of these changes.

2.   See Chandler (1965) Part III.

3.   *Survey of Current Business,* April 1969, 14.

4.   Kaysen and Turner, 26-37.

5.   Scherer, 40.

6.   Means, *Economic Concentration,* 15-19, 281-84.

7.   M.A. Adelman, "Concept and Statistical Measurement of Vertical Integration," *Business Concentration and Price Policy,* 281-83.

8.   For example, if there were three competitors in each of five regional markets, local concentration would be high but national concentration would be much lower. If the market grows to national size and mergers and competition eliminate all but three firms, national concentration will have risen though concentration in any regional market may well be unchanged.

9.   Gort (1962), 80-82.

10.   Adelman, 308-11.

11.   Laffer (1969).

12.   H.F. Houghton, *Economic Concentration,* 157.

13.   Appendix by P. Glenn Porter and Harold C. Livesay to Chandler (1969), 297-98. Also, *Business Week,* 1/16/78.

14.   Scherer, 41. Marx, *Capital,* I, 836. Swezy (1942), 254-69.

15.   Coase (1934).

References

Chapter 1

Adelman, M.A. *The World Petroleum Market*. Baltimore: Johns Hopkins University Press. 1972.

American Petroleum Institute. *Petroleum Facts and Figures*. Irregular. Washington: API. 1928-.

Andrews, K.R. *The Concept of Corporate Strategy*. Homewood, Ill.: Dow Jones-Irwin, Inc. 1971.

Bain, J.S. "Economies of Scale, Concentration, and the Condition of Entry in Twenty Manufacturing Industries," *American Economic Review*, March 1954, pp. 15-39.

_____. *Barriers to New Competition*. Cambridge: Harvard University Press. 1956.

_____. *Industrial Organization*. New York: Wiley and Sons. 1959.

Berle, A.A. and Means, G. *The Modern Corporation and Private Property*. New York: Macmillan. 1932.

Bower, J.L. *Managing the Resource Allocation Process*. Division of Research, Graduate School of Business Administration, Boston: Harvard University. 1969.

British Petroleum. *BP Statistical Review of the World Oil Industry*. Annual. London.

Business Week. 1/16/78.

Caves, R.E. *American Industry: Structure, Conduct, and Performance*. 3d ed. Englewood Cliffs, N.J.: Prentice-Hall. 1972.

_____. "Industrial Organization, Corporate Strategy and Structure." *Journal of Economic Literature*. March 1980.

Chamberlin, E.H. *The Theory of Monopolistic Competition*. Cambridge: Harvard University Press. 1933.

Chandler, A.D., Jr. *Strategy and Structure: Chapters in the History of Industrial Enterprise*. Cambridge: The MIT Press. 1962.

_____. *The Railroads: The Nation's First Big Business*. New York: Harcourt, Brace, and World. 1965.

_____. "The Structure of American Industry in the Twentieth Century: A Historical Overview, *Business History Review*. Autumn 1969.

Chase Manhattan Bank. *Capital Investments of the World Petroleum Industry*, annual. New York.

_____. *Financial Analysis of a Group of Petroleum Companies*, annual. New York.

Christensen, R.C., Andrews, K.R., and Bower, J.L. *Business Policy: Text and Cases*. Homewood, Ill.: Richard D. Irwin. 1973.

Coase, R.H. "The Nature of the Firm," *Economica*—November 1937.

Cohen, K.J. and Cyert, R.M. *Theory of the Firm: Resource Allocation in a Market Economy*. Englewood Cliffs, N.J.: Prentice-Hall. 1965.

Cournot, A.A. *Researches into the Mathematical Principles of the Theory of Wealth*. 1838. Translated by Nathaniel T. Bacon. New York: Macmillan. 1927.

Cyert, R.M. and March, J.G. *A Behavorial Theory of the Firm.* Englewood Cliffs, N.J.: Prentice Hall. 1963.

Europe Documents.

Europe Energy.

Federal Trade Commission, *The International Petroleum Cartel,* 1952.

Fellner, W. *Competition Among the Few.* New York: Knopf. 1949.

Friedman, M. *Essays in Positive Economics,* Part 1, "The Methodology of Positive Economics." The University of Chicago Press. 1953.

Galbraith, John K., *The New Industrial State.* New York: Houghton Mifflin. 1971. 2nd ed.

Gort, M. *Diversification and Integration in American Industry.* Princeton: Princeton University Press. 1962.

Graaff, J. de V. *Theoretical Welfare Economics.* Cambridge (England): Cambridge University Press. 1957.

Graham, E.M. "Oligopolistic Imitation and European Direct Investment in the United States." DBA thesis, Harvard Business School. 1974.

Hidy, R.W. and Hidy, M.E. *Pioneering in Big Business, 1882-1911: History of Standard Oil Company (New Jersey)* vol. 1. New York: Harper and Brothers. 1955.

Hunt, M.S. "Competition in the Major Home Appliance Industry 1960-1970." Ph.D. thesis, Harvard University. 1972.

Kaysen, C. and Turner, D.F. *Antitrust Policy.* Cambridge: Harvard University Press. 1959.

Laffer, A.B. "Vertical Integration by Corporations, 1929-1965," *Review of Economics and Statistics.* February 1969.

Landeau, J.F. "Strategies of the Independent Oil Companies." DBA thesis, Harvard Business School. 1977.

Marx, K. *Capital.* (trans. by Ernest Untermann) Chicago: Kerr. 1912.

Mason, E.S. "The Current State of the Monopoly Problem in the United States," *Harvard Law Review.* June 1949, pp. 1265-85.

McLean, J.G. and Haigh, R.W. *The Growth of Integrated Oil Companies.* Division of Research, Graduate School of Business Administration. Boston: Harvard University. 1954.

National Bureau of Economic Research. *Business Concentration and Price Policy.* Princeton: Princeton University Press. 1955.

Newman, H.H. "Strategic Groups and the Structure-Performance Relationship." Ph.D. thesis, Harvard University. 1973.

Petroleum Intelligence Weekly.

Platt's OILGRAM Price Service.

Porter, M.E. *Interbrand Choice, Strategy and Bilateral Market Power.* Cambridge: Harvard University Press. 1976.

――――. *Competitive Strategy: Techniques for Analyzing Industries and Competitors.* New York: Macmillan. 1980.

Rumelt, R.P. "Strategy, Structure, and Economic Performance." DBA thesis, Harvard Business School. 1972.

Sampson, A. *The Seven Sisters: The Great Oil Companies and the World They Shaped.* New York: The Viking Press. 1975.

Samuelson, P.A. "The Monopolistic Competition Revolution," in Robert C. Merton, ed. *The Collected Scientific Papers of Paul A. Samuelson.* Vol. III. Cambridge: The M.I.T. Press. 1972, pp. 18-56.

Saving, T.R. "Estimation of Optimum Size of Plant by the Survivor Technique," *Quarterly Journal of Economics.* November 1961, pp.569-607.

Scherer, R.M. *Industrial Market Structure and Economic Performance.* Chicago: Rand McNally and Company. 1970 (1st ed.), 1980 (2d ed.).

Scott, B.R. "Stages of Corporate Development," unpublished paper, Harvard Business School. 1970.

Smith, A. *The Wealth of Nations*. New York: Random House. 1937.

Stigler, G.J. "The Division of the Labor is Limited by the Extent of the Market," *Journal of Political Economy*. June 1951, pp. 185-192.

———. "Perfect Competition, Historically Contemplated," *Journal of Political Economy*, February 1957, pp. 1-17.

———. "The Economies of Scale," *Journal of Law and Economics*. October 1958, pp. 54-71.

Stigler, G.J. and Boulding, K.E., eds. *Readings in Price Theory*. Homewood, Ill: Irwin. 1952.

Swezy, P.M. *The Theory of Capitalist Development*. New York: Monthly Review Press. 1942.

U.S. Department of Commerce. *Survey of Current Business*.

U.S. Senate, Committee on Foreign Relations, Subcommittee on Multinational Corporations, *Hearings, 1974*. Sen. Frank Church, Chmn.

U.S. Senate, Committee on the Judiciary, Subcommittee on Antitrust and Monopoly, Hearings, *Economic Concentration*, Washington, 1964.

Vernon, R. *Manager in the International Economy*. Englewood Cliffs, N.J.: Prentice-Hall, Inc. 1972.

Von Neumann, J. and Morgenstern, O. *Theory of Games and Economic Behavior*. Princeton, 1944.

von Stackelberg, H. *Marktform und Gleichgewicht*. Berlin: Springer. 1934.

Wilkins, M. *The Emergence of Multinational Enterprise: American Business Abroad from the Colonial Era to 1914*. Cambridge: Harvard University Press. 1970.

———. *The Maturing of Multinational Enterprise: American Business Abroad from 1914 to 1970*. Cambridge: Harvard University Press. 1974.

Williamson, H.F. and Daum, A.R. *The American Petroleum Industry*, vol. 1, *The Age of Illumination 1959-99*. Evanston, Ill.: Northwestern University Press. 1959.

Williamson, O.E. *Corporate Control and Business Behavior*. Englewood Cliffs, N.J.: Prentice-Hall. 1970.

Chapters 2 and 3

American Petroleum Institute. *Petroleum Facts and Figures*. Washington, D.C.: API. 3d ed., 1930.

Business Week, February 28, 1977.

Chandler, Alfred D. Jr. *Strategy and Structure: Chapters in the History of Industrial Enterprise*. Cambridge: The MIT Press. 1962. "Standard Oil and the Early Development of the American Oil Industry." Inter-Collegiate Case Clearing House, Harvard Business School (BH120).

Hidy, Ralph W. and Hidy, Muriel E. *Pioneering in Big Business, 1881-1911*, vol. 1 of the *History of Standard Oil Company (New Jersey)*. New York: Harper and Brothers, 1955.

McLean, John G. and Haigh, Robert Wm. *The Growth of Integrated Oil Companies*. Division of Research, Graduate School of Business Administration, Cambridge, Massachusetts: Harvard University. 1954.

Mellon, W.L. *Judge Mellon's Sons*. Privately printed, 1948.

Sampson, Anthony. *The Seven Sisters*. New York: Viking. 1975.

U.S. Consular Reports, "Petroleum and Kerosene Oil in Foreign Countries," no. 37 (January 1884).

White, Gerald T. *Formative Years in the Far West*. New York: Appleton Century-Crofts. 1962.

Wilkins, Mira. *The Emergence of Multinational Enterprise*. Cambridge: Harvard University Press, 1970.

Williamson, Harold F. and Daum, Arnold R. *The American Petroleum Industry*. Northwestern University Press, Evanston, Ill., 1959.

Chapter 4

British Petroleum. *BP Statistical Review of the World Oil Industry*. Annual. London.
Chandler, Alfred D. *Strategy and Structure*. Cambridge: M.I.T. Press, 1962.
Forbes, "Exxon's Small Businesses," 12 June 1978; "Does Exxon Have a Future"? 15 August 1977.
Gibb, George S. and Knowlton, Evelyn H. *The Resurgent Years, 1911-27* in *History of Standard Oil Company (New Jersey)* Vol. 2. New York: Harper and Brothers, 1956.
Hidy, Ralph and Hidy, Muriel E. *Pioneering in Big Business, 1882-1911* in *History of Standard Oil Company (New Jersey)* Vol. 1. Harper and Brothers, New York, 1955.
House Judiciary Committee on Monopolies and Commercial Law. Submission of Exxon Company, U.S.A. 9/11/75.
Kaufman, Burton I. "Oil and Antitrust: The Oil Cartel Case and the Cold War," *Business History Review*, Spring 1977, pp. 35-36.
Larson, Henrietta M., Knowlton, Evelyn H., and Popple, Charles S. *New Horizons, 1927-50* in *History of Standard Oil Company (New Jersey)* Vol. 3. Harper and Row, New York, 1971.
McLean, John G. and Haigh, Robt. Wm. *The Growth of Integrated Oil Companies*. Division of Research, Graduate School of Business Administration, Harvard University, Cambridge, 1954.
New York Times, 15 December 1912.
Sampson, Anthony. *The Seven Sisters*. New York: The Viking Press, 1975.
Wall, Bennett H. and Gibb, George S. *Teagle of Jersey Standard*. New Orleans: Tulane University Press, 1974.
Wilkins, Mira. *The Maturing of Multinational Enterprise*. Cambridge: Harvard University Press, 1974.

Chapter 5

Adelman, M.A. *The World Petroleum Market*. Johns Hopkins Univ. Press. Baltimore, 1972.
American Petroleum Institute. *Petroleum Facts and Figures*. API: Washington, irregular.
Anderson, J.R.L. *East of Suez*. Hodder and Stoughton. London, 1969.
Annual Reports.
British Petroleum. *BP Statistical Review of the World Oil Industry*. Annual. London.
Business Week. 25 November 1961; 26 July 1976; 31 January 1977.
Chisolm, Archibald. *The First Kuwait Oil Concession*. Frank Cass. London, 1975.
Federal Trade Commission. *The International Petroleum Cartel*. FTC: Washington 1952.
Forbes. 15 October 1977; 26 May 1978.
Gibb, G.S. and Knowlton, E.H. *The Resurgent Years, 1911-27*. Harper and Row. New York, 1956.
Hidy, R.W. and Hidy, M.E. *Pioneering in Big Business, 1882-1911*. Harper and Brothers. New York, 1955.
Kaufman, Burton I., "Oil and Antitrust: The Oil Cartel and the Cold War," *Business History Review*, Spring 1977, pp. 35-56.
McCloy, John J., Nathan W. Pearson and Beverly Matthews. *The Great Oil Spill, Report of the Special Review Committee of the Board of Directors of Gulf Oil Corporation*. Chelsea House Publishers. New York, 1976.
McLean J.G. and Haigh, R.W. *The Growth of Integrated Oil Companies*. Division of Research, Graduate School of Business Administration, Harvard University. Cambridge, MA. 1954.
Mellon, W.L. *Judge Mellon's Sons*. Privately printed, 1948.
Mikdashi, Zuhayr. *A Financial Analysis of Middle Eastern Oil Concessions: 1901-65*. Praeger. New York, 1966.
Sampson, Anthony. *The Seven Sisters*. The Viking Press. New York, 1975.
Schoenleber, Alvin. *Doctors in Oil*. Privately printed by SONJ. 1950.

Stegner, W. *Discovery! The Search for Arabian Oil.* Middle East Export Press. Beirut, Lebanon, 1971.

Thompson, Craig. *Since Spindletop.* Gulf Oil Corporation, 1952. A history commemorating Gulf's first 50 years.

Wilkins, Mira. *The Maturing of Multinational Enterprise.* Cambridge: Harvard University Press, 1974.

Chapter 6

Anderson, Irvine H. *The Standard-Vacuum Oil Company and United States East Asian Policy, 1933-1941.* Princeton, N.J.: Princeton University Press, 1975.

British Petroleum. *BP Statistical Review of the World Oil Industry.* Annual. London.

Business Week. 13 June 1977.

Federal Trade Commission. *The International Petroleum Cartel.* Washington. 1952.

Harrington, E.I. "General Petroleum Corporation," *California World and Petroleum Industry.* 5 July 1937 and three subsequent issues.

Henriques. *Marcus Samuel.* London: Barrie and Rockliff, 1960.

Larson, Henrietta M., Evelyn H. Knowlton, and Charles S. Popple. *New Horizons, 1927-1950.* in *History of Standard Oil Company (New Jersey)* Vol. 3. New York: Harper and Row, 1971.

Lenzen, Theodore L. "Inside International Oil." Manuscript 1972.

McLean, John G. and Robt. Wm. Haigh. *The Growth of Integrated Oil Companies.* Division of Research, Graduate School of Business Administration, Harvard University, Cambridge, MA. 1954.

"Mobil I." unpublished company history covering 1866-1962.

"Mobil II." unpublished company history covering 1866-1938.

Mobil World. July-August 1976.

Ross, Irwin. "Public Relations Isn't Kid-Glove Stuff at Mobil," *Fortune.* September 1976.

U.S. Senate. Committee on Foreign Relations. *Multinational Corporations and United States Foreign Policy.* Hearings before Subcommittee on Multinational Corporations. 93rd Congress, 2nd session. 1974 (called *Multinational Hearings* in text).

White, Gerald T. *Formative Years in the Far West.* New York: Appleton Century-Crofts. 1962.

Wilkins, Mira. *The Maturing of Multinational Enterprise.* Cambridge, MA: Havard University Press, 1974.

Chapter 7

Adams, Nathan M. "A Tankful of Gas," *Reader's Digest* (condensation), May, 1976.

American Petroleum Institute. *Petroleum Facts and Figures.* API: Washington, irregular.

Business Week 10 June 1961, 7 December 1974.

Federal Trade Commission. *The International Petroleum Cartel.* Washington. 1952.

Feis, Herbert. *Seen from EA.* New York, 1947.

Forbes 1 February 1977.

Gibb, George S. and Evelyn H. Knowlton. *The Resurgent Years, 1911-27* in *History of Standard Oil Company (New Jersey)* Vol. 2. New York: Harper and Brothers, 1956.

Larson, Henrietta M., Evelyn H. Knowlton, and Charles S. Popple. *New Horizons, 1927-50* in *History of Standard Oil Company (New Jersey)* Vol. 3. New York: Harper and Row, 1971.

Lenzen, Theodore L. "Inside International Oil." manuscript 1972.

Longhurst, Henry. *Adventure in Oil: The Story of British Petroleum.* Sidgwick and Jackson, London, 1959.

McLean, John G. and Robt. Wm. Haigh. *The Growth of Integrated Oil Companies*. Division of Research, Graduate School of Business Administration, Harvard University, Cambridge, MA 1954.

Sampson, Anthony. *The Seven Sisters: The Great Oil Companies and the World They Shaped*. New York: The Viking Press, 1975.

Standard Oil Company (New Jersey). "Background Memorandum." (pamphlet) 1954.

Stegner, Wallace. *Discovery! The Search for Arabian Oil*. Beirut: Middle East Export Press, Inc., 1971.

U.S. Senate. Committee on Foreign Relations. *Multinational Corporations and United States Foreign Policy*. Hearings before Sub-committee on Multinational Corporations. 93rd Congress, 2nd session, 1974 (called *Multinational Hearings* in notes).

White, Gerald T. *Formative Years in the Far West: A History of Standard Oil Company of California and Predecessors Through 1919*. New York: Appleton-Century-Crofts. 1962.

Wilkins, Mira. *The Maturing of Multinational Enterprise*. Cambridge, MA: Harvard University Press, 1974.

Chapter 8

Annual Reports and Statistical Supplements.

Forbes, "Rebuilding the House that Long Built," 17 April 1978.

James, Marquis. *The Texaco Story: The First Fifty Years 1902-52*, The Texas Company, 1953.

King, John O. *Joseph Stephen Cullinan*, Nashville, 1970.

McLean, John G. and Haigh, Robert William. *The Growth of Integrated Oil Companies*, Division of Research, Graduate School of Business Administration, Harvard University, 1954.

Sampson, Anthony. *The Seven Sisters*. New York: The Viking Press. 1975.

The Texaco Star, January 1977.

Chapter 9

Barran, Sir David. "How Shell Works: Management Techniques in a Large International Group of Companies," a paper presented to the British Institute of Management, Shell Centre, London, 30 November 1967.

Beaton, Kendall. *Enterprise in Oil: A History of Shell Oil Company in the United States 1912-1955*. New York: Appleton-Century-Crofts, 1957.

Deterding, Sir Henri. *An International Oilman* (as told to Stanley Naylor). Ivor Nicholson and Watson, London, 1934.

Gerretson, F.C. *The History of the Royal Dutch Petroleum Company*. (Four volumes) Leiden: E.J. Brill, 1953-57.

Gibb, George S. and Knowlton, Evelyn H. *The Resurgent Years, 1911-27* in *History of Standard Oil Company (New Jersey)* Vol. 2. New York: Harper and Brothers, 1956.

Henriques, Robert. *Marcus Samuel, First Viscount Bearsted and Founder of the 'Shell' Transport and Trading Company 1853-1927*. London: Barrie and Rockliff, 1960.

Godber, Lord Frederick. "I Remember ..." *Shell World* 1957, pp. 158-60.

Kessler, J.B. August. "I Remember ..." *Shell World* 1957, pp. 161-63.

Larson, Henrietta M., Knowlton, Evelyn H. and Popple, Charles S. *New Horizons, 1927-50* in *History of Standard Oil Company (New Jersey)* Vol. 3. New York: Harper and Row, 1971.

Royal Dutch/Shell. *A Short History of the Royal Dutch/Shell Group of Companies*. London: Shell Printing Limited, 1975.

Sampson, Anthony. *The Seven Sisters: the Great Oil Companies and the World they Shaped*. New York: The Viking Press, 1975.

Shell Briefing Service. "Chemicals Digest" November, 1976; "The Lubricants Business" September, 1977.

The "Shell" Transport and Trading Company Limited. *Survey of Activities of the Shell and Royal Dutch Group of Companies 1957*. London: Alabaster Passmore and Sons, Limited, 1957.

The Shell Magazine, March–October 1934.

Wilkins, Mira. *The Emergence of Multinational Enterprise: A History of American Business Abroad from Colonial Times until World War I*. Cambridge, MA: Harvard University Press, 1970.

———. *The Maturing of Multinational Enterprise: American Business Abroad from 1914 to 1970*. Cambridge, MA: Harvard University Press, 1974.

Chapter 10

Acheson, Dean. *Present at the Creation*. New York: Signet. 1974.

Adelman, M.A. *The World Petroleum Market*. Baltimore: Johns Hopkins University Press. 1972.

Anderson, J.R.L. *East of Suez, A Study of Britain's Greatest Trading Enterprise*. London: Hodder and Stoughton. 1969.

Barron's. 24 March 1975.

British Petroleum. Annual reports and statements.

Business Week 6 October 1975, 8 December 1975.

Federal Trade Commission. *The International Petroleum Cartel*. Washington 1952.

Forbes 15 March 1977.

Gibb, G.S. and Knowlton, E.H. *The Resurgent Years, 1911-27*. New York: Harper and Row. 1956.

Hewins, Ralph. *Mr. Five Percent, the Biography of Calouste Gulbenkian*. London. 1957.

Lenzen, T.L. "Inside International Oil." manuscript. 1972.

Longhurst, Henry. *Adventure in Oil, the Story of British Petroleum*. London: Sidgwick and Jackson. 1959.

Longrigg, S.H. *Oil in the Middle East, Its Discovery and Development*. 3d edition. London: Oxford University Press. 1968.

Mikdashi, Zuhayr. *A Financial Analysis of Middle Eastern Oil Concesions: 1901-65*. New York: Praeger. 1966.

Rowland, John and Cadman, Basil. *Ambassador for Oil: The Life of John, First Baron Cadman*. London: Herbert Jenkins. 1960.

Sampson, Anthony. *The Seven Sisters*. New York: Viking Press. 1975.

Wilkins, Mira. *The Maturing of Multinational Enterprise: American Business Abroad from 1914 to 1970*. Cambridge: Harvard University Press. 1974.

Appendices

References are included with those of Chapter 1 above.

Index